A Handbook of Small Data Sets

A Handbook of Small Data Sets

Edited by

D.J. Hand,
F. Daly,
A.D. Lunn,
K.J. McConway
and
E. Ostrowski

CHAPMAN & HALL
London · Glasgow · New York · Tokyo · Melbourne · Madras

Published by Chapman & Hall, 2–6 Boundary Row, London SE1 8HN, UK

Chapman & Hall, 2–6 Boundary Row, London SE1 8HN, UK

Blackie Academic & Professional, Wester Cleddens Road, Bishopbriggs, Glasgow G64 2NZ, UK

Chapman & Hall Inc., One Penn Plaza, 41st Floor, New York NY 10119, USA

Chapman & Hall Japan, Thomson Publishing Japan, Hirakawacho Nemoto Building, 6F, 1-7-11 Hirakawa-cho, Chiyoda-ku, Tokyo 102, Japan

Chapman & Hall Australia, Thomas Nelson Australia, 102 Dodds Street, South Melbourne, Victoria 3205, Australia

Chapman & Hall India, R. Seshadri, 32 Second Main Road, CIT East, Madras 600 035, India

First edition 1994

© 1994 D.J. Hand, F. Daly, A.D. Lunn, K.J. McConway and E. Ostrowski

The editors and publisher would like to acknowledge the kind permission to reproduce each data set given by the individual copyright holders. We have made every effort to contact the copyright holder for each data set and would be grateful if any errors were brought to the attention of the publisher for correction at a later printing.

Printed in Great Britain by Clays Ltd, St. Ives plc, Bungay, Suffolk

ISBN 0 412 39920 2

Printed on permanent acid-free text paper, manufactured in accordance with ANSI/NISO Z39.48-1992 and ANSI/NISO Z39.48-1984 (Permanence of Paper).

CONTENTS

INTRODUCTION

During our work as teachers of statistical methodology we have often been in the position of trying to find suitable data sets to illustrate techniques or phenomena or to use in examination questions. In common with many other teachers, we have often fabricated numbers to fill the role. But this is far from ideal for several reasons:

- Obviously unreal data sets ('In a country called Randomania, the Grand Vizier wanted to know the average number of sheep per household') do not convey to the students the importance and relevance of the discipline of statistics. If the technique being taught is as important as the teacher claims, how is it that he/she has been unable to find a real example?
- If data purporting to come from some real domain are invented ('Suppose a researcher wanted to find out if women scored higher than men on the WAIS-R test') there is the risk of misleading – it is in fact quite difficult to create realistic artificial data sets unless one is very familiar with the application area. One needs to be sure that the means are in the right range, that the dispersion is realistic, and so on.
- Inventing data serves to reinforce the misconception that statistics is a science of calculation, instead of a science of problem solving. To avoid this risk it is necessary to present real problems along with the statistical solutions.

Since artificial data sets have a number of drawbacks, real ones must be found. And this is often not easy. Many subject matter journals do not give the raw data, but only the results of statistical analyses, typically insufficient to allow reconstruction of the data. One can spend hours browsing through books and journals to locate a suitable set. We have spent many hours so doing, and we are certain that many other teachers of statistics share our experience.

For this reason we decided that a source book, a volume containing a large number of small data sets suitable for teaching, would be valuable. This book is the result. In what follows, about 500 real small data sets, with brief descriptions and details of their sources, are given.

Of course, a book such as this will only realize its potential if users can locate a data set to illustrate the sort of technique that they wish to use. This obviously must be achieved through an index, but it is perhaps not as easy as it might appear. Data sets can be analysed in many different ways. A contingency table can be used to illustrate chi-squared tests, for log-linear modelling, and for correspondence analysis, but it might also be used for less obvious purposes. It might be used to illustrate methods of outlier detection, the dangers of collapsing tables, Simpson's paradox, methods for estimating small probabilities, problems with structural zeros, sampling inadequacies, or doubtless a whole host of other things we have not thought of. So indexing the data sets by possible statistical technique, while in a sense ideal, seemed impracticable.

An alternative was to index the data sets by their properties, so that users could find a data set which had the sort of structure they needed to demonstrate whatever it was that they were interested in.

This was the strategy we finally adopted. The book has two indexes:

(i) a **data structure** index,
(ii) a **subject** index.

The second of these is straightforward. It simply contains keywords describing the application domain – the technical area and the problem from which the data arose.

The first is more difficult. There are various theories of data which we could have used to produce a taxonomy through which to classify the data sets in this volume (for example, Coombs, C.H. (1964) *A theory of data.* New York: John Wiley & Sons). However, none of them seemed to provide the right mix of simplicity and power for our purposes. We needed an approach which could handle most of the data sets, but which was not excessively complicated and difficult to grasp. It was not critical if particularly unorthodox data sets had to be handled by supplementary comments, provided this did not occur too frequently.

The approach we adopted was to describe the data sets in terms of:

(a) two numbers, the first representing the number of independent units described in the set and the second the number of measurements taken on each unit;
(b) a categorization of the variables measured;
(c) an optional supplementary word or phrase describing the structure in familiar terms.

Of course, such descriptions are not always unambiguous. For example, they tell us nothing about any grouping structure beyond that contained in terms such as **nominal**, **categorical**, and **binary** in (b) above. However, to have included such descriptions would have led to substantially greater complexity of description.

Also, there is often more than one way of describing any given data set. In particular, the description of a data set will depend on the objective of the analysis. As a simple example, consider responses given as percentages of items correct in each of six tests taken by two people. If the aim is to compare people, then one could describe this as two units, each with six scores. Alternatively, if the aim is to compare tests, then one could describe it as six units, each with two scores.

Such disadvantages have to be weighed against the merits of keeping the descriptive scheme short and this is the spirit in which our data structure index was treated. It is not intended to lead the user to **the single data set** which will do the job but to several which might be suitable and from which a choice can be made. And, a point to which we return below, it is not intended to take the place of casual browsing through the data sets.

The terms used in (i) above are as follows:

- A grouping of subjects has been represented by a **nominal, binary** (if two groups), or **categorical** variable, though the table might show the groups separately rather than give an explicit grouping variable.
- **Nominal** represents a variable with unordered response categories, and **categorical** represents a variable with ordered response categories.
- The term **numeric** has been used to indicate measurement on an interval or ratio scale.
- Values expressed as 'parts per million' could sensibly be regarded as **proportions, ratios, counts, rates**, or simply **numeric**. We adopted whatever seemed most sensible to us in the context of the example (which, of course, need not seem sensible to you, though we hope it does).
- Other terms have been used occasionally, where we thought them desirable, such as **maxima**, if the values represent the maximum values observed in some process, **count**, if the values are counts, and so on.
- The description of the numbers and types of the variables is sometimes followed by an overall description of the data set. These occur in square brackets. Examples are:

   [**survival**] to indicate that the data show survival times (and will often be censored). Censoring is indicated by a binary variable in the description of the data set, though again it may not appear explicitly in the data (it may appear as asterisks against appropriate values – the descriptions of each data set will make things clear).

   [**spatial**] to indicate that there is a spatial component to the data. (Such data would normally require a complex descriptive scheme to describe them adequately, which would be contrary to our aim of simplicity and could not be justified for a mere handful of data sets.)

   [**time series**]: the fact that the data are a time series will be apparent from the descriptions of the variables – having the form *n m rep(r)*, with *rep(r)* signifying repeated *r* times. Nevertheless, we thought it helpful to flag such data sets explicitly.

   [**latin square**].

[**transition matrix**].

[**dissimilarity matrix**].

[**correlation matrix**].

Some examples of data descriptions are:

40 3 numeric(2), binary [survival]

which means that there are 40 cases, each measured on three variables, two of which are numeric and the third binary (if there is censoring in the survival data, this is indicated by this binary variable). The final term indicates that it is survival data.

1 26 rep(26)count [time series]

which indicates that a single object is measured 26 times, producing a single count on each occasion.

9 x 9 [correlation matrix]

is a correlation matrix of size 9.

13 22 rep(11)(numeric, rate)

shows that 13 objects each produce a numeric score and a rate on each of 11 occasions.

We stress again that this does not remove all ambiguity. For example, if two counts are given, with one necessarily being part of the other (e.g. number of children in a family and number of female children in a family) then it may be described as **count(2)** or **count, proportion**. Similarly, very large counts might arguably be treated as **numeric**. It is thus possible that our way of describing a data set may not be the way you would have chosen. While it may be possible to define a formal language, free from ambiguity, to describe all conceivable data structures, the complexity of such a language would be out of place here. We hope and expect that users of this volume will browse through it. In any case, it is worth noting that many of the data sets have intrinsic interest in their own right and are informative, educational, or even amusing.

The data sets have been drawn from a very wide range of sources and application domains and we have also tried to provide material which can be used to illustrate a correspondingly wide range of statistical methodology. However, we are all too aware of the enormous size of the discipline of statistics. If you feel that our coverage of some subdomain of statistics is too weak, then please let us know – we can try to rectify the inadequacy in any future edition that may be produced.

Similarly, while we have made every effort to ensure the accuracy of the figures, given the number of digits reproduced it is likely that some inaccuracies exist. We apologise in advance should this prove to be the case and would appreciate being informed of any inaccuracies that you spot. At least the presence of the data disk will remove the risk of further data entry errors beyond those we may have introduced!

In this connection, the filename of each data set, as used on the data disk, is indicated in the data structure index.

We hope that the data sets collected here will be of value to both teachers and students of statistics. And that both teachers and students will enjoy analysing them.

David J. Hand
Fergus Daly
A. Dan Lunn
Kevin J. McConway
Elizabeth Ostrowski

The Open University, July 1993

HOW TO USE THE DISK

Each of the data sets in this book is provided in electronic form on the accompanying data disk, which is in DOS format. The files are in the sub-directory called \ DATASETS. There is a separate plain text (ASCII) file for each data set. The names of the files are given in the Data Structure Index.In each file,the data are laid out much as they are on the pages of the book, except that data in different columns are separated by a single tab space. (Two tab spaces directly following each other in a row indicate that there is an empty column in that row of the data set.) You should be able to read all the files with any text editor or word processor on an IBM-compatible PC. Different statistical packages require their input data in different formats, and we hope that ASCII files will prove fairly universal. Details of how to read ASCII files with tab separators in any particular package will be given in the documentation for the package in question. If there are difficulties it may be necesary to reformat the data, for example by using a word processor.

THE DATA
SETS

1. Germinating seeds

Chatfield, C. (1982) Teaching a course in applied statistics, *Applied Statistics,* **31**, 272–89.

The data come from an experiment to study the effect of different amounts of water on the germination of seeds. For each amount of water, four identical boxes were sown with 100 seeds each, and the number of seeds having germinated after two weeks was recorded. The experiment was repeated with the boxes covered to slow evaporation. There were six levels of watering, coded 1 to 6, with higher codes corresponding to more water. Chatfield used the data as an example of how preparing appropriate graphical representations is an essential for appropriate interpretation of data.

Number of seeds germinating per box:

Uncovered boxes

		Amount of water			
1	**2**	**3**	**4**	**5**	**6**
22	41	66	82	79	0
25	46	72	73	68	0
27	59	51	73	74	0
23	38	78	84	70	0

Covered boxes

		Amount of water			
1	**2**	**3**	**4**	**5**	**6**
45	65	81	55	31	0
41	80	73	51	36	0
42	79	74	40	45	0
43	77	76	62	*	0

* No result was recorded for this box

2. Guessing lengths

Hills, M. and the M345 Course Team (1986) M345 *Statistical Methods, Unit 1: Data, distributions and uncertainty,* Milton Keynes: The Open University. Tables 2.1 and 2.4.

Shortly after metric units of length were officially introduced in Australia, each of a group of 44 students was asked to guess, to the nearest metre, the width of the lecture hall in which they were sitting. Another group of 69 students in the same room was asked to guess the width in feet, to the nearest foot. The true width of the hall was 13.1 metres (43.0 feet). The data were collected by Professor T. Lewis.

Guesses in metres:

8	9	10	10	10	10	10	10	11	11	11	11	12	12	13
13	13	14	14	14	15	15	15	15	15	15	15	15	16	16
16	17	17	17	17	18	18	20	22	25	27	35	38	40	

Guesses in feet:

24	25	27	30	30	30	30	30	30	32	32	33	34	34	34
35	35	36	36	36	37	37	40	40	40	40	40	40	40	40
40	41	41	42	42	42	42	43	43	44	44	44	45	45	45
45	45	45	46	46	47	48	48	50	50	50	51	54	54	54
55	55	60	60	63	70	75	80	94						

3. Darwin's cross-fertilized and self-fertilized plants

Darwin, C. (1876) *The Effect of Cross- and Self-fertilization in the Vegetable Kingdom*, 2nd edition, London: John Murray.

These data are from Charles Darwin's study of cross- and self-fertilization. Pairs of seedlings of the same age, one produced by cross-fertilization and the other by self-fertilization, were grown together so that the members of each pair were reared under nearly identical conditions. The aim was to demonstrate the greater vigour of the cross-fertilized plants. The data are the final heights of each plant after a fixed period of time. Darwin consulted Galton about the analysis of these data, and they were discussed further in Fisher's *Design of Experiments*.

Final heights of plants (inches):

Pair	Cross-fertilized	Self-fertilized
1	23.5	17.4
2	12.0	20.4
3	21.0	20.0
4	22.0	20.0
5	19.1	18.4

6	21.5	18.6
7	22.1	18.6
8	20.4	15.3
9	18.3	16.5
10	21.6	18.0
11	23.3	16.3
12	21.0	18.0
13	22.1	12.8
14	23.0	15.5
15	12.0	18.0

4. Intervals between cars on the M1 motorway

Lewis, T. and the M345 Course Team (1986) *M345 Statistical Methods, Unit 2: Basic Methods: Testing and Estimation*, Milton Keynes: The Open University, 16.

The data were collected by Professor Toby Lewis as a sample of event times in a point process which it *may* be appropriate to model as a Poisson process. They are the times that 41 successive vehicles travelling northwards along the M1 motorway in England passed a fixed point near Junction 13 in Bedfordshire on Saturday 23 March 1985. The times are recorded on the 24-hour clock to the nearest second.

Vehicle	Time	Vehicle	Time	Vehicle	Time
1	22 34 38	15	22 36 44	29	22 38 48
2	50	16	50	30	49
3	52	17	37 01	31	54
4	58	18	09	32	57
5	35 00	19	37	33	39 11
6	19	20	43	34	16
7	24	21	47	35	19
8	58	22	52	36	23
9	36 02	23	53	37	28
10	03	24	38 11	38	29
11	07	25	20	39	32
12	15	26	25	40	48
13	22	27	26	41	50
14	23	28	47		

5. Tearing factor for paper

Williams, E.J. (1959) *Regression Analysis,* New York: John Wiley & Sons, 17.

The data come from an experiment in which five different pressures were applied during the sheet-pressing phase in the manufacture of paper. The aim was to investigate the effect of pressure on the *tear factor* of the paper, which is the percentage of a standard force which is necessary to tear the paper. Four sheets of paper were selected and tested from each of five batches made at different pressures. The pressures are equally spaced on a log scale, suggesting that the investigator expected equal changes in the ratio of pressures to produce equal changes in tearing factor. A log transformation of pressure may thus be appropriate.

Pressure applied during pressing	Tear factor			
35.0	112	119	117	113
49.5	108	99	112	118
70.0	120	106	102	109
99.0	110	101	99	104
140.0	100	102	96	101

6. Abrasion loss

Davies, O.L. and Goldsmith, P.L. (eds.) (1972) *Statistical Methods in Research and Production,* 4th Edition, Edinburgh: Oliver and Boyd, 239.

The data come from an experiment to investigate how the resistance of rubber to abrasion is affected by the hardness of the rubber and its tensile strength. Each of 30 samples of rubber was tested for hardness (in degrees Shore; the larger the number, the harder the rubber) and for tensile strength (measured in kg per square centimetre), and was then subjected to steady abrasion for a fixed time. The weight loss due to abrasion was measured in grams per hour. The data could be analysed by regression with two explanatory variables.

Abrasion loss (g/h)	Hardness (degrees S)	Tensile strength (kg/cm^2)
372	45	162
206	55	233
175	61	232
154	66	231
136	71	231

112	71	237
55	81	224
45	86	219
221	53	203
166	60	189
164	64	210
113	68	210
82	79	196
32	81	180
228	56	200
196	68	173
128	75	188
97	83	161
64	88	119
249	59	161
219	71	151
186	80	165
155	82	151
114	89	128
341	51	161
340	59	146
283	65	148
267	74	144
215	81	134
148	86	127

7. Mortality and water hardness

Hills, M. and the M345 Course Team (1986) *M345 Statistical Methods, Unit 3: Examining Straight-line Data*, Milton Keynes: The Open University, 28. Data provided by Professor M.J. Gardner, Medical Research Council Environmental Epidemiology Research Unit, Southampton.

These data were collected in an investigation of environmental causes of disease. They show the annual mortality rate per 100 000 for males, averaged over the years 1958–1964, and the calcium concentration (in parts per million) in the drinking water supply for 61 large towns in England and Wales. (The higher the calcium concentration, the harder the water.) Towns at least as far north as Derby are identified in the data table. How are mortality and water hardness related, and is there a geographical factor in the relationship?

Town	Mortality per 100 000	Calcium (ppm)	Town	Mortality per 100 000	Calcium (ppm)
Bath	1247	105	•Blackpool	1609	18
•Birkenhead	1668	17	•Bolton	1558	10
Birmingham	1466	5	•Bootle	1807	15
•Blackburn	1800	14	Bournemouth	1299	78

•Bradford	1637	10	Plymouth	1486	5
Brighton	1359	84	Portsmouth	1456	90
Bristol	1392	73	•Preston	1696	6
Cardiff	1519	21	Reading	1236	101
•Burnley	1755	12	•Rochdale	1711	13
Coventry	1307	78	•Rotherham	1444	14
Croydon	1254	96	•St Helens	1591	49
•Darlington	1491	20	•Salford	1987	8
•Derby	1555	39	•Sheffield	1495	14
•Doncaster	1428	39	Southampton	1369	68
East Ham	1318	122	Southend	1257	50
Exeter	1260	21	•Southport	1587	75
•Gateshead	1723	44	•South Shields	1713	71
•Grimsby	1379	94	•Stockport	1557	13
•Halifax	1742	8	•Stoke	1640	57
•Huddersfield	1574	9	•Sunderland	1709	71
•Hull	1569	91	Swansea	1625	13
Ipswich	1096	138	•Wallasey	1625	20
•Leeds	1591	16	Walsall	1527	60
Leicester	1402	37	West Bromwich	1627	53
•Newcastle	1702	44	West Ham	1486	122
Newport	1581	14	•Liverpool	1772	15
Northampton	1309	59	•Manchester	1828	8
Norwich	1259	133	•Middlesbrough	1704	26
•Nottingham	1427	27	Wolverhampton	1485	81
•Oldham	1724	6	•York	1378	71
Oxford	1175	107			

• denotes Derby and towns north of Derby

8. Tensile strength of cement

Hald, A. (1952) *Statistical Theory with Engineering Applications*, New York: John Wiley & Sons, 451.

The tensile strength of cement depends on (among other things) the length of time for which the cement is dried or 'cured'. In an experiment, different batches of cement were tested for tensile strength after different curing times. The relationship between curing time and strength is non-linear. Hald regresses log tensile strength on the reciprocal of curing time.

Curing time (days)	Tensile strength (kg/cm^2)		
1	13.0	13.3	11.8
2	21.9	24.5	24.7

3	29.8	28.0	24.1	24.2	26.2
7	32.4	30.4	34.5	33.1	35.7
28	41.8	42.6	40.3	35.7	37.3

9. Weight gain in rats

Snedecor, G.W. and Cochran, G.C. (1967) *Statistical Methods*, 6th edition, Ames, Iowa: Iowa State University Press, 347.

The data come from an experiment to study the gain in weight of rats fed on four different diets, distinguished by amount of protein (low and high) and by source of protein (beef and cereal). The design of the experiment is completely randomized with ten rats on each of the four treatments (which have a complete factorial structure).

Weight gains of rats:

Protein source	Beef	Beef	Cereal	Cereal
Protein amount	Low	High	Low	High
	90	73	107	98
	76	102	95	74
	90	118	97	56
	64	104	80	111
	86	81	98	95
	51	107	74	88
	72	100	74	82
	90	87	67	77
	95	117	89	86
	78	111	58	92

10. Weight of chickens

Snee, R.D. (1985) Graphical display of results of three-treatment randomized block experiments. *Applied Statistics*, **34**, 71–7.

A randomized blocks experiment was carried out to investigate a drug added to the feed of chicks in an attempt to promote growth. The comparison is between three treatments: standard feed (control), standard feed plus low dose of drug, standard feed plus high dose of drug. The experimental unit is a group of chicks, reared and fed together in the birdhouse. The experimental units are grouped three to a block, with physically adjacent units going in the same block. The response is the average weight per bird at maturity for the group of birds in each experimental unit.

Average weight of birds in pounds:

Block	Control	Low dose	High dose
1	3.93	3.99	3.96
2	3.78	3.96	3.94
3	3.88	3.96	4.02
4	3.93	4.03	4.06
5	3.84	4.10	3.94
6	3.75	4.02	4.09
7	3.98	4.06	4.17
8	3.84	3.92	4.12

11. Flicker frequency

Hedges, A., Hills, M., Maclay, W.P., Newman-Taylor, A.J. and Turner, P. (1971) Some central and peripheral effects of meclastine, a new antihistaminic drug, in man. *Journal of Clinical Pharmacology*, **2**, 112–119

An undesirable side-effect of some antihistamines is drowsiness, which is a consequence of the effect of the drugs on the central nervous system. These data come from an experiment to compare a placebo and two antihistaminic drugs in their effect on the central nervous system. This can be done by measuring what is called the *flicker frequency* in volunteers who have taken the treatments. In the experiment there were 9 subjects, who each took one of the three treatments on each of three days. The experimental unit was thus a subject-day combination. The design of the experiment was based on three 3×3 latin squares in which the columns (days) are permuted randomly separately for each square, and the rows (subjects) are permuted using a random permutation of 1, 2,..., 9. The three treatments are labelled A (meclastine, then a new antihistaminic drug), B (placebo) and C (promethazine, an established antihistaminic drug known to cause drowsiness).

Flicker frequency values 6 hours after drug administration:

Subject number	Day 1	Day 2	Day 3
1	31.25 (A)	31.25 (C)	33.12 (B)
2	25.87 (C)	26.63 (A)	26.00 (B)
3	23.75 (C)	26.13 (B)	24.87 (A)
4	28.75 (A)	29.63 (B)	29.87 (C)
5	24.50 (C)	28.63 (A)	28.37 (B)
6	31.25 (B)	30.63 (A)	29.37 (C)
7	25.50 (B)	23.87 (C)	24.00 (A)
8	28.50 (B)	27.87 (C)	30.12 (A)
9	25.13 (A)	27.00 (B)	24.63 (C)

12. Effect of ammonium chloride on yield

Davies, O.L. (ed.) (1956) *The Design and Analysis of Industrial Experiments*, 2nd Edition, London: Longman, 375.

The data come from an experiment to investigate the yield of an organic chemical from a process which used ammonium chloride. The main interest lay in the comparison between the use of finely ground and coarse ammonium chloride; the fine grade cost more. The experimenters were also interested in the effect of adding an extra 10% of ammonium chloride above the usual amount. There were two production units at the plant, and it was thought that the unit used would have an effect on yield, but that interactions between the unit and the two factors involving ammonium chloride were unlikely. The experiment thus had a 2^3 factorial treatment structure. Experimental runs were carried out in blocks, each block using a blend of input materials, and it was not desirable for these blends to be too large. Therefore a complete randomized block design was not considered appropriate; instead, blocks of size 4 were used and the 3-way interaction was confounded with blocks.

Yield (pounds):

Ammonium chloride quality	low	high	low	high	low	high	low	high
Ammonium chloride amount	low	low	high	high	low	low	high	high
Production unit	1	1	1	1	2	2	2	2
Block								
1	155	—	—	157	—	150	152	—
2	—	162	168	—	156	—	—	161
3	—	171	175	—	161	—	—	173
4	164	—	—	171	—	153	162	—

13. Comparing dishwashing detergents

John, P.W.M. (1961) An application of a balanced incomplete block design, *Technometrics*, **3**, 51–54.

An experiment was carried out to compare dishwashing detergents. Standard plates were soiled with standard dirt, and an operator then washed the plates in a detergent solution one at a time. The response variable was the number of plates washed before the foam disappeared. Three basins were set up, with an operator for each, and the three operators took care to wash at the same rate. A different detergent was used in each basin. Thus the experimental unit was a single washing-up session in one basin, and the units were grouped in blocks of three (being the washing-up sessions carried out simultaneously). Since there were nine detergents to compare in blocks of size three, a balanced incomplete block design was used.

Number of plates washed before disappearance of foam:

| | | | | | Detergent | | | | |
Block	1	2	3	4	5	6	7	8	9
1	19	17	11	—	—	—	—	—	—
2	—	—	—	6	26	23	—	—	—
3	—	—	—	—	—	—	21	19	28
4	20	—	—	7	—	—	20	—	—
5	—	17	—	—	26	—	—	19	—
6	—	—	15	—	—	23	—	—	31
7	20	—	—	—	26	—	—	—	31
8	—	16	—	—	—	23	21	—	—
9	—	—	13	7	—	—	—	20	—
10	20	—	—	—	—	24	—	19	—
11	—	17	—	6	—	—	—	—	29
12	—	—	14	—	24	—	21	—	—

14. Software system failures

Musa, J.D., Iannino, A. and Okumoto, K. (1987) *Software reliability: measurement, prediction, application*, New York: McGraw-Hill, 305. Data originally from Musa, J.D. (1979) *Software reliability data*, Data and Analysis Center for Software, Rome Air Development Center, Rome, NY.

These data give the failure times (in CPU seconds, measured in terms of execution time) of a real-time command and control software system. The data can be used as an example for fitting various reliability models. Musa *et al.* recommend fitting a nonhomogeneous Poisson process with linearly or exponentially decreasing failure intensity.

Time of failure (CPU sec):

3	2676	7843	16185	35338	53443
33	3098	7922	16229	36799	54433
146	3278	8738	16358	37642	55381
227	3288	10089	17168	37654	56463
342	4434	10237	17458	37915	56485
351	5034	10258	17758	39715	56560
353	5049	10491	18287	40580	57042
444	5085	10625	18568	42015	62551
556	5089	10982	18728	42045	62651
571	5089	11175	19556	42188	62661
709	5097	11411	20567	42296	63732
759	5324	11442	21012	42296	64103
836	5389	11811	21308	45406	64893
860	5565	12559	23063	46653	71043
968	5623	12559	24127	47596	74364
1056	6080	12791	25910	48296	75409
1726	6380	13121	26770	49171	76057
1846	6477	13486	27753	49416	81542
1872	6740	14708	28460	50145	82702
1986	7192	15251	28493	52042	84566
2311	7447	15261	29361	52489	88682
2366	7644	15277	30085	52875	
2608	7837	15806	32408	53321	

End of test occurred at 91208 CPU sec.

15. Piston-ring failures

Davies, O.L. and Goldsmith, P.L. (eds.) (1972) *Statistical Methods in Research and Production*, 4th Edition, Edinburgh: Oliver and Boyd, 324.

These are data on the number of failures of piston-rings in each of three legs in each of four steam-driven compressors located in the same building. The compressors have identical design and are oriented in the same way. Questions of interest are whether the probability of failure varies between compressors or between different legs, and whether the pattern of the location of failures is different for different compressors. Davies and Goldsmith investigate these questions using chi-squared tests; log-linear analysis might be an alternative.

Numbers of failures:

Leg

Compressor no.	North	Centre	South	total
1	17	17	12	46
2	11	9	13	33
3	11	8	19	38
4	14	7	28	49
Total	53	41	72	166

16. Strength of chemical pastes

Davies, O.L. and Goldsmith, P.L. (eds.) (1972) *Statistical Methods in Research and Production*, 4th Edition, Edinburgh: Oliver and Boyd, 136.

Davies and Goldsmith use the following data to demonstrate the use of analysis of variance in a hierarchical situation where the aim is to estimate variance components. A chemical paste product is delivered in casks, and there are thought to be variations in the mean strength of the paste between delivery batches. When a batch arrives, the paste is tested by sampling casks from the batch and then analysing material from the casks. Further sources of error arise in the sampling and in the analysis. The data come from ten randomly chosen delivery batches; from each, three casks are selected at random, and two analyses are carried out on the contents of each selected cask. The response is the percentage paste strength in the analysed sample.

Batch	Cask 1		Cask 2		Cask 3	
1	62.8	62.6	60.1	62.3	62.7	63.1
2	60.0	61.4	57.5	56.9	61.1	58.9
3	58.7	57.5	63.9	63.1	65.4	63.7
4	57.1	56.4	56.9	58.6	64.7	64.5
5	55.1	55.1	54.7	54.2	58.8	57.5
6	63.4	64.9	59.3	58.1	60.5	60.0
7	62.5	62.6	61.0	58.7	56.9	57.7
8	59.2	59.4	65.2	66.0	64.8	64.1
9	54.8	54.8	64.0	64.0	57.7	56.8
10	58.3	59.3	59.2	59.2	58.9	56.6

17. Human age and fatness

Mazess, R.B., Peppler, W.W., and Gibbons, M. (1984) Total body composition by dual-photon (^{153}Gd) absorptiometry. *American Journal of Clinical Nutrition*, **40**, 834–839.

The data come from a study investigating a new method of measuring body composition, and give the body fat percentage (% fat), age and sex for 18 normal adults aged between 23 and 61 years. How are age and % fat related, and is there any evidence that the relationship is different for males and females?

The data are also analysed in Altman, D.G. (1991) *Practical statistics for medical research*, London: Chapman and Hall, 286.

age	% fat	sex
23	9.5	M
23	27.9	F
27	7.8	M
27	17.8	M
39	31.4	F
41	25.9	F
45	27.4	M
49	25.2	F
50	31.1	F
53	34.7	F
53	42.0	F
54	29.1	F
56	32.5	F
57	30.3	F
58	33.0	F
58	33.8	F
60	41.1	F
61	34.5	F

18. Motion sickness

Burns, K.C. (1984) Motion sickness incidence: distribution of time to first emesis and comparison of some complex motion conditions. *Aviation Space and Environmental Medicine*, **56**, 521–527.

Experiments were performed as part of a research programme investigating motion sickness at sea. Human subjects were placed in a cubical cabin mounted on a hydraulic piston and subjected to vertical motion for two hours. The length of time until each subject first vomited was recorded. Censoring occurred because some subjects requested an early stop to the experiment, while several others survived the whole two hours without vomiting.   The data presented here come from two experiments, one with 21 subjects experiencing motion at a frequency of 0.167 Hz and acceleration 0.111g, and the other with 28 subjects moving at frequency 0.333 Hz and acceleration 0.222g. As well as estimating survival curves in each case, the equality of the curves can be tested, for example using the logrank test.

These data are also analysed in Altman, D.G. (1991) *Practical statistics for medical research*, London: Chapman and Hall, 369–371.

Experiment 1: 0.167Hz, 0.111g		Experiment 2: 0.333Hz, 0.222g	
Subject no.	**Survival time (min)**	**Subject no.**	**Survival time (min)**
1	30	1	5
2	50	2	6*
3	50*	3	11
4	51	4	11
5	66*	5	13
6	82	6	24
7	92	7	63
8	120*	8	65
9	120*	9	69
10	120*	10	69
11	120*	11	79
12	120*	12	82
13	120*	13	82
14	120*	14	102
15	120*	15	115
16	120*	16	120*
17	120*	17	120*
18	120*	18	120*
19	120*	19	120*
20	120*	20	120*
21	120*	21	120*
		22	120*
		23	120*
		24	120*
		25	120*
		26	120*
		27	120*
		28	120*

* censored data

19. Plum root cuttings

Bartlett, M.S (1935) Contingency table interactions, *Journal of the Royal Statistical Society Supplement*, **2**, 248–252.

In an experiment to investigate the effect of cutting length (at two levels) and planting time (at two levels) on the survival of plum root cuttings, 240 cuttings were planted for each of the 2 × 2 combinations of these factors, and their survival was later recorded. The data were used by Bartlett to illustrate a method of testing for no three-way interaction in a contingency table.

Numbers of cuttings:

Length	Time	Survival Dead	Alive	Total
long	at once	84	156	240
	in spring	156	84	240
short	at once	133	107	240
	in spring	209	31	240

20. Spectacle wearing and delinquency

Weindling, A.M., Bamford, F.N. and Whittall, R.A. (1986) Health of juvenile delinquents. *British Medical Journal*, **292**, 447.

The following data come from a study comparing the health of juvenile delinquent boys and a non-delinquent control group. They relate to the subset of the boys who failed a vision test, and show the numbers who did and did not wear glasses. Are delinquents with poor eyesight more or less likely to wear glasses than are non-delinquents with poor eyesight? There is insufficient data for a chi-squared test, but Fisher's exact test is possible.

These data are also analysed in Altman, D.G. (1991) *Practical statistics for medical research*, London: Chapman and Hall, 254.

Spectacle wearers		Juvenile delinquents	Non-delinquents	Total
	Yes	1	5	6
	No	8	2	10
	Total	9	7	16

21. Yield of isatin derivative

Davies, O.L. (ed.) (1956) *The Design and Analysis of Industrial Experiments,* 2nd Edition, London: Longman, 275.

The data come from a complete, unreplicated, 2^4 factorial experiment to investigate the effect of the following four factors on the yield of an isatin derivative.

(i)   Acid strength (at levels 87% and 93%)
(ii)   Time of reaction (at levels 15 minutes and 30 minutes)
(iii)   Amount of acid (at levels 35 ml and 45 ml)
(iv)   Temperature of reaction (at levels 60°C and 70°C)

The response is the yield of isatin (in g per 100 g of base material). Analysis using anova is straightforward; however, since there is no replication one must think about what error term to use. Davies suggests that, on technical grounds, the existence of three- and four-factor interactions is unlikely.

		Temperature of reaction			
		60°C		70°C	
		Amount of acid			
Acid strength	Reaction time	35 ml	45 ml	35 ml	45 ml
87%	15 min	6.08	6.31	6.79	6.77
87%	30 min	6.53	6.12	6.73	6.49
93%	15 min	6.04	6.09	6.68	6.38
93%	30 min	6.43	6.36	6.08	6.23

22. Fecundity of fruitflies

Sokal, R.R. and Rohlf, F.J. (1981) *Biometry,* 2nd edition, San Francisco: W.H. Freeman, 239.

The data (collected by R.R. Sokal) give the per diem fecundity (number of eggs laid per female per day for the first 14 days of life) for 25 females of each of three genetic lines of the fruitfly *Drosophila melanogaster.* The lines labelled RS and SS were selectively bred for resistance and for susceptibility to DDT, respectively, and the NS line is a nonselected control strain. The aim was to test two hypotheses. First, did the two selected lines (RS and SS) differ in fecundity from the nonselected line? Second, did the line selected for resistance differ in fecundity from the line selected for susceptibility?

Resistant (RS)	Susceptible (SS)	Nonselected (NS)
12.8	38.4	35.4
21.6	32.9	27.4
14.8	48.5	19.3
23.1	20.9	41.8
34.6	11.6	20.3
19.7	22.3	37.6
22.6	30.2	36.9
29.6	33.4	37.3
16.4	26.7	28.2
20.3	39.0	23.4
29.3	12.8	33.7
14.9	14.6	29.2
27.3	12.2	41.7
22.4	23.1	22.6
27.5	29.4	40.4
20.3	16.0	34.4
38.7	20.1	30.4
26.4	23.3	14.9
23.7	22.9	51.8
26.1	22.5	33.8
29.5	15.1	37.9
38.6	31.0	29.5
44.4	16.9	42.4
23.2	16.1	36.6
23.6	10.8	47.4

23. Butterfat

Sokal, R.R. and Rohlf, F.J. (1981) *Biometry*, 2nd edition, San Francisco: W.H. Freeman, 368.

These data come from Canadian records of pure-bred dairy cattle. They give average butterfat percentages for random samples of 10 mature ($\geq$ 5 years old) and 10 two-year-old cows from each of five breeds. The data can be subjected to analysis of variance to investigate questions about the effects of maturity and breed on butterfat.

Breed

Ayrshire		Canadian		Guernsey		Holstein-Fresian		Jersey	
Mature	2-yr	Mature	2-yr	Mature	2-yr	Mature	2-yr	Mature	2-yr
3.74	4.44	3.92	4.29	4.54	5.30	3.40	3.79	4.80	5.75
4.01	4.37	4.95	5.24	5.18	4.50	3.55	3.66	6.45	5.14
3.77	4.25	4.47	4.43	5.75	4.59	3.83	3.58	5.18	5.25
3.78	3.71	4.28	4.00	5.04	5.04	3.95	3.38	4.49	4.76
4.10	4.08	4.07	4.62	4.64	4.83	4.43	3.71	5.24	5.18
4.06	3.90	4.10	4.29	4.79	4.55	3.70	3.94	5.70	4.22
4.27	4.41	4.38	4.85	4.72	4.97	3.30	3.59	5.41	5.98
3.94	4.11	3.98	4.66	3.88	5.38	3.93	3.55	4.77	4.85
4.11	4.37	4.46	4.40	5.28	5.39	3.58	3.55	5.18	6.55
4.25	3.53	5.05	4.33	4.66	5.97	3.54	3.43	5.23	5.72

24. Snoring and heart disease

Norton, P.G. and Dunn, E.V. (1985) Snoring as a risk factor for disease: an epidemiological survey, *British Medical Journal*, **291**, 630–632.

The data come from a report of a survey which investigated whether snoring was related to various diseases. Those surveyed were classified according to the amount they snored, on the basis of reports from their spouses. These particular data relate to the presence or absence of heart disease. One question for the analysis is how to take account of the ordered nature of the snoring categories. The authors used scores 1, 3, 5 and 6 respectively for the four snoring groups and used a chi-squared test for trend.

Numbers of subjects:

Heart disease	Non-snorers	Occasional snorers	Snore nearly every night	Snore every night	Total
Yes	24	35	21	30	110
No	1355	603	192	224	2374
Total	1379	638	213	254	2484

25. Trees' nearest neighbours

Digby, P.G.N and Kempton, R.A. (1987) *Multivariate analysis of ecological communities*, London: Chapman and Hall, 150.

These data describe the spatial association between tree species in Lansing Wood, Michigan, measured by the number of times that each species occurs as the nearest neighbour of each other species. This measure of association is not symmetric, and therefore cannot be analysed by methods like principal co-ordinates. Digby and Kempton suggest other approaches including biplots, correspondence analysis and a method due to Gower involving partitioning the matrix into symmetric and skew-symmetric parts.

Number of occurrences as nearest neighbour

Species	Red oak	White oak	Black oak	Hickory	Maple	Other	Total trees
Red oak	104	59	14	95	64	10	346
White oak	62	138	20	117	95	16	448
Black oak	12	20	27	51	25	0	135
Hickory	105	108	48	355	71	16	703
Maple	74	70	21	79	242	28	514
Other	11	14	0	25	30	25	105

26. Air pollution in US cities

Sokal, R.R. and Rohlf, F.J. (1981) *Biometry*, 2nd edition, San Francisco: W.H. Freeman, 239.

These data relate to air pollution in 41 US cities. There is a single dependent variable, the annual mean concentration of sulphur dioxide, in micrograms per cubic metre. These data generally relate to means for the three years 1969–71. The values of six explanatory variables are also recorded, two of which relate to human ecology, and four to climate. The data were collated by Sokal and Rohlf from several US government publications. Multiple regression can be used to investigate hypotheses about determinants of pollution.

The variables are as follows:

Y SO_2 content of air in micrograms per cubic metre
X_1 Average annual temperature in °F
X_2 Number of manufacturing enterprises employing 20 or more workers
X_3 Population size (1970 census); in thousands
X_4 Average annual wind speed in miles per hour
X_5 Average annual precipitation in inches
X_6 Average number of days with precipitation per year

Cities	Y	X_1	X_2	X_3	X_4	X_5	X_6
Phoenix	10	70.3	213	582	6.0	7.05	36
Little Rock	13	61.0	91	132	8.2	48.52	100
San Francisco	12	56.7	453	716	8.7	20.66	67
Denver	17	51.9	454	515	9.0	12.95	86
Hartford	56	49.1	412	158	9.0	43.37	127
Wilmington	36	54.0	80	80	9.0	40.25	114
Washington	29	57.3	434	757	9.3	38.89	111
Jacksonville	14	68.4	136	529	8.8	54.47	116
Miami	10	75.5	207	335	9.0	59.80	128
Atlanta	24	61.5	368	497	9.1	48.34	115
Chicago	110	50.6	3344	3369	10.4	34.44	122
Indianapolis	28	52.3	361	746	9.7	38.74	121
Des Moines	17	49.0	104	201	11.2	30.85	103
Wichita	8	56.6	125	277	12.7	30.58	82
Louisville	30	55.6	291	593	8.3	43.11	123
New Orleans	9	68.3	204	361	8.4	56.77	113
Baltimore	47	55.0	625	905	9.6	41.31	111
Detroit	35	49.9	1064	1513	10.1	30.96	129
Minneapolis–St. Paul	29	43.5	699	744	10.6	25.94	137
Kansas City	14	54.5	381	507	10.0	37.00	99
St. Louis	56	55.9	775	622	9.5	35.89	105

Omaha	14	51.5	181	347	10.9	30.18	98
Albuquerque	11	56.8	46	244	8.9	7.77	58
Albany	46	47.6	44	116	8.8	33.36	135
Buffalo	11	47.1	391	463	12.4	36.11	166
Cincinnati	23	54.0	462	453	7.1	39.04	132
Cleveland	65	49.7	1007	751	10.9	34.99	155
Columbus	26	51.5	266	540	8.6	37.01	134
Philadelphia	69	54.6	1692	1950	9.6	39.93	115
Pittsburgh	61	50.4	347	520	9.4	36.22	147
Providence	94	50.0	343	179	10.6	42.75	125
Memphis	10	61.6	337	624	9.2	49.10	105
Nashville	18	59.4	275	448	7.9	46.00	119
Dallas	9	66.2	641	844	10.9	35.94	78
Houston	10	68.9	721	1233	10.8	48.19	103
Salt Lake City	28	51.0	137	176	8.7	15.17	89
Norfolk	31	59.3	96	308	10.6	44.68	116
Richmond	26	57.8	197	299	7.6	42.59	115
Seattle	29	51.1	379	531	9.4	38.79	164
Charleston	31	55.2	35	71	6.5	40.75	148
Milwaukee	16	45.7	569	717	11.8	29.07	123

27. Acacia ants

Sokal, R.R. and Rohlf, F.J. (1981) *Biometry*, 2nd edition, San Francisco: W.H. Freeman, 740.

These data record the results of an experiment with acacia ants. All but 28 trees of two species of acacia were cleared from an area in Central America, and the 28 trees were cleared of ants using insecticide. Sixteen colonies of a particular species of ant were obtained from other trees of species A. The colonies were placed roughly equidistant from the 28 trees and allowed to invade them. In the resulting 2 × 2 table, both margins are fixed, because the numbers of trees of the two species are fixed, and because each of the 16 ant colonies will invade a tree (and two colonies never invade the same tree). The question of interest is whether the invasion rate differs for the two species of tree; analysis by chi-squared or Fisher's exact test indicated that the ants really prefer Species A.

Acacia species	Not invaded	Invaded	Total
A	2	13	15
B	10	3	13
Total	12	16	28

28. Vaccination

Mead, R. (1988) *The design of experiments: statistical principles for practical application*, Cambridge: Cambridge University Press, 289.

A vaccination study was carried out in three different areas, using four different batches of vaccine and two vaccination methods. One of the methods was used with two different types of needle. After vaccination, the individuals involved were tested to see if the vaccination had succeeded; the response is therefore binary (positive or negative test). The data can be analysed using logistic regression; there are some minor complications because of the non-orthogonality of the estimates of the treatment effects.

Area	Method of vaccination	Type of needles	Vaccine batch	Number tested	Number positive
Staffordshire	Multiple puncture	Fixed	1	228	223
	Multiple puncture	Detachable	1	221	210
	Multiple puncture	Fixed	2	230	218
	Intradermal	—	4	240	238
Cardiff	Multiple puncture	Fixed	2	221	181
	Multiple puncture	Detachable	2	213	158
	Multiple puncture	Fixed	3	200	160
	Intradermal	—	4	214	186
Sheffield	Multiple puncture	Fixed	1	223	198
	Multiple puncture	Detachable	3	228	189
	Multiple puncture	Fixed	3	216	177
	Intradermal	—	4	224	206

29. Peppers in glasshouses

Mead, R. (1988) *The design of experiments: statistical principles for practical application*, Cambridge: Cambridge University Press, 148.

An experiment was carried out over a two-year period to find the best treatment for growing peppers in glasshouses. Three factors were investigated, each at two levels:
    Heating: standard (0) or supplementary (1)
    Lighting: standard (0) or supplementary (1)
    Carbon dioxide: control (0) or added CO_2 (1).
Each treatment combination requires a glasshouse compartment, and 12 compartments, divided into two blocks of 6, are available. In the first year of the experiment, all 8 treatment combinations were used. In the second year, the 5 most successful treatments from the first year were retained, and one treatment was replicated in each block.

The responses are a measure of the excess of yield over costs.

Heating	0	0	0	0	1	1	1	1
Lighting	0	0	1	1	0	0	1	1
CO_2	0	1	0	1	0	1	0	1
Year 1 Block 1	11.4	13.2	10.4	-	13.7	-	12.0	12.5
Year 1 Block 2	-	8.4	6.5	6.1	10.8	9.4	-	9.1
Year 2 Block 1	-	13.7	-	-	14.6	16.5	12.8	12.9
						15.4		
Year 2 Block 2	-	10.7	-	-	10.9	10.9	9.0	10.2
							10.1	

30. A clinical trial in lymphoma

Ezdinli, E., Pocock, S., Berard, C.W., *et al.* (1976) Comparison of intensive versus moderate chemotherapy of lymphocytic lymphomas: a progress report. *Cancer*, **38**, 1060–1068.

These data come from a clinical trial of cytoxan + prednisone (CP) and BCNU + prednisone (BP) in lymphocytic lymphoma. The outcome variable is the response of the tumour in each patient, measured on a qualitative scale from 'Complete response' (best) to 'Progression' (worst). Do the treatments differ in their efficacy? The data are also analysed in Pocock, S. (1983) *Clinical trials*, Chichester: John Wiley & Sons, 191.

Numbers of patients:

	Treatment		
	BP	CP	total
Complete response	26	31	57
Partial response	51	59	110
No change	21	11	32
Progression	40	34	74
Total no. of patients	138	135	273

31. Danish do-it-yourself

Edwards, D.E. and Kreiner, S. (1983) The analysis of contingency tables by graphical models. *Biometrika*, **70**, 553–565.

The data come from a sample of employed men aged between 18 and 67, who were asked whether, in the preceding year, they had carried out work on their home which they would previously have employed a craftsman to do. The response variable is the answer (yes/no) to that question. There are four categorical explanatory variables:
   Age: under 30, 31–45, over 45
   Accommodation type: apartment or house
   Tenure: rent or own
   Work of respondent: skilled, unskilled, office.
The data thus form a 5-way contingency table. Edwards and Kreiner (and also Whittaker, J. (1990), *Graphical models in applied statistics*, Chichester: John Wiley & Sons, 336) analyse the data using graphical models.

Numbers of respondents:

			Accommodation type					
			apartment			house		
			age			age		
work	tenure	response	<30	31-45	46+	<30	31-45	46+
skilled	rent	yes	18	15	6	34	10	2
		no	15	13	9	28	4	6
	own	yes	5	3	1	56	56	35
		no	1	1	1	12	21	8
unskilled	rent	yes	17	10	15	29	3	7
		no	34	17	19	44	13	16
	own	yes	2	0	3	23	52	49
		no	3	2	0	9	31	51
office	rent	yes	30	23	21	22	13	11
		no	25	19	40	25	16	12
	own	yes	8	5	1	54	191	102
		no	4	2	2	19	76	61

32. Testing cement

Davies, O.L. and Goldsmith, P.L. (eds.) (1972) *Statistical Methods in Research and Production*, 4th Edition, Edinburgh: Oliver and Boyd, 154.

An experiment was carried out to investigate sources of variability in testing the strength of Portland cement. A sample of cement was divided into small samples for testing. The cement was 'gauged', or mixed with water and worked for a fixed time, by three different gaugers, and then it was cast into cubes. Three testers or 'breakers' later tested the cubes for compressive strength. Each gauger gauged 12 cubes, which were then divided into three sets of four, and each breaker tested one set of four cubes from each gauger. All the tests were carried out on the same machine; it was thought that individual breakers might differ in the way they set up the machine, and the general aim of the study was to investigate and quantify the relative importance of the variability in test results due to personal differences between gaugers and between breakers. (Breakers and gaugers are people, not machines.) Analysis of variance can be applied, and since there is replication the breaker-gauger interaction can be investigated. The investigators' interest was in these particular breakers and gaugers; in other words the breaker and gauger effects are fixed. Davies and Goldsmith give the data in 'working units' derived by subtracting 5000 from the measurements below and then dividing by 10. The measurements given here are in the original units, pounds per square inch.

	Breaker 1		Breaker 2		Breaker 3	
Gauger 1	5280	5520	4340	4400	4160	5180
	4760	5800	5020	6200	5320	4600
Gauger 2	4420	5280	5340	4880	4180	4800
	5580	4900	4960	6200	4600	4480
Gauger 3	5360	6160	5720	4760	4460	4930
	5680	5500	5620	5560	4680	5600

33. Irises

Fisher, R.A. (1936) The use of multiple measurements in taxonomic problems. *Annals of Eugenics*, **7**, 179–184.

The data give measurements of four flower parts (sepal length, sepal width, petal length and petal width, in centimetres) on 50 specimens of each of three species of iris. They were collected (but not published in full) by E. Anderson, and most famously analysed by R.A. Fisher. His interest was in developing a methodology (discriminant analysis) for discriminating between the species on the basis of the four measurements. The data can also be used for demonstrating other multivariate techniques such as principal components analysis.

Iris setosa				*Iris versicolor*				*Iris virginica*			
Sepal length	Sepal width	Petal length	Petal width	Sepal length	Sepal width	Petal length	Petal width	Sepal length	Sepal width	Petal length	Petal width
5.1	3.5	1.4	0.2	7.0	3.2	4.7	1.4	6.3	3.3	6.0	2.5
4.9	3.0	1.4	0.2	6.4	3.2	4.5	1.5	5.8	2.7	5.1	1.9
4.7	3.2	1.3	0.2	6.9	3.1	4.9	1.5	7.1	3.0	5.9	2.1

4.6	3.1	1.5	0.2	5.5	2.3	4.0	1.3	6.3	2.9	5.6	1.8
5.0	3.6	1.4	0.2	6.5	2.8	4.6	1.5	6.5	3.0	5.8	2.2
5.4	3.9	1.7	0.4	5.7	2.8	4.5	1.3	7.6	3.0	6.6	2.1
4.6	3.4	1.4	0.3	6.3	3.3	4.7	1.6	4.9	2.5	4.5	1.7
5.0	3.4	1.5	0.2	4.9	2.4	3.3	1.0	7.3	2.9	6.3	1.8
4.4	2.9	1.4	0.2	6.6	2.9	4.6	1.3	6.7	2.5	5.8	1.8
4.9	3.1	1.5	0.1	5.2	2.7	3.9	1.4	7.2	3.6	6.1	2.5
5.4	3.7	1.5	0.2	5.0	2.0	3.5	1.0	6.5	3.2	5.1	2.0
4.8	3.4	1.6	0.2	5.9	3.0	4.2	1.5	6.4	2.7	5.3	1.9
4.8	3.0	1.4	0.1	6.0	2.2	4.0	1.0	6.8	3.0	5.5	2.1
4.3	3.0	1.1	0.1	6.1	2.9	4.7	1.4	5.7	2.5	5.0	2.0
5.8	4.0	1.2	0.2	5.6	2.9	3.6	1.3	5.8	2.8	5.1	2.4
5.7	4.4	1.5	0.4	6.7	3.1	4.4	1.4	6.4	3.2	5.3	2.3
5.4	3.9	1.3	0.4	5.6	3.0	4.5	1.5	6.5	3.0	5.5	1.8
5.1	3.5	1.4	0.3	5.8	2.7	4.1	1.0	7.7	3.8	6.7	2.2
5.7	3.8	1.7	0.3	6.2	2.2	4.5	1.5	7.7	2.6	6.9	2.3
5.1	3.8	1.5	0.3	5.6	2.5	3.9	1.1	6.0	2.2	5.0	1.5
5.4	3.4	1.7	0.2	5.9	3.2	4.8	1.8	6.9	3.2	5.7	2.3
5.1	3.7	1.5	0.4	6.1	2.8	4.0	1.3	5.6	2.8	4.9	2.0
4.6	3.6	1.0	0.2	6.3	2.5	4.9	1.5	7.7	2.8	6.7	2.0
5.1	3.3	1.7	0.5	6.1	2.8	4.7	1.2	6.3	2.7	4.9	1.8
4.8	3.4	1.9	0.2	6.4	2.9	4.3	1.3	6.7	3.3	5.7	2.1
5.0	3.0	1.6	0.2	6.6	3.0	4.4	1.4	7.2	3.2	6.0	1.8
5.0	3.4	1.6	0.4	6.8	2.8	4.8	1.4	6.2	2.8	4.8	1.8
5.2	3.5	1.5	0.2	6.7	3.0	5.0	1.7	6.1	3.0	4.9	1.8
5.2	3.4	1.4	0.2	6.0	2.9	4.5	1.5	6.4	2.8	5.6	2.1
4.7	3.2	1.6	0.2	5.7	2.6	3.5	1.0	7.2	3.0	5.8	1.6
4.8	3.1	1.6	0.2	5.5	2.4	3.8	1.1	7.4	2.8	6.1	1.9
5.4	3.4	1.5	0.4	5.5	2.4	3.7	1.0	7.9	3.8	6.4	2.0
5.2	4.1	1.5	0.1	5.8	2.7	3.9	1.2	6.4	2.8	5.6	2.2
5.5	4.2	1.4	0.2	6.0	2.7	5.1	1.6	6.3	2.8	5.1	1.5
4.9	3.1	1.5	0.2	5.4	3.0	4.5	1.5	6.1	2.6	5.6	1.4
5.0	3.2	1.2	0.2	6.0	3.4	4.5	1.6	7.7	3.0	6.1	2.3
5.5	3.5	1.3	0.2	6.7	3.1	4.7	1.5	6.3	3.4	5.6	2.4
4.9	3.6	1.4	0.1	6.3	2.3	4.4	1.3	6.4	3.1	5.5	1.8
4.4	3.0	1.3	0.2	5.6	3.0	4.1	1.3	6.0	3.0	4.8	1.8
5.1	3.4	1.5	0.2	5.5	2.5	4.0	1.3	6.9	3.1	5.4	2.1
5.0	3.5	1.3	0.3	5.5	2.6	4.4	1.2	6.7	3.1	5.6	2.4
4.5	2.3	1.3	0.3	6.1	3.0	4.6	1.4	6.9	3.1	5.1	2.3
4.4	3.2	1.3	0.2	5.8	2.6	4.0	1.2	5.8	2.7	5.1	1.9
5.0	3.5	1.6	0.6	5.0	2.3	3.3	1.0	6.8	3.2	5.9	2.3
5.1	3.8	1.9	0.4	5.6	2.7	4.2	1.3	6.7	3.3	5.7	2.5
4.8	3.0	1.4	0.3	5.7	3.0	4.2	1.2	6.7	3.0	5.2	2.3
5.1	3.8	1.6	0.2	5.7	2.9	4.2	1.3	6.3	2.5	5.0	1.9
4.6	3.2	1.4	0.2	6.2	2.9	4.3	1.3	6.5	3.0	5.2	2.0
5.3	3.7	1.5	0.2	5.1	2.5	3.0	1.1	6.2	3.4	5.4	2.3
5.0	3.3	1.4	0.2	5.7	2.8	4.1	1.3	5.9	3.0	5.1	1.8

34. Water voles

Corbet, G.B., Cummins, J., Hedges, S.R. and Krzanowski, W.J. (1970) The taxonomic status of British water voles, genus *Arvicola*. *Journal of Zoology*, **161**, 301–316.

The original data (not given here) record the presence or absence of 13 characteristics in about 300 water vole skulls divided into samples from 14 populations from Britain and the rest of Europe. A similarity matrix was computed as follows. First, the percentage incidence x of each characteristic in each population was transformed to $y = \arcsin(1 - 2x/100)$. For a pair of samples of sizes n_1 and n_2, the dissimilarity was calculated by finding the squared Euclidean distance between the y values, dividing by 13, subtracting a bias correction $(1/n_1 + 1/n_2)$ and taking the square root. (The bias correction led to three small negative squared dissimilarities: their absolute values were substituted.)

The main aim of the study which used these data was to compare British populations of water voles with other European ones, to investigate whether more than one species might be present in Britain. The British populations are numbered 1–6 below. The non-British populations are from two species, *Arvicola terrestris* (populations 7–11) and *Arvicola sapidus* (populations 12–14), and a pre-existing hypothesis was that both these species were present in Britain. Corbet *et al.* used metric multi-dimensional scaling (principal co-ordinates) to analyse the data; other scaling methods might be applied.

The populations are: 1 Surrey, 2 Shropshire, 3 Yorkshire, 4 Perthshire, 5 Aberdeen, 6 Eilean Gamhna, 7 Alps, 8 Yugoslavia, 9 Germany, 10 Norway, 11 Pyrenees I, 12 Pyrenees II, 13 North Spain, 14 South Spain. Details of the 13 skull characteristics are not given here.

34. Water voles

Percentage incidence of 13 characteristics in each of 14 samples of water vole skulls:

Char.	Populations													
	1	2	3	4	5	6	7	8	9	10	11	12	13	14
1	48.5	67.7	51.7	42.9	18.1	65.0	57.1	26.7	38.5	33.3	47.6	60.0	53.8	29.2
2	89.2	67.0	84.8	50.0	79.6	81.8	76.2	53.1	67.9	83.3	92.9	90.9	88.1	74.0
3	7.9	2.0	0.0	0.0	4.1	9.1	21.4	23.5	17.9	27.8	26.7	13.6	7.1	16.0
4	42.1	23.0	31.3	50.0	44.9	31.8	38.1	38.2	21.4	29.4	10.0	68.2	33.3	46.0
5	92.1	93.0	88.2	77.3	79.6	81.8	66.7	44.1	82.1	86.1	36.7	40.9	88.1	86.0
6	100.0	100.0	100.0	100.0	100.0	100.0	97.6	94.1	100.0	100.0	100.0	100.0	100.0	100.0
7	100.0	86.0	94.1	90.9	77.6	59.1	14.3	11.8	60.7	63.9	50.0	18.2	19.0	18.0
8	35.3	44.0	18.8	36.4	16.7	20.0	23.5	11.8	35.7	53.8	14.3	100.0	85.7	88.0
9	11.4	14.0	25.0	59.1	37.1	30.0	9.5	18.2	24.0	18.8	7.4	5.0	9.8	16.3
10	100.0	99.0	100.0	100.0	100.0	100.0	100.0	100.0	100.0	100.0	100.0	80.0	73.8	72.0
11	71.9	97.0	83.3	100.0	90.4	100.0	91.4	94.9	91.7	83.3	86.4	90.0	72.2	80.4
12	31.6	31.0	33.3	38.9	9.8	5.0	11.8	12.5	37.5	8.3	90.9	50.0	73.7	69.6
13	2.8	17.0	5.9	0.0	0.0	9.1	17.5	5.9	0.0	34.3	3.3	0.0	2.4	4.0
Sample size	19	50	17	11	49	11	21	17	14	18	16	11	21	25

Matrix of squared dissimilarities for water vole populations:

Populations

	1	2	3	4	5	6	7	8	9	10	11	12	13
2	0.099												
3	0.033	0.022											
4	0.183	0.114	0.042										
5	0.148	0.224	0.059	0.068									
6	0.198	0.039	0.053	0.085	0.051								
7	0.462	0.266	0.322	0.435	0.268	0.025							
8	0.628	0.442	0.444	0.406	0.240	0.129	0.014						
9	0.113	0.070	0.046	0.047	0.034	0.002	0.106	0.129					
10	0.173	0.119	0.162	0.331	0.177	0.039	0.089	0.237	0.071				
11	0.434	0.419	0.339	0.505	0.469	0.390	0.315	0.349	0.151	0.430			
12	0.762	0.633	0.781	0.700	0.758	0.625	0.469	0.618	0.440	0.538	0.607		
13	0.530	0.389	0.482	0.579	0.597	0.498	0.374	0.562	0.247	0.383	0.387	0.084	
14	0.586	0.435	0.550	0.530	0.552	0.509	0.369	0.471	0.234	0.346	0.456	0.090	0.038

35. Facilities in East Jerusalem

Gabriel, K.R. (1971) The biplot graphical display of matrices with application to principal components analysis. *Biometrika*, **58**, 453–467.

The data give the percentages of households with various facilities and equipment in nine areas of East Jerusalem in 1967. The aim of the analysis is to investigate similarities and differences between the different areas, and to see how patterns of access to facilities differ between them. Gabriel used the data to demonstrate the use of the biplot technique.

Percentage of households possessing:	Old city quarters			
	Christian	**Armenian**	**Jewish**	**Moslem**
Toilet	98.2	97.2	97.3	96.9
Kitchen	78.8	81.0	65.6	73.3
Bath	14.4	17.6	6.0	9.6
Electricity	86.2	82.1	54.5	74.7
Water*	32.9	30.3	21.1	26.9
Radio	73.0	70.4	53.0	60.5
TV set	4.6	6.0	1.5	3.4
Refrigerator**	29.2	26.3	4.3	10.5

Percentage of households possessing:	Modern		Other		Rural
	American Colony Sh. Jarah	**Shaafat Bet-Hanina**	**A-Tur Isawiye**	**Silwan Abu-Tor**	**Sur-Bahar Bet-Safafa**
Toilet	97.6	94.4	90.2	94.0	70.5
Kitchen	91.4	88.7	82.2	84.2	55.1
Bath	56.2	69.5	31.8	19.5	10.7
Electricity	87.2	80.4	68.6	65.5	26.1
Water*	80.1	74.3	46.3	36.2	9.8
Radio	81.2	78.0	67.9	64.8	57.1
TV set	12.7	23.0	5.6	2.7	1.3
Refrigerator**	52.8	49.7	21.7	9.5	1.2

* In dwelling. ** Electric.

36. Yields of winter wheat

Lyons, R. (1980) *A review of multidimensional scaling*, Unpublished M.Sc. dissertation, University of Reading.

The data give the yields of winter wheat in each of the years 1970-1973 at twelve different sites in England. Can general conclusions be drawn about the pattern of the yields in different years and at different sites? Lyons used the data to illustrate metric multidimensional unfolding. This analysis is also presented on pages 144–145 of Krzanowski, W.J. (1988) *Principles of multivariate analysis*, Oxford: Oxford University Press. Many other graphical displays are possible.

Yields of winter wheat (kg per unit area):

| | *Year* | | | |
Site	1970	1971	1972	1973
Cambridge	46.81	39.40	55.64	32.61
Cockle Park	46.49	34.07	45.06	41.02
Harpers Adams	44.03	42.03	40.32	50.23
Headley Hall	52.24	36.19	47.03	34.56
Morley	36.55	43.06	38.07	43.17
Myerscough	34.88	49.72	40.86	50.08
Rosemaund	56.14	47.67	43.48	38.99
Seale-Hayne	45.67	27.30	45.48	50.32
Sparsholt	42.97	46.87	38.78	47.49
Sutton Bonington	54.44	49.34	24.48	46.94
Terrington	54.95	52.05	50.91	39.13
Wye	48.94	48.63	31.69	59.72

37. WISC blocks

Aitkin, M., Anderson, D., Francis, B. and Hinde, J. (1989) *Statistical modelling in GLIM*, Oxford: Oxford University Press, 70.

The data were collected from a sample of 24 primary school children in Sydney, Australia. Each child completed the Embedded Figures Test (EFT), which measures 'field dependence', i.e. the extent to which a person can abstract the logical structure of a problem from its context. Then the children were allocated to one of two experimental groups. They were timed as they constructed a 3×3 pattern from nine coloured blocks, taken from the Wechsler Intelligence Scale for Children (WISC). The two groups differed in the instructions they were given for the task: the 'row group' were told to start with a row of three blocks, and the 'corner group' were told to start with a corner of three blocks. The experimenter was interested in whether the different instructions produced any change in the average time to complete the pattern and in whether this time was affected by field dependence.

Completion times in seconds, and EFT scores:

Row group:

Time	317	464	525	298	491	196	268	372	370	739	430	410
EFT	59	33	49	69	65	26	29	62	31	139	74	31

Corner group:

Time	342	222	219	513	295	285	408	543	298	494	317	407
EFT	48	23	9	128	44	49	87	43	55	58	113	7

38. Byssinosis

Higgins, J.E. and Koch, G.G. (1977) Variable selection and generalized chi-square analysis of categorical data applied to a large cross-sectional occupational health survey. *International Statistical Review*, **45**, 51–62.

The data come from a survey of workers in the US cotton industry, and record whether they were suffering from the lung disease byssinosis, as well as the values of five categorical explanatory variables: the race, sex and smoking status of the worker, the length of employment and the dustiness of the workplace. Primary interest is in how the incidence of byssinosis is related to the dustiness of the workplace, but the other variables must be taken into account.

Numbers of workers in different categories suffering from (*yes*) and not suffering from (*no*) byssinosis:

Yes	*No*	*Dust*	*Race*	*Sex*	*Smoking*	*Emp. length*
3	37	1	1	1	1	1
0	74	2	1	1	1	1
2	258	3	1	1	1	1
25	139	1	2	1	1	1
0	88	2	2	1	1	1
3	242	3	2	1	1	1
0	5	1	1	2	1	1
1	93	2	1	2	1	1
3	180	3	1	2	1	1
2	22	1	2	2	1	1
2	145	2	2	2	1	1
3	260	3	2	2	1	1
0	16	1	1	1	2	1
0	35	2	1	1	2	1

0	134	3	1	1	2	1
6	75	1	2	1	2	1
1	47	2	2	1	2	1
1	122	3	2	1	2	1
0	4	1	1	2	2	1
1	54	2	1	2	2	1
2	169	3	1	2	2	1
1	24	1	2	2	2	1
3	142	2	2	2	2	1
4	301	3	2	2	2	1
8	21	1	1	1	1	2
1	50	2	1	1	1	2
1	187	3	1	1	1	2
8	30	1	2	1	1	2
0	5	2	2	1	1	2
0	33	3	2	1	1	2
0	0	1	1	2	1	2
1	33	2	1	2	1	2
2	94	3	1	2	1	2
0	0	1	2	2	1	2
0	4	2	2	2	1	2
0	3	3	2	2	1	2
2	8	1	1	1	2	2
1	16	2	1	1	2	2
0	58	3	1	1	2	2
1	9	1	2	1	2	2
0	0	2	2	1	2	2
0	7	3	2	1	2	2
0	0	1	1	2	2	2
0	30	2	1	2	2	2
1	90	3	1	2	2	2
0	0	1	2	2	2	2
0	4	2	2	2	2	2
0	4	3	2	2	2	2
31	77	1	1	1	1	3
1	141	2	1	1	1	3
12	495	3	1	1	1	3
10	31	1	2	1	1	3
0	1	2	2	1	1	3
0	45	3	2	1	1	3
0	1	1	1	2	1	3
3	91	2	1	2	1	3
3	176	3	1	2	1	3
0	1	1	2	2	1	3
0	0	2	2	2	1	3
0	2	3	2	2	1	3
5	47	1	1	1	2	3
0	39	2	1	1	2	3
3	182	3	1	1	2	3
3	15	1	2	1	2	3
0	1	2	2	1	2	3
0	23	3	2	1	2	3
0	2	1	1	2	2	3

3	187	2	1	2	2	3
2	340	3	1	2	2	3
0	0	1	2	2	2	3
0	2	2	2	2	2	3
0	3	3	2	2	2	3

Meaning of explanatory variable codes:

Dust: dustiness of workplace (1 high, 2 medium, 3 low)

Race: ethnic group of worker (1 white, 2 other)

Sex: (1 male, 2 female)

Smoking: smoking status (1 smoker, 2 non-smoker)

Emp. length: length of employment in years (1 less than 10 years, 2 10–20 years, 3 over 20 years)

39. A tomato crossing experiment

MacArthur, J.W. (1931) Linkage studies with the tomato. III. Fifteen factors in six groups, *Transactions of the Royal Canadian Institute*, **18**, 1–19.

The data come from an genetics experiment involving a dihybrid cross where the expected ratio of the four phenotypes listed is 9:3:3:1 (assuming no linkage). Analysis, using for example a chi-squared test, shows no reason to doubt this hypothesis. The data are analysed in Sokal, R.S. and Rohlf, F.J. (1981) *Biometry*, 2nd edition, San Francisco: W.H. Freeman, 697.

Numbers of offspring:

Phenotype	Observed frequency
Tall cut-leaf	926
Tall, potato-leaf	288
Dwarf, cut-leaf	293
Dwarf, potato-leaf	104
Total	1611

40. Coronary heart disease

Ku, H.H. and Kullback, S. (1974) Loglinear models in contingency table analysis. *American Statistician*, **28**, 115–122.

The data come from an American study of 1329 men. The subjects are classified by blood pressure and by serum cholesterol, each measured as a four-value categorical variable. In each cell of the table are given the number of men r with coronary heart disease (CHD) and the total number of men examined n. The aim is to investigate how the proportion of men suffering from CHD is related to blood pressure and serum cholesterol levels.

		Blood pressure in mm Hg							
		<127		127–146		147–166		>166	
		r	*n*	*r*	*n*	*r*	*n*	*r*	*n*
Serum	<200	2	119	3	124	3	50	4	26
cholesterol	200–219	3	88	2	100	0	43	3	23
in mg/100cc	220–259	8	127	11	220	6	74	6	49
	>259	7	74	12	111	11	57	11	44

41. Toxaemia of pregnancy

Brown, P.J., Stone, J. and Ord-Smith, C. (1983) Toxaemic signs during pregnancy. *Applied Statistics*, **32**, 69–72.

The data were collected in Bradford, England, between 1968 and 1977, and relate to 13 384 women giving birth to their first child. The women were classified according to social class (five categories on the Registrar General's scale, I–V) and according to the number of cigarettes smoked per day during pregnancy (on a three-level categorization: 1 means no smoking, 2 means 1–19 cigarettes per day, and 3 means 20 or more cigarettes per day). The data for each category consist of counts of women showing toxaemic signs (hypertension and/or proteinuria) during pregnancy. The question of interest is how the toxaemic signs vary with social class and smoking status.

Signs of toxaemia

Class	Smoking	Hypertension and proteinuria	Proteinuria only	Hypertension only	Neither sign exhibited
1	1	28	82	21	286
1	2	5	24	5	71
1	3	1	3	0	13
2	1	50	266	34	785
2	2	13	92	17	284
2	3	0	15	3	34
3	1	278	1101	164	3160
3	2	120	492	142	2300
3	3	16	92	32	383
4	1	63	213	52	656
4	2	35	129	46	649
4	3	7	40	12	163
5	1	20	78	23	245
5	2	22	74	34	321
5	3	7	14	4	65

42. Starting positions in horse racing

New York Post, August 30 1955, 42. Reprinted in Siegel, S. and Castellan, N.J. (1988) *Nonparametric statistics for the behavioral sciences*, 2nd edition, New York: McGraw-Hill, 47.

These data were collected in the US to investigate the hypothesis that a horse's chances of winning a race on a circular track are affected by its position in the starting line-up. They relate to eight-horse races. Starting position 1 is closest to the rail on the inside of the track. The data give the starting position of each of 144 winners.

	Starting position							
	1	2	3	4	5	6	7	8
Number of wins	29	19	18	25	17	10	15	11

43. Saltiness judgements

Kroeze, J.H.A. (1982) The influence of relative frequencies of pure and mixed stimuli on mixture suppression in taste. *Perception and psychophysics*, **31**, 276–278.

Subjects tasted a mixture of salt and sucrose, in an experiment to investigate how judged saltiness depended on salt and sucrose concentration. The data presented here relate to just one concentration. The question of interest here is the shape of the distribution of saltiness judgements. In Example 4.4 of Siegel, S. and Castellan, N.J. (1988) *Nonparametric statistics for the behavioral sciences*, 2nd edition, New York: McGraw-Hill, these data are used to illustrate a test of distribution symmetry.

13.53	28.42	48.11	48.64	51.40	59.91	67.98	79.13	103.05

44. Oral socialization and explanations of illness

Whiting, J.W.M and Child, I.L. (1953) *Child training and personality*, New Haven: Yale University Press, 156.

Whiting and Child studied the relationship between child-rearing practices and customs related to illness in several non-literate cultures. On the basis of ethnographical reports, 39 societies were each given a rating for the degree of typical oral socialization anxiety, a concept derived from psychoanalytic theory relating to the severity and rapidity of oral socialization practice in child-rearing. For each of the societies, a judgement was also made (by an independent set of judges) of whether oral explanations of illness were present. Do the data support the hypothesis that oral explanations of illness are more likely to be present in societies with high levels of oral socialization anxiety? (If measurement-theoretic principles are important to you in choosing an appropriate technique, you should note that the oral socialization anxiety scores are on an ordinal scale at best.)

Oral socialization anxiety scores:

Societies where oral explanations of illness are absent

6 7 7 7 7 7 8 8 9 10 10 10 10 12 12 13

Societies where oral explanations of illness are present

6 8 8 10 10 10 11 11 12 12 12 12 13 13 13 14 14 14 15 15 15 16 17

45. Dopamine and schizophrenia

Sternberg, D.E., Van Kammen, D.P and Bunney, W.E. (1982) Schizophrenia: dopamine *b*-hydroxylase activity and treatment response. *Science*, **216**, 1423–1425.

Many theories about the causation of schizophrenia involve changes in the activity of a substance called dopamine in the central nervous system. In this study, 25 hospitalized schizophrenic patients were treated with antipsychotic medication, and after a period of time were classified as psychotic or nonpsychotic by hospital staff. Samples of cerebrospinal fluid were taken from each patient and assayed for the dopamine *b*-hydroxylase (DBH) activity. (DBH is an enzyme.) The data are given below: the units are nmol/(ml)(h)/(mg) of protein. How does DBH activity differ between the two groups of patients?

Judged nonpsychotic

.0104 .0105 .0112 .0116 .0130 .0145 .0154 .0156 .0170 .0180 .0200
.0200 .0210 .0230 .0252

Judged psychotic

.0150 .0204 .0208 .0222 .0226 .0245 .0270 .0275 .0306 .0320

46.   US cancer mortality

Selvin, S. (1991) *Statistical analysis of epidemiological data*, New York: Oxford University Press, Table 1.21.

These are US cancer mortality data by age for white males in 1940 and 1960. The corresponding population sizes are also given. Selvin uses the data to exemplify age-adjustment procedures.

Age	1960 Deaths	1960 Population	1940 Deaths	1940 Population
<1	141	1,784,033	45	906,897
1–4	926	7,065,148	201	3,794,573
5–14	1,253	15,658,730	320	10,003,544
15–24	1,080	10,482,916	670	10,629,526
25–34	1,869	9,939,972	1,126	9,465,330
35–44	4,891	10,563,872	3,160	8,249,558
45–54	14,956	9,114,202	9,723	7,294,330
55–64	30,888	6,850,263	17,935	5,022,499
65–74	41,725	4,702,482	22,179	2,920,220
75–84	26,501	1,874,619	13,461	1,019,504
85+	5,928	330,915	2,238	142,532
Total	130,158	78,367,152	71,058	59,448,513

47.   Cholesterol and behaviour type

Selvin, S. (1991) *Statistical analysis of epidemiological data*, New York: Oxford University Press, Table 2.1.

The data come from the Western Collaborative Group Study, which was carried out in California in 1960–61 and studied 3,154 middle-aged men to investigate the relationship between behaviour pattern and the risk of coronary heart disease. These particular data were obtained from the 40 heaviest men in the study (all weighing at least 225 pounds) and record cholesterol measurements (mg per 100 ml), and behaviour type on a twofold categorization. In general terms, type A behaviour is characterized by urgency, aggression and ambition, while type B behaviour is relaxed, non-competitive and less hurried. The question of interest is whether, in heavy middle-aged men, cholesterol level is related to behaviour type.

Type A behaviour: cholesterol levels

233	291	312	250	246	197	268	224	239	239
254	276	234	181	248	252	202	218	212	325

Type B behaviour: cholesterol levels

344	185	263	246	224	212	188	250	148	169
226	175	242	252	153	183	137	202	194	213

48. Aflatoxin in peanuts

Draper, N.R. and Smith, H. (1981) *Applied regression analysis*, 2nd edition, New York: John Wiley & Sons, 63.

The data give, for 34 batches of peanuts, the average level of aflatoxin (parts per billion) in a mini-lot sample of 120 pounds of peanuts (X) and the percentage of noncontaminated peanuts in the batch (Y). The aim is to investigate the relationship between the two variables, and to predict Y from X.

Y	X	Y	X	Y	X
99.971	3.0	99.942	18.8	99.863	46.8
99.979	4.7	99.932	18.9	99.811	46.8
99.982	8.3	99.908	21.7	99.877	58.1
99.971	9.3	99.970	21.9	99.798	62.3
99.957	9.9	99.985	22.8	99.855	70.6
99.961	11.0	99.933	24.2	99.788	71.1
99.956	12.3	99.858	25.8	99.821	71.3
99.972	12.5	99.987	30.6	99.830	83.2
99.889	12.6	99.958	36.2	99.718	83.6
99.961	15.9	99.909	39.8	99.642	99.5
99.982	16.7	99.859	44.3	99.658	111.2
99.975	18.8				

49. Anscombe's correlation data

Anscombe, F.J. (1973) Graphs in statistical analysis. *American Statistician*, **27**, 17–21.

Anscombe invented these data to demonstrate the importance of graphs in correlation and regression. There are four different data sets; the correlation coefficients and the regression lines for all four data sets are the same, but their scatter diagrams look very different. (Data sets 1–3 all have the same X values.)

Data set:	1–3	1	2	3	4	4
Variable:	X	Y	Y	Y	X	Y
	10	8.04	9.14	7.46	8	6.58
	8	6.95	8.14	6.77	8	5.76
	13	7.58	8.74	12.74	8	7.71
	9	8.81	8.77	7.11	8	8.84
	11	8.33	9.26	7.81	8	8.47
	14	9.96	8.10	8.84	8	7.04
	6	7.24	6.13	6.08	8	5.25
	4	4.26	3.10	5.39	8	5.56
	12	10.84	9.13	8.15	8	7.91
	7	4.82	7.26	6.42	8	6.89
	5	5.68	4.74	5.73	19	12.50

50. Caffeine and finger tapping

Draper, N.R. and Smith, H. (1981) *Applied regression analysis*, 2nd edition, New York: John Wiley & Sons, 425.

A double-blind experiment was carried out to investigate the effect of the stimulant caffeine on performance on a simple physical task. Thirty male college students were trained in finger tapping. They were then divided at random into three groups of 10 and the groups received different doses of caffeine (0, 100 and 200 mg). Two hours after treatment, each man was required to do finger tapping and the number of taps per minute was recorded. Does caffeine affect performance on this task? If it does, can you describe the effect?

0 ml caffeine

242   245   244   248   247   248   242   244   246   242

100 ml caffeine

248   246   245   247   248   250   247   246   243   244

200 ml caffeine

246   248   250   252   248   250   246   248   245   250

51. Jackal mandible lengths

Manly, B.F.J. (1991) *Randomization and Monte Carlo methods in biology*, London: Chapman and Hall, 4.

These data give the mandible lengths in millimetres for 10 male and 10 female golden jackals (*Canis aureus*) in the collection of the British Museum (Natural History). They form part of a larger data set collected by Higham *et al.* (Higham, C.F.W., Kijngam, A. and Manly, B.F.J. (1980) An analysis of prehistoric canid remains from Thailand, *Journal of Archaeological Science*, **7**, 149–165). Is there evidence that average mandible length differs between the sexes in this species? Manly uses the data to exemplify a randomization test, but other analyses are of course possible.

Males

| 120 | 107 | 110 | 116 | 114 | 111 | 113 | 117 | 114 | 112 |

Females

| 110 | 111 | 107 | 108 | 110 | 105 | 107 | 106 | 111 | 111 |

52. Birds in paramo vegetation

Manly, B.F.J. (1991) *Randomization and Monte Carlo methods in biology*, London: Chapman and Hall, 107.

These data were originally derived from a study by Vuilleumier (Vuilleumier, F. (1970) Insular biogeography in continental regions. I. The northern Andes of South America, *American Naturalist*, **104**, 373–388), which investigated numbers of bird species in isolated 'islands' of paramo vegetation in the northern Andes. The aim is to investigate how the number of species (N) is related to the four explanatory variables AR (area of 'island' in thousands of square km), EL (elevation in thousands of m), DEc (distance from Ecuador in km) and DNI (distance to nearest 'island' in km).

'Island'	N	AR	EL	DEc	DNI
Chiles	36	0.33	1.26	36	14
Las Papas-Coconuco	30	0.50	1.17	234	13
Sumapaz	37	2.03	1.06	543	83
Tolima-Quindio	35	0.99	1.90	551	23
Paramillo	11	0.03	0.46	773	45
Cocuy	21	2.17	2.00	801	14
Pamplona	11	0.22	0.70	950	14
Cachira	13	0.14	0.74	958	5

Tama	17	0.05	0.61	995	29
Batallon	13	0.07	0.66	1065	55
Merida	29	1.80	1.50	1167	35
Perija	4	0.17	0.75	1182	75
Santa Marta	18	0.61	2.28	1238	75
Cende	15	0.07	0.55	1380	35

53. Extinction of marine genera

Manly, B.F.J. (1991) *Randomization and Monte Carlo methods in biology*, London: Chapman and Hall, 174–175.

These data give the estimated percentages of marine genera which became extinct in 48 geological ages. The letter and number codes refer to the geological stages in the Harland time scale, and the times shown relate to the ends of the stages in millions of years before the present (MYBP). The aim of the study is to investigate whether there is a trend in extinction rate over time. Another question of interest is to identify the periods at which 'mass extinctions' (abnormally high extinction rates) took place.

Period		Time (MYBP)	% extinct
Permian	1 A	265	22
	2 K	258	23
	3 G	253	61
	4 D	248	60
Triassic	5 S	243	45
	6 A	238	29
	7 L	231	23
	8 C	225	40
	9 N	219	28
	10 R	213	46
Jurassic	11 H	201	7
	12 S	200	14
	13 P	194	26
	14 T	188	21
	15 A	181	7
	16 B	175	22
	17 B	169	16
	18 C	163	19

	19 O	156	18
	20 K	150	15
	21 T	144	30
Cretaceous	22 B	138	7
	23 V	131	14
	24 H	125	10
	25 B	119	11
	26 A	113	18
	27 A1	108	7
	28 A2	105	9
	29 A3	98	11
	30 C	91	26
	31 T	88	13
	32 C	87	8
	33 S	83	11
	34 C	73	13
	35 M	65	48
Tertiary	36 D	60	9
	37 T	55	6
	38 E1	50	7
	39 E2	42	13
	40 E3	38	16
	41 O1	33	6
	42 O2	25	5
	43 M1	21	4
	44 M1	16	3
	45 M2	11	11
	46 M3	5	6
	47 P	2	7
	48 R	0	2

54. Household expenditures

Aitchison, J. (1986) *The statistical analysis of compositional data*, London: Chapman and Hall, 362.

The data come from a survey of household expenditure and give the expenditure of 20 single men (M) and 20 single women (W) on four commodity groups. The units of expenditure are Hong Kong dollars, and the commodity groups are as follows.

1   Housing, including fuel and light
2   Foodstuffs, including alcohol and tobacco
3   Other goods, including clothing, footwear and durable goods
4   Services, including transport and vehicles

The aim is to investigate how the division of household expenditure between the four commodity groups depends on the total expenditure, and to find out whether (and if so, how) this relationship differs for men and women.

House-hold no.	Commodity group 1	2	3	4	House-hold no.	Commodity group 1	2	3	4
M1	497	591	153	291	W1	820	114	183	154
M2	839	942	302	365	W2	184	74	6	20
M3	798	1308	668	584	W3	921	66	1686	455
M4	892	842	287	395	W4	488	80	103	115
M5	1585	781	2476	1740	W5	721	83	176	104
M6	755	764	428	438	W6	614	55	441	193
M7	388	655	153	233	W7	801	56	357	214
M8	617	879	757	719	W8	396	59	61	80
M9	248	438	22	65	W9	864	65	1618	352
M10	1641	440	6471	2063	W10	845	64	1935	414
M11	1180	1243	768	813	W11	404	97	33	47
M12	619	684	99	204	W12	781	47	1906	452
M13	253	422	15	48	W13	457	103	136	108
M14	661	739	71	188	W14	1029	71	244	189
M15	1981	869	1489	1032	W15	1047	90	653	298
M16	1746	746	2662	1594	W16	552	91	185	158
M17	1865	915	5184	1767	W17	718	104	583	304
M18	238	522	29	75	W18	495	114	65	74
M19	1199	1095	261	344	W19	382	77	230	147
M20	1524	964	1739	1410	W20	1090	59	313	177

55. Cork deposits

Rao, C.R. (1948) Tests of significance in multivariate analysis. *Biometrika*, **35**, 58–79.

These data give the weights of cork deposits (in centigrams) of 28 trees, in each of the four directions north, east, south and west. As well as simply describing the pattern of variation, one might test the hypothesis that the mean deposit weight is the same all round a tree, or one might investigate appropriate sets of contrasts.

N	E	S	W
72	66	76	77
60	53	66	63
56	57	64	58
41	29	36	38
32	32	35	36
30	35	34	26
39	39	31	27
42	43	31	25
37	40	31	25
33	29	27	36
32	30	34	28
63	45	74	63
54	46	60	52
47	51	52	43
91	79	100	75
56	68	47	50
79	65	70	61
81	80	68	58
78	55	67	60
46	38	37	38
39	35	34	37
32	30	30	32
60	50	67	54
35	37	48	39
39	36	39	31
50	34	37	40
43	37	39	50
48	54	57	43

56. Creatinine kinase and heart attacks

Sackett, D.L., Haynes, R.B., Guyatt, G.H. and Tugwell, P. (1991) *Clinical epidemiology*, 2nd edition, Boston: Little, Brown and Company, 71.

The data come from a study in which the level of an enzyme, creatinine kinase (CK), was measured in patients who were suspected of having had a myocardial infarction (heart attack). The aim was to investigate whether measuring the level of this enzyme on admission to hospital was a useful diagnostic indicator for whether patients had really had a heart attack.   The enzyme was measured in 360 patients on admission, and later an expert clinician reviewed the records of these patients to decide with hindsight which of them had actually had a heart attack. The data give the numbers of patients with and without a confirmed heart attack whose CK level fell into each of 13 ranges, as shown below (measured in International Units). In each case the lower endpoint of the range is included in the range.

These data were originally collected in Edinburgh Royal Infirmary; the original study was published as Smith, A.F. (1967) Diagnostic value of serum-creatinine-kinase in a coronary care unit. *Lancet*, **2**, 178.

CK range	Patients with heart attack	Patients without heart attack
Below 40	2	88
40–80	13	26
80–120	30	8
120–160	30	5
160–200	21	0
200–240	19	1
240–280	18	1
280–320	13	1
320–360	19	0
360–400	15	0
400–440	7	0
440–480	8	0
480 and over	35	0

57. Petrol expenditure

The Open University (1993) *MDST242 Statistics in Society Unit A0: Introduction*, 2nd edition, Milton Keynes: The Open University, Table 1.2.

These data were collected by Dr G. Moss, who was trying to keep track of the running costs of his car, a Vauxhall Cavalier 1.6L Estate. Every time he put petrol into the tank, he continued until the flow stopped automatically, showing the tank was nearly full. Then he ran the pump on gently until, if possible, the cost on the meter reached the next whole number of pounds (to avoid change). He tried to record the date, the odometer reading, the price of petrol and how much he spent. Some data are missing. Most of the prices are in £/gallon; a few are in pence/litre (where 1 gallon = 4.54609 litres). The data can be used to exemplify the problems that arise in 'cleaning' a real data set, and to investigate hypotheses about seasonal variations in petrol consumption rates.

Date	Odometer reading (miles)	Petrol price (£/gallon)	Expenditure (£)
12 Jan 1990	68644	1.891	10.00
3 Feb 1990	68952	1.891	19.00

13 Feb 1990	69261	1.891	17.50
23 Feb 1990	69543	1.891	16.00
—	69871	1.864	19.50
23 Mar 1990	70024	1.891	10.60
26 Mar 1990	70338	1.896	16.00
6 Apr 1990	70603	1.896	16.00
22 Apr 1990	70863	1.964	14.00
4 May 1990	71131	1.964	15.00
10 May 1990	71394	1.982	14.75
23 May 1990	71596	1.887	12.00
23 May 1990	71772	2.014	10.00
1 Jun 1990	71984	1.932	11.00
4 Jun 1990	72254	2.041	18.00
11 Jun 1990	72448	2.018	—
15 Jun 1990	72613	2.018	10.50
27 Jun 1990	73053	1.882	21.00
16 Jul 1990	73268	—	14.00
4 Aug 1990	73483	1.964	13.00
10 Aug 1990	73690	2.087	11.00
8 Sep 1990	73989	2.132	18.50
20 Sep 1990	74156	2.255	12.00
9 Oct 1990	74328	50.3 p/litre	—
17 Oct 1990	74494	2.296	12.00
26 Oct 1990	74692	50.5 p/litre	12.00
8 Nov 1990	74919	2.118	14.79
15 Nov 1990	75277	2.128	21.00
30 Nov 1990	75520	2.028	17.00
14 Dec 1990	75745	1.964	15.00
23 Dec 1990	75878	1.928	9.00
7 Jan 1991	76047	1.928	10.00

— indicates missing data.

58. North Buckinghamshire moths

The Open University (1993) *MDST242 Statistics in Society Unit A0: Introduction*, 2nd edition, Milton Keynes: The Open University, 17.

These data are the numbers of moths caught in an ultraviolet trap at a location in North Buckinghamshire, England on 24 consecutive nights in September 1990. One might investigate the form of the distribution of counts, or one might use time series methods to look for features such as autocorrelation.

47	21	16	39	24	34	21	34	49	20	37
65	67	21	37	46	29	41	47	24	22	19
54	71									

59. Crowds and threatened suicide

Mann, L. (1981) The baiting crowd in episodes of threatened suicide. *Journal of Personality and Social Psychology*, **41**, 703–709.

A study was carried out to investigate the causes of jeering or baiting behaviour by a crowd when a person is threatening to commit suicide by jumping from a high structure. A hypothesis is that baiting is more likely to occur in warm weather. Mann classified 21 accounts of threatened suicide according to two factors, the time of year and whether or not baiting occurred. Is the hypothesis supported? (The data come from the northern hemisphere, so June-September are the warm months.)

	Crowd	
Months	**Baiting**	**Nonbaiting**
June–September	8	4
October–May	2	7

60. Putting the shot

The Open University (1993) *MDST242 Statistics in Society Unit A0: Introduction*, 2nd edition, Milton Keynes: The Open University, Table 2.3.

These are the distances in metres of throws by the 20 senior male athletes who competed in the shot-put in English area championships in June 1991. The quoted source uses them in exercises on graphical presentation of data.

17.79	17.21	16.47	16.27	15.53	14.92	14.72	14.63	14.23	13.70
14.52	14.46	13.22	11.75	16.31	15.81	14.98	14.30	14.05	13.66

61. Wooden toy prices

The Open University (1993) *MDST242 Statistics in Society Unit A0: Introduction*, 2nd edition, Milton Keynes: The Open University, Table 3.1.

The following are the prices (in £) of the 31 different children's wooden toys on sale in a Suffolk craft shop in April 1991. The quoted source uses them in exercises on graphical presentation of data.

4.20	1.12	1.39	2.00	3.99	2.15	1.74	5.81	1.70	2.85
0.50	0.99	11.50	5.12	0.90	1.99	6.24	2.60	3.00	12.20
7.36	4.75	11.59	8.69	9.80	1.85	1.99	1.35	10.00	0.65
1.45									

62. Food prices

The Open University (1993) *MDST242 Statistics in Society Unit A0: Introduction*, 2nd edition, Milton Keynes: The Open University, Table 4.1.

These data are the prices of ten different types of food, bought in a particular area of the UK, first at a particular time during 1987, and then again exactly one year later at the corresponding time in 1988. They are used in the quoted source to demonstrate simple techniques for displaying and summarizing bivariate data.

Food (quantity)	1987 price (pence)	1988 price (pence)
Fish (5 lb)	856	943
Milk (20 pints)	496	516
Cheese (5 lb)	663	732
Bread (20 loaves)	478	511
Breakfast cereal (10 lb)	753	802
Potatoes, old (70 lb)	659	617
Apples (20 lb)	685	711
Frozen peas (20 lb)	750	891
Beef (5 lb)	823	896
Margarine (5 lb)	407	427

63. UK earnings ratios

The Open University (1993) *MDST242 Statistics in Society Unit A2: Earnings*, 2nd edition, Milton Keynes: The Open University, Table 1.4.

These data were collated as part of an investigation of equality of pay between men and women. They are derived from data from the British Government's annual New Earnings Survey, using the distributions of earnings of men and women in full-time employment on adult rates of pay. The survey reports give the median, quartiles and highest and lowest deciles of weekly earnings for men and for women. These earnings ratios give the quantiles for women as a percentage of those for men. The data are given at two-year intervals between 1970 and 1990, together with data for 1975, the year in which the Equal Pay Act of 1970 was first required to be implemented in full.

Earnings ratios F/M (%):

Year	Highest decile	Upper quartile	Median	Lower quartile	Lowest decile
1970	56	54	53	53	54
1972	57	56	55	55	55
1974	57	57	56	56	57

1975	63	60	61	61	61
1976	67	64	64	63	62
1978	64	63	63	64	65
1980	63	63	63	64	66
1982	65	64	64	65	67
1984	64	66	66	68	71
1986	65	67	66	68	72
1988	67	68	67	68	72
1990	67	70	68	70	73

64. Census data on 10 towns

The Open University (1993) *MDST242 Statistics in Society Unit A4: Relationships*, 2nd edition, Milton Keynes: The Open University, Tables 1.2 and 1.6.

These are data from the 1981 Great Britain census. The administrative areas involved are local authority districts chosen as a stratified sample from non-metropolitan areas of England. The source uses the data to illustrate basic methods of displaying bivariate data, and to introduce the notion of a correlation coefficient. The variables are the percentage of males aged between 16 and 64 who were unemployed at the census date, the percentage of households that did not have the use of a car, the percentage of households who owned their dwelling, and the percentage of employed residents who worked in manufacturing industry.

District	% males unemployed	% households with no car	% in owner-occupied dwellings	% in manufacturing industry
Bromsgrove	9.3	23.4	66.7	34.4
Vale Royal	10.2	28.7	63.3	35.3
Rotherham	13.6	45.9	45.9	33.5
Alnwick	8.9	39.0	40.3	8.9
Rutland	5.1	24.1	57.1	24.3
West Dorset	6.6	26.7	57.9	15.5
Norwich	12.5	46.1	35.5	28.2
Bracknell	5.7	22.9	47.4	23.6
Rother	8.5	32.8	71.8	12.9
Mole Valley	4.0	22.2	66.9	19.0

65. Expenditure on food

Central Statistical Office (1989), *Regional trends 24*, London: HMSO, Table 11.10.

These data show the average expenditure per household per week in the 11 main regions of the UK, and the average percentage of that total that is spent on food. The data were originally derived from the UK *Family Expenditure Survey*. They are averages for the years 1986 and 1987. These data can be used to illustrate basic correlation and regression; but there are potential difficulties over influential points.

Region	Average expenditure per household per week (£)	Average percentage spent on food
England		
North	150.2	20.8
Yorkshire and Humberside	157.3	20.3
East Midlands	169.8	20.3
East Anglia	188.2	18.1
South East	219.2	17.8
South West	189.5	19.0
West Midlands	166.4	20.4
North West	172.0	19.9
Wales	163.6	21.7
Scotland	161.8	20.5
Northern Ireland	178.5	21.4

66. Anaerobic threshold

Bennett, G.W. (1988) Determination of anaerobic threshold. *Canadian Journal of Statistics*, **16**, 307–310.

These data were collected in an experiment in kinesiology. A subject performed a standard exercise task at a gradually increasing level. The two variables are the oxygen uptake and the expired ventilation, which is related to the rate of exchange of gases in the lungs. The object is to describe the (non-linear) relationship between these variables.

Oxygen uptake	Expired ventilation	Oxygen uptake	Expired ventilation
574	21.9	2577	46.3
592	18.6	2766	55.8
664	18.6	2812	54.5
667	19.1	2893	63.5
718	19.2	2957	60.3
770	16.9	3052	64.8
927	18.3	3151	69.2
947	17.2	3161	74.7
1020	19.0	3266	72.9
1096	19.0	3386	80.4
1277	18.6	3452	83.0
1323	22.8	3521	86.0
1330	24.6	3543	88.9
1599	24.9	3676	96.8
1639	29.2	3741	89.1
1787	32.0	3844	100.9
1790	27.9	3878	103.0
1794	31.0	4002	113.4
1874	30.7	4114	111.4
2049	35.4	4152	119.9
2132	36.1	4252	127.2
2160	39.1	4290	126.4
2292	42.6	4331	135.5
2312	39.9	4332	138.9
2475	46.2	4390	143.7
2489	50.9	4393	144.8
2490	46.5		

67. Road casualties on Fridays

Department of Transport (1987) *Road accidents in Great Britain 1986: the casualty report*, London: HMSO, Table 28.

These data give the number of casualties in Great Britain due to road accidents on Fridays in 1986, for each hour of the day. How can the relationship between casualty numbers and time be simply described?

Time of day (24hr clock)	Casualties
0–1	938
1–2	621
2–3	455
3–4	207
4–5	138

5–6	215
6–7	526
7–8	1933
8–9	3377
9–10	2045
10–11	2078
11–12	2351
12–13	3015
13–14	2966
14–15	2912
15–16	4305
16–17	4923
17–18	4427
18–19	3164
19–20	2950
20–21	2601
21–22	2420
22–23	2557
23–0	4319

68. UK sprinters

The Open University (1993) *MDST242 Statistics in Society Unit A4: Relationships*, 2nd edition, Milton Keynes: The Open University, Table 2.1.

These data are for male UK sprinters in 1988. They include the best times of all men who ran 200 metres in under 21.20 seconds without wind assistance and who also recorded a time without wind assistance for 100 metres that year. Times are in seconds. The data were originally supplied by the (British) National Union of Track Statisticians.

Athlete	200 m best	100 m best
L Christie	20.09	9.97
J Regis	20.32	10.31
M Rosswess	20.51	10.40
A Carrott	20.76	10.56
T Bennett	20.90	10.92
A Mafe	20.94	10.64
D Reid	21.00	10.54
P Snoddy	21.14	10.85
L Stapleton	21.17	10.71
C Jackson	21.19	10.56

69. Unemployment and expenditure on motoring

Central Statistical Office (1989) *Regional trends 24*, London: HMSO, Tables 10.17 and 11.10.

These data show the percentage male unemployment rate in 1987 in the 11 main regions of the UK, and the average percentage of household weekly expenditure on motoring and travel fares. These data can be used to illustrate basic correlation and regression, but there are potential difficulties over Northern Ireland, which is an outlier.

Region	Unemployment rate 1987 (%)	Average percentage spent on motoring and fares (1986-87)
England		
North	14.0	12.8
Yorkshire and Humberside	11.3	14.5
East Midlands	9.0	15.4
East Anglia	6.8	15.0
South East	7.1	15.0
South West	8.2	15.3
West Midlands	11.1	14.6
North West	12.7	14.2
Wales	12.5	14.4
Scotland	13.0	14.1
Northern Ireland	17.6	15.5

70. Cloud seeding

Woodley, W.L., Simpson, J., Biondini, R. and Berkeley, J. (1977) Rainfall results 1970–75: Florida Area Cumulus Experiment. *Science*, **195**, 735–742.

These data were collected in the summer of 1975 from an experiment to investigate the use of silver iodide in cloud seeding to increase rainfall. In the experiment, which was conducted in an area of Florida, 24 days were judged suitable for seeding on the basis that a measured suitability criterion, denoted $S - Ne$, was not less than 1.5. On each day, the decision to seed was made randomly. The response variable Y is the amount of rain (in cubic metres $\times 10^7$) that fell in the target area for a 6 hour period on each suitable day. As well as $S - Ne$, the following explanatory variables were also recorded on each suitable day.

A   Action: an indicator of whether seeding action occurred (1 yes, 0 no)

T   Time: number of days after the first day of the experiment (June 1, 1975)

C   Echo coverage: the percentage cloud cover in the experimental area, measured using radar

P   Pre-wetness: the total rainfall in the target area 1 hour before seeding (in cubic metres $\times 10^7$)

E   Echo motion: an indicator showing whether the radar echo was moving (1) or stationary (2)

The aim is to set up a model to investigate how Y is related to the explanatory variables. There are several difficulties; for instance, the second day seems untypical in several ways. These data are extensively analysed by R.D. Cook and S. Weisberg (1982, *Residuals and influence in regression*, New York: Chapman and Hall).

A	T	S – Ne	C	P	E	Y
0	0	1.75	13.40	0.274	2	12.85
1	1	2.70	37.90	1.267	1	5.52
1	3	4.10	3.90	0.198	2	6.29
0	4	2.35	5.30	0.526	1	6.11
1	6	4.25	7.10	0.250	1	2.45
0	9	1.60	6.90	0.018	2	3.61
0	18	1.30	4.60	0.307	1	0.47
0	25	3.35	4.90	0.194	1	4.56
0	27	2.85	12.10	0.751	1	6.35
1	28	2.20	5.20	0.084	1	5.06
1	29	4.40	4.10	0.236	1	2.76
1	32	3.10	2.80	0.214	1	4.05
0	33	3.95	6.80	0.796	1	5.74
1	35	2.90	3.00	0.124	1	4.84
1	38	2.05	7.00	0.144	1	11.86
0	39	4.00	11.30	0.398	1	4.45
0	53	3.35	4.20	0.237	2	3.66
1	55	3.70	3.30	0.960	1	4.22
0	56	3.80	2.20	0.230	1	1.16
1	59	3.40	6.50	0.142	2	5.45
1	65	3.15	3.10	0.073	1	2.02
0	68	3.15	2.60	0.136	1	0.82
1	82	4.01	8.30	0.123	1	1.09
0	83	4.65	7.40	0.168	1	0.28

71. Calculator random digits

Dunsmore, I.R., Daly, F. and the M345 Course Team (1987) *M345 Statistical Methods, Unit 9: Categorical data*, Milton Keynes: The Open University, Table 1.5.

A sequence of 300 pseudo-random digits was generated on a Casio fx-3600p calculator. The data give the number of times each of the digits 0, 1, ... , 9 occurred. Do these data look like a sample from a discrete uniform distribution on 0, 1, ... , 9?

Digit	0	1	2	3	4	5	6	7	8	9
Frequency	25	28	29	35	35	31	27	33	32	25

72. Captopril and blood pressure

MacGregor, G.A., Markandu, N.D., Roulston. J.E. and Jones, J.C. (1979) Essential hypertension: effect of an oral inhibitor of angiotensin-converting enzyme. *British Medical Journal*, **2**, 1106–1109.

These data give the supine systolic and diastolic blood pressures (mm Hg) for 15 patients with moderate essential hypertension, immediately before and two hours after taking a drug, captopril. The interest is in investigating the response to the drug treatment. The data are analysed as Example E in D.R. Cox and E.J. Snell (1981), *Applied Statistics*, London: Chapman and Hall.

Patient number	Systolic before	Systolic after	Diastolic before	Diastolic after
1	210	201	130	125
2	169	165	122	121
3	187	166	124	121
4	160	157	104	106
5	167	147	112	101
6	176	145	101	85
7	185	168	121	98
8	206	180	124	105
9	173	147	115	103
10	146	136	102	98
11	174	151	98	90
12	201	168	119	98
13	198	179	106	110
14	148	129	107	103
15	154	131	100	82

73. Food prices and house prices

The Open University (1993) *MDST242 Statistics in Society Unit A4: Relationships*, 2nd edition, Milton Keynes: The Open University, Table 5.1.

These data give the values of a food price index and a house price measure for the UK for each year from 1971 to 1989. The food price index is based on the Food group of the official UK Retail Prices Index. The house prices are the average prices of new dwellings for which mortgages were approved, as reported by the Building Societies Association. How do the two price measures change over time, and how are the changes related?

Year	Food price index (base date January 1962)	Average price of new dwellings (in £100)
1971	155.6	60
1972	169.4	79
1973	194.9	107
1974	230.0	113
1975	288.9	124
1976	346.5	134
1977	412.4	148
1978	441.6	177
1979	494.7	227
1980	554.5	272
1981	601.3	280
1982	648.6	285
1983	669.2	317
1984	706.7	342
1985	728.8	373
1986	752.6	436
1987	775.6	513
1988	802.4	646
1989	847.7	750

74. UNICEF data on child mortality

The Open University (1993) *MDST242 Statistics in Society, Unit A5: Review*, 3rd edition, Milton Keynes: The Open University, Tables 3.1–3.3.

The United Nations Children's Fund, UNICEF, publishes an annual report called *The State of the World's Children*. The second part of this report consists of statistical tables. In the 1992 report, these give the values of 96 different variables, related to child health and the status of women and children, for 129 countries. (There is also limited information on 41 smaller countries.) The following data were drawn as small samples from this data set to illustrate correlation and regression techniques. The three variables involved are the under 5 mortality rate (annual deaths of children under five years of age per 1000 live births, measured in 1990 or the latest available year before then), the percentage of the population with access to safe water (1988–90), and the adult literacy rate for females (the percentage of females aged 15 or over who can read or write, 1990). The first table gives data for 14 countries on all three variables, drawn as a systematic sample with a random start from a list of the countries in order of under 5 mortality. It demonstrates one major problem with these data: the values for many of the key development variables are missing for most of the developed countries. The other two tables give under 5 mortality and adult female literacy for two groups of countries, in Central and South America and in the Middle East.

Country	Under 5 mortality rate per 1000 live births	% of population with access to safe water	Adult female literacy rate (%)
Guinea-Bissau	246	25	24
Rwanda	198	50	37
Sudan	172	46	12
Laos	152	29	
Lesotho	129	48	
Guatemala	94	61	47
Turkey	80	78	71
Viet Nam	65	42	84
Oman	49	47	
USSR	31		
Costa Rica	22	92	93
Czechoslovakia	13		
Belgium	9		
Canada	9		

Country	Under 5 mortality rate per 1000 live births	Adult female literacy rate (%)
Bolivia	160	71
Peru	116	79
Guatemala	94	47
El Salvador	87	70
Honduras	84	71
Ecuador	83	84
Brazil	83	80
Nicaragua	78	
Paraguay	60	88
Colombia	50	86
Venezuela	43	90
Argentina	35	95
Panama	31	88
Chile	27	93
Uruguay	25	96
Costa Rica	22	93

Country	Under 5 mortality rate per 1000 live births	Adult female literacy rate (%)
Yemen	187	21
Saudi Arabia	91	48
Iraq	86	49
Syria	59	51
Iran	59	43
Lebanon	56	73
Jordan	52	70
Oman	49	
United Arab Emirates	30	38
Kuwait	19	67
Israel	11	

75. Length of stay on a psychiatric observation ward

Lunn, A.D., McConway, K. and the M345 Course Team (1987) *M345 Statistical Methods, Unit 15: Excursions in data analysis*, Milton Keynes: The Open University, Table 1.5.

These data come from a study on length of stay on a psychiatric observation ward. They relate to 336 patients who went on from the ward to another ward in the same hospital or to another mental hospital. The patients are classified into four categories by sex and status (certified, i.e. compulsorily admitted, and voluntary). Do the different categories of patient differ in the time they spend on the ward, and if so how? There are *a priori* reasons for expecting the time spent on the ward to be shorter for male certified patients than for the other categories. (Would you keep potentially violent patients any longer than you had to?)

This study was reported by J. Hoenig and I.M. Crotty in the *International Journal of Social Psychiatry,* **3** (1958), 260–77, but the data shown below, provided by Professor Hoenig, do not appear in that source.

		Type of patient				
		Male voluntary	Female voluntary	Male certified	Female certified	Total
	1	5	9	4	11	29
	2	16	25	18	18	77
	3	20	34	20	28	102
Number of	4	10	17	6	8	41
days in ward	5	5	15	1	12	33
	6	3	8	0	5	16
	7	3	7	0	5	15
	≥8	5	11	1	6	23
Total		67	126	50	93	336

76. Hodgkin's disease

Dunsmore, I.R., Daly, F. and the M345 Course Team (1987) *M345 Statistical Methods, Unit 9: Categorical data*, Milton Keynes: The Open University, 18.

The following data were recorded during a study of Hodgkin's disease, a cancer of the lymph nodes. (The study is described in Hancock, B.W. *et al.* (1979) *Clinical Oncology*, **5**, 283–297.) Each of 538 patients with the disease was classified by histological type, and by their response to treatment three months after it had begun. The histological types are LP = lymphocyte predominance, NS = nodular sclerosis, MC = mixed cellularity, and LD = lymphocyte depletion. What, if any, is the relationship between histological type and response to treatment?

		Response			Total
		Positive	Partial	None	
	LP	74	18	12	104
Histological	NS	68	16	12	96
type	MC	154	54	58	266
	LD	18	10	44	72
Total		314	98	126	538

77. A Church Assembly vote

The Daily Telegraph, 4 July 1967.

In 1967 the Church Assembly of the Church of England voted on a motion that individual women who felt called to exercise 'the office and work of a priest in the church' should now be considered, on the same basis as men, as candidates for Holy Orders. The voting figures are given separately for Bishops, Clergy and Laity. How did the voting patterns of the three groups differ? (An interesting question is the extent to which different flavours of inferential statistics have anything to say about this question. Does it make sense, for instance, to think of these data as random samples from some hypothetical populations?)

	Vote		
	Aye	No	Abstained
House of Bishops	1	8	8
House of Clergy	14	96	20
House of Laity	45	207	52

78. A vandalized experiment

Rayner, A.A. (1969) *A first course in biometry for agriculture students*, Pietermaritzburg: University of Natal Press.

Six varieties of turnip were grown in 36 plots arranged in a latin square design. The response variable is the fresh weight (roots plus tops) of turnips in pounds per plot (15ft × 15ft). Three plots in one corner of the experiment had been attacked by vandals and therefore did not yield any usable data. Do the varieties of turnip differ in mean weight per plot; and if so, how do they differ? The data below are laid out in the pattern of the experiment. The letters denote the varieties, A to F.

E, 29.0	F, 14.5	D, 20.5	A, 22.5	B, 16.0	C, 6.5
B, 17.5	A, 29.5	E, 12.0	C, 9.0	D, 33.0	F, 12.5
F, 17.0	B, 30.0	C, 13.0	D, 29.0	A, 27.0	E, 12.0
A, 31.5	D, 31.5	F, 24.0	E, 19.5	C, 10.5	B, 21.0
D, 25.0	C, 13.0	B, 31.0	F, 26.0	E, 19.5	A, —
C, 12.2	E, 13.0	A, 34.0	B, 20.0	F, —	D, —

79. Chest, waist and hips measurements

Hills, M., McConway, K. and the M345 Course Team (1987) *M345 Statistical Methods, Unit 13: Multivariate data*, Milton Keynes: The Open University, 16.

These data were collected by Dr M. Hills, and give the chest, hips and waist measurements (in inches) of each of 20 individuals. They are used in the source to demonstrate the method of principal components, and could be used to demonstrate other multivariate techniques. How can the variability between the individuals in terms of these three measurements be summarized? Ten of the individuals were male and ten were female: how can you tell from the data which are which?

Individual	Chest	Waist	Hips
1	34	30	32
2	37	32	37
3	38	30	36
4	36	33	39
5	38	29	33
6	43	32	38
7	40	33	42
8	38	30	40
9	40	30	37
10	41	32	39
11	36	24	35
12	36	25	37
13	34	24	37
14	33	22	34
15	36	26	38
16	37	26	37
17	34	25	38
18	36	26	37
19	38	28	40
20	35	23	35

80. Morse code mistakes

Rothkopf, E.Z. (1957) A measure of stimulus similarity and errors in some paired-associate learning tasks. *Journal of Experimental Psychology*, **53**, 94–101.

As part of an investigation into the nature of mistakes made in the perception of Morse codes, a group of 598 untrained subjects listened to pairs of single-digit numbers transmitted in Morse code. The Morse codes for the single digits are as follows.

```
0     - - - - -
1     . - - - -
2     . . - - -
3     . . . - -
4     . . . . -
5     . . . . .
6     - . . . .
7     - - . . .
8     - - - . .
9     - - - - .
```

The data below give, for each possible pair of digits, the percentage of times that the two codes in the pair were declared to be the same by the subjects. What types of confusion are being made?

These data are analysed in Chapter 14 of Mardia, K.V., Kent, J.T. and Bibby, J.M. (1979) *Multivariate analysis*, London: Academic Press. The Rothkopf paper gives data for the 26 letters as well.

	1	*2*	*3*	*4*	*5*	*6*	*7*	*8*	*9*	*0*
1	84									
2	62	89								
3	16	59	86							
4	6	23	38	89						
5	12	8	27	56	90					
6	12	14	33	34	30	86				
7	20	25	17	24	18	65	85			
8	37	25	16	13	10	22	65	88		
9	57	28	9	7	5	8	31	58	91	
0	52	18	9	7	5	18	15	39	79	94

81. Distribution of minimum temperatures

Barnett, V.D. and Lewis, T. (1967) A study of low-temperature probabilities in the context of an industrial problem. *Journal of the Royal Statistical Society, Series A*, **130**, 177–206.

As part of a study of how the performance of fuel and lubricating oils is affected by climate and weather, Barnett and Lewis investigated annual minimum temperatures at 16 locations in Britain. The data below are the annual minimum temperatures (in degrees Fahrenheit) at Kew, Manchester Airport and Plymouth for several years up to 1964. (There is a much longer run of data for Plymouth than for the other sites.) What do these data tell us about the distribution of minimum temperatures? Barnett and Lewis fitted a distribution with c.d.f. $F(t) = 1 - \exp(-\exp(-[at + b]))$, where a (< 0) and b are constants.

81. Distribution of minimum temperatures

	1964	63	62	61	60	59	58	57	56	55	54	53	52	51	50	49	48	47	46	45
Kew	22	15	19	29	25	23	23	29	17	22	22	24	23	25	23	22	23	15	23	19
Manchester Airport	19	11	14	24	24	18	16	23	14	16	18	18	14	8	21	19	22	12	15	7
Plymouth	23	20	24	26	25	25	21	27	17	23	21	25	24	25	22	31	22	16	25	18

	1944	43	42	41	40	39	38	37	36	35	34	33	32	31	30	29	28	27	26	25
Kew	27	28	18	21	—	—	—	—	—	—	—	—	—	—	—	—	—	—	—	—
Manchester Airport	19	23	—	—	—	—	—	—	—	—	—	—	—	—	—	—	—	—	—	—
Plymouth	25	30	21	23	17	24	29	29	25	29	24	28	27	29	29	24	29	28	29	31

	1924	23	22	21	20	19	18	17	16
Plymouth	30	31	28	25	27	26	23	21	28

82. Rat skeletal muscle

Lunn, A.D., McConway, K. and the M345 Course Team (1987) *M345 Statistical Methods, Unit 15: Excursions in data analysis*, Milton Keynes: The Open University, Table 6.1.

These data were collected by M. Khan and M. Khan. They are counts of fibres in rat skeletal muscle. A group of fibres is called a *fascicle*. Any fascicle can contain fibres of two types, Type I and Type II. Type I fibres are subdivided into three categories, reticulated, punctate, and both reticulated and punctate. The fibres of all these categories were counted in 25 different fascicles. The aim is to set up a model which relates the number of Type II fibres to the counts of the three different types of Type I fibre.

Fascicle number	Number of Type I fibres			Number of Type II fibres
	Reticulated	Punctate	Both	
1	1	13	5	15
2	2	8	4	12
3	9	27	16	46
4	4	5	2	12
5	2	12	7	24
6	2	31	16	66
7	1	13	15	45
8	8	27	16	50
9	1	5	5	18
10	1	8	5	15
11	1	2	2	4
12	1	11	3	17
13	1	8	6	15
14	2	17	5	30
15	1	14	4	17
16	1	11	3	16
17	2	12	2	19
18	1	8	7	14
19	1	8	4	14
20	0	4	3	7
21	1	18	5	26
22	0	11	10	26
23	2	15	7	24
24	0	0	4	5
25	0	4	3	6

83. Ground cover under apple trees

Pearce, S.C. (1983) *The agricultural field experiment*, Chichester: John Wiley & Sons, 284.

These data, which were first published by Professor Pearce in 1953, come from an experiment to study the best way of forming ground cover in an apple plantation. Treatment O represents what was the usual treatment, keeping the land clear during the growing season but letting the weeds grow up towards the end. Treatments A, B, C, D and E represent the growing of various permanent crops under the trees. There were four randomized blocks. The response Y was the total crop weight in pounds over a four-year period after the treatments were begun. The trees were old and their crop sizes would be likely to vary considerably from one tree to the next. However, records were available of cropping before the experiment began. These were used to provide a covariate X, the total volume of crop in bushels over a four-year period before the new treatments began. It is instructive to compare the results of analysis with and without using the covariate.

| | Blocks | | | | | | | |
| | 1 | | 2 | | 3 | | 4 | |
Treatments	X	Y	X	Y	X	Y	X	Y
A	8.2	287	9.4	290	7.7	254	8.5	307
B	8.2	271	6.0	209	9.1	243	10.1	348
C	6.8	234	7.0	210	9.7	286	9.9	371
D	5.7	189	5.5	205	10.2	312	10.3	375
E	6.1	210	7.0	276	8.7	279	8.1	344
O	7.6	222	10.1	301	9.0	238	10.5	357

84. Lung cancer and occupation

Office of Population Censuses and Surveys (1978) *Occupational mortality: the Registrar General's decennial supplement for England and Wales, 1970–72*, Series DS, no. 1, London: HMSO, 149.

These data give, for males in England and Wales in 1970–72, the standardized mortality ratio for deaths from lung cancer for each of 25 'occupation orders' or broad groups of jobs. The population used for standardization was the male population of the whole of England and Wales. Also given are smoking ratios for each occupation order. The smoking ratio is a measure of cigarette consumption for an occupational order, again calculated using indirect standardization, so that in an order with smoking ratio 100, the men would smoke (per man per day) the same number of cigarettes that one would expect on the basis of its age structure and national age-specific smoking rates. What is the relationship between lung cancer SMR and smoking ratio?

Occupation order		Smoking ratio	Lung cancer SMR
I	Farmers, foresters, fishermen	77	84
II	Miners and quarrymen	137	116
III	Gas, coke and chemical makers	117	123
IV	Glass and ceramics makers	94	128
V	Furnace, forge, foundry, rolling mill workers	116	155
VI	Electrical and electronic workers	102	101
VII	Engineering and allied trades not included elsewhere	111	118
VIII	Woodworkers	93	113
IX	Leather workers	88	104
X	Textile workers	102	88
XI	Clothing workers	91	104
XII	Food, drink and tobacco workers	104	129
XIII	Paper and printing workers	107	86
XIV	Makers of other products	112	96
XV	Construction workers	113	144
XVI	Painters and decorators	110	139
XVII	Drivers of stationary engines, cranes, etc.	125	113
XVIII	Labourers not included elsewhere	133	146
XIX	Transport and communications workers	115	128
XX	Warehousemen, storekeepers, packers, bottlers	105	115
XXI	Clerical workers	87	79
XXII	Sales workers	91	85
XXIII	Service, sport and recreation workers	100	120
XXIV	Administrators and managers	76	60
XXV	Professional, technical workers, artists	66	51

85. House insulation: Whitburn

The Open University (1983) *MDST242 Statistics in Society, Unit C5: Review*, Milton Keynes: The Open University, Figure 2.2.

In the mid-1970s an experiment on insulation was carried out on an estate of housing in public ownership in Whitburn, Scotland. The control houses and flats were insulated to the usual standard at that time: no insulation in the wall cavities, and 25mm of insulation in the roof. The experimental houses had cavity insulation, and 100mm of roof insulation. The houses were heated by electricity. These data give the total electricity consumption in kWh for the year from June 1975 for the 15 two-bedroom houses which were constantly occupied throughout that period. How much energy did the extra insulation save?

Control houses	Experimental houses	Control houses	Experimental houses
10225	9708	8451	8017
10689	6700	12086	8162
14643	4307	12467	8022
6584	10315	12669	

86. House insulation: Bristol

The Open University (1983) *MDST242 Statistics in Society, Unit A0: Introduction*, 1st edition, Milton Keynes: The Open University, Figure 2.9.

These data were originally provided by the Electricity Council and give the total winter energy consumption in MWh of ten houses in the Fishponds area of Bristol, for a winter before the installation of cavity-wall insulation, and the winter energy consumption for the same ten houses for another winter after insulation was installed. How much energy does the insulation save? Is the saving proportional to the amount of energy used?

Before insulation	After insulation	Before insulation	After insulation
12.1	12.0	12.2	13.6
11.0	10.6	12.8	12.6
14.1	13.4	9.9	8.8
13.8	11.2	10.8	9.6
15.5	15.3	12.7	12.4

87. Engine capacity and fuel consumption

The Open University (1983) *MDST242 Statistics in Society, Unit A3: Relationships*, 1st edition, Milton Keynes: The Open University, Figure 1.2.

These data give the engine capacity in cc and the overall fuel consumption in miles per (imperial) gallon for nine models of car produced by the manufacturer Peugeot. They were collated from the *Motor* magazine for 8 May 1982. The 305GRD and 604D models had diesel engines; the others were petrol-driven.

Model	Engine capacity (cc)	Overall fuel consumption (mpg)
104ZL	954	33.2
104SL	1124	30.6
104ZS	1360	34.0
305SR	1472	25.8
305S	1472	29.2
305GRD	1548	37.6
505STi	1995	23.0
604SL	2664	18.0
604D	2304	26.4

88. House insulation: Whiteside's data

The Open University (1984) *MDST242 Statistics in Society, Unit A5: Review*, 2nd edition, Milton Keynes: The Open University, Figures 2.5 and 2.6.

These data were collected in the 1960s by Derek Whiteside of the UK Building Research Station. He recorded (among other things) the weekly gas consumption (in 1000 cubic feet) and the average outside temperature (in degrees Celsius) at his own house in south-east England, for 26 weeks before and 30 weeks after cavity-wall insulation had been installed. The house thermostat was set at 20°C throughout. The data are not given in chronological order. How is gas consumption related to outside temperature, and how did this relationship change after insulation? How much energy does the insulation save, allowing for the temperature difference between the inside and outside of the house?

Before insulation		*After insulation*	
Average outside temperature (°C)	*Gas consumption* (*1000 cubic feet*)	*Average outside temperature* (°C)	*Gas consumption* (*1000 cubic feet*)
−0.8	7.2	−0.7	4.8
−0.7	6.9	0.8	4.6
0.4	6.4	1.0	4.7
2.5	6.0	1.4	4.0
2.9	5.8	1.5	4.2
3.2	5.8	1.6	4.2
3.6	5.6	2.3	4.1
3.9	4.7	2.5	4.0
4.2	5.8	2.5	3.5
4.3	5.2	3.1	3.2
5.4	4.9	3.9	3.9
6.0	4.9	4.0	3.5
6.0	4.3	4.0	3.7
6.0	4.4	4.2	3.5
6.2	4.5	4.3	3.5
6.3	4.6	4.6	3.7
6.9	3.7	4.7	3.5
7.0	3.9	4.9	3.4
7.4	4.2	4.9	3.7
7.5	4.0	4.9	4.0
7.5	3.9	5.0	3.6
7.6	3.5	5.3	3.7
8.0	4.0	6.2	2.8
8.5	3.6	7.1	3.0
9.1	3.1	7.2	2.8
10.2	2.6	7.5	2.6
		8.0	2.7
		8.7	2.8
		8.8	1.3
		9.7	1.5

89. 1981 coffee prices

The Open University (1983) *MDST242 Statistics in Society, Unit A1: Prices*, 1st edition, Milton Keynes: The Open University, Figure 1.1.

These data are the prices in pence of a 100 g pack of a particular brand of instant coffee, on sale in 15 different shops in Milton Keynes on the same day in 1981. The source uses the data to exemplify methods of summarizing data.

100	109	101	93	96	104	98	97	95	107
102	104	101	99	102					

90. House temperatures: Neath Hill

The Open University (1984) *MDST242 Statistics in Society, Unit A5: Review*, 2nd edition, Milton Keynes: The Open University, Figure 2.13.

These data were collected by the Open University's Energy Research Group in the early 1980s. For each of 15 houses in the Neath Hill district of Milton Keynes, they recorded over a period of time the average temperature difference (in °C) between the inside and the outside of the house, and the average daily gas consumption (in kWh). How are these two quantities related?

Temperature difference (°C)	Daily gas consumption (kWh)
10.3	69.81
11.4	82.75
11.5	81.75
12.5	80.38
13.1	85.89
13.4	75.32
13.6	69.81
15.0	78.54
15.2	81.29
15.3	99.20
15.6	86.35
16.4	110.23
16.5	106.55
17.0	85.50
17.1	90.02

91. Sex differences at school

The Open University (1983) *MDST242 Statistics in Society, Unit B3: Education: does sex matter?*, Milton Keynes: The Open University, Figure 1.4.

These data were derived from figures given in Table 8 of the UK Department of Education and Science publication *Statistics of Education 1980, Volume 2, School-leavers*. They classify a sample of 749 students leaving schools in England in 1979/80 by sex and by achievement in public examinations in two subjects, Mathematics and French. At that time, there were two separate public examinations taken at around age 16, the Certificate of Secondary Education (CSE) and the General Certificate of Education Ordinary Level (O-level). O-levels were more challenging and more academic than CSEs, but a grade 1 pass at CSE was considered equivalent to an O-level pass. The four categories of achievement for each subject are as follows.

(a)   Did not attempt CSE or O-level.
(b)   Attempted CSE/O-level but did not pass.
(c)   Passed CSE at grades 2–5 or O-level at grades D–E.
(d)   Passed CSE at grade 1 or O-level at grades A–C.

What is the relationship, if any, between sex and achievement in each of the two subjects?

Mathematics

| | Achievement category | | | | |
	(a)	(b)	(c)	(d)	Total
Males	82	13	176	112	383
Females	74	23	184	85	366
Total	156	36	360	197	749

French

| | Achievement category | | | | |
	(a)	(b)	(c)	(d)	Total
Males	289	10	44	40	383
Females	221	8	74	63	366
Total	510	18	118	103	749

92. Destination of school leavers

The Open University (1983) *MDST242 Statistics in Society, Unit B3: Education: does sex matter?*, Milton Keynes: The Open University, Figure 3.7.

These data were derived from figures given in Table C12 of the UK Department of Education and Science publication *Statistics of Education 1980, Volume 2, School-leavers*. They classify a random sample of 115 students leaving schools in England in 1979/80 with at least one pass in the General Certificate of Education Advanced Level (A-level) public examinations (taken usually around age 18) by what they did after leaving school and by their number of A-level passes. What is the relationship, if any, between the destination of school leavers and their performance at A-level?

| | **Number of A-level passes** | | | | |
	1	**2**	**3**	**4 or more**	**Total**
Degree course	1	10	29	16	56
Other course	7	7	4	1	19
Employment	14	12	11	3	40
Total	22	29	44	20	115

93. House insulation: Pennyland

The Open University (1984) *MDST242 Statistics in Society, Unit B5: Review*, 2nd edition, Milton Keynes: The Open University, Figures 2.3, 2.7 and 2.8.

The Open University (1983) *MDST242 Statistics in Society, Unit C5: Review*, Milton Keynes: The Open University, Figure 3.1.

These four data sets were all collected from a large-scale experiment on energy conservation in domestic houses, carried out in the early 1980s in the Pennyland district of Milton Keynes. An estate of around 180 houses had been built. About half of them had a standard (*S*) level of insulation (70 mm of roof insulation and 50 mm of wall cavity insulation). The others had extra (*E*) insulation (120 mm of roof insulation, 100 mm of wall cavity insulation, double glazing and under-floor insulation). Energy consumption in the houses was monitored over a period of several years. As well as the differences in insulation, most of the houses had features designed to make more use of the sun's heat; most notably, they faced generally south and had most of their windows on the south side. These were known as single-aspect houses. Others had a more conventional design with more evenly spread window patterns. These were known as dual-aspect houses. The data give annual gas consumptions in kWh for various samples of Pennyland houses. In each case the object is to see how large a difference, if any, the design and insulation variations make to energy use.

Unrestricted random samples of houses with standard (S) and extra (E) insulation:

S	11400	13900	13900	14000	15300	18000	18000	18100	18900	19000
	19000	21700								

E	8300	11700	12700	13000	13400	13600	13700	13700	13800	14600
	15300	15600	16000	18800						

Random samples of single-aspect houses with standard (S) and extra (E) insulation:

S	12300	13300	13700	13800	14900	15600	15900	16300	16500	17200
	17500	17600	17800	17900	18000	19900				

E	10500	11300	11400	12600	13000	14500	15200	15700	15700	17600
	19000									

Paired samples of houses with standard (S) and extra (E) insulation, paired such that houses in the same pair have the same number of inhabitants:

S	E
13800	15100
17800	13900
18000	15900
17300	17200
16900	15200
19900	13800
13600	11300
17600	19000
15900	15400
17900	13200
12300	18800
11700	18900
18600	17100
16200	17500
14500	16400
18000	14000

Random samples of dual-aspect and single-aspect houses, all with standard (S) insulation

Dual	13670	15930	17180	15790	15490	14810	13310	17190	13730	17200
	22890	19800	12240							

Single	36340	12280	15740	14270	13960	13610	10520	31910	14570	11780
	17860	17710	15420	15630	18660	20710	8770			

94. Wave energy device mooring

The Open University (1984) *MDST242 Statistics in Society, Unit B5: Review*, 2nd edition, Milton Keynes: The Open University, Figure 2.5.

These data were collected as part of a design study for a device to generate electricity from wave power at sea. The study was carried out on scale models in a wave tank. The aim of the experiment was to establish how the choice of mooring method for the system affected the bending stress produced in part of the device. The wave tank could simulate a wide range of sea states. The model system was subjected to the same sample of 18 sea states with each of two mooring methods, one of which was considerably cheaper than the other. The resulting data (root mean square bending moment in Newton metres) were as follows.

Sea state	Mooring method 1	Mooring method 2
1	2.23	1.82
2	2.55	2.42
3	7.99	8.26
4	4.09	3.46
5	9.62	9.77
6	1.59	1.40
7	8.98	8.88
8	0.82	0.87
9	10.83	11.20
10	1.54	1.33
11	10.75	10.32
12	5.79	5.87
13	5.91	6.44
14	5.79	5.87
15	5.50	5.30
16	9.96	9.82
17	1.92	1.69
18	7.38	7.41

95. The effect of light on mustard root growth

The Open University (1983) *MDST242 Statistics in Society, Unit C2: Scientific experiments*, Milton Keynes: The Open University, Figure 3.7.

These data come from an experiment to investigate the effect of light on root growth in mustard seedlings. Two groups of seedlings were grown in identical conditions, except that one was kept in the dark while the other had daylight during the day. After germination the stems were cut off some of the seedlings, to allow for the possibility that light affected the vigour of the whole plant through the stem and leaves.

Later the root lengths (in mm) of all the seedlings were measured. Does light affect root growth; and does this effect depend on whether the stem is cut?

| Grown in the light | | Grown in the dark | |
Stems cut	Stems not cut	Stems cut	Stems not cut
21	27	22	21
39	21	16	39
31	26	20	20
13	12	14	24
52	11	32	20
39	8	28	
55		36	
50		41	
29		17	
17		22	

96. Height and weight of eleven-year-old girls

The Open University (1983) *MDST242 Statistics in Society, Unit C3: Is my child normal?*, Milton Keynes: The Open University, Figure 3.12.

The heights (cm) and weights (kg) of 30 eleven-year-old girls attending Heaton Middle School, Bradford, were measured. How are the two variables related?

Height (cm)	Weight (kg)	Height (cm)	Weight (kg)
135	26	133	31
146	33	149	34
153	55	141	32
154	50	164	47
139	32	146	37
131	25	149	46
149	44	147	36
137	31	152	47
143	36	140	33
146	35	143	42
141	28	148	32
136	28	149	32
154	36	141	29
151	48	137	34
155	36	135	30

97. Consulting the doctor about a child

Mapes, R. and Dajda, R. (1976) Children and general practitioners, the General Household Survey as a source of information. *Sociological Review Monographs*, **M22**, 99–110.

These data come from a study in Britain of possible influences on the extent to which parents consult a doctor when their child is ill. The authors used data from a survey, which asked parents how often each child had been ill in the past two weeks, and how often each child had visited the doctor, to calculate the percentage of each child's illnesses that were reported to the doctor. One of the potential influences considered was the child's age. Does age have an effect on the rate of reporting illness to doctors?

Child's age	0	1	2	3	4	5	6	7	8	9	10	11	12	13	14
% of illnesses reported to doctor	70	76	51	62	67	48	50	51	65	70	60	40	55	45	38

98. House heating: Great Linford

The Open University (1983) *MDST242 Statistics in Society, Unit C5: Review*, Milton Keynes: The Open University, Figure 3.2.

The Open University Energy Research Group ran a constant-heating experiment in an unoccupied test house in Great Linford, Milton Keynes. The house was well-insulated, faces south and has most of its glazing on the south side. For a ten-week period from February to May 1982, the house was heated to a constant 21°C using thermostatically controlled electric heaters. Among other things, the researchers measured the following three variables.

E The electrical energy (kWh per day) required to keep the house at 21°C.

S Energy from solar radiation (kWh per square metre per day) falling on a south-facing vertical surface at the house site.

TD Temperature difference (°C) between the inside and outside of the house.

The following data are weekly averages of each of these variables. How is E determined by the other two variables?

Week	S	TD	E
1	1.36	15.1	74.5
2	2.77	16.8	60.5
3	2.79	16.1	63.2
4	2.70	16.0	55.6

5	3.27	15.6	45.0
6	2.00	12.4	47.7
7	2.67	15.7	52.8
8	3.43	13.5	31.1
9	2.30	12.0	33.4
10	2.68	14.4	48.9

99. Births in Basel

Walser, P. (1969) Untersuchung über die Verteilung der Geburtstermine bei der mehrgebärenden Frau, *Helvetica Paediatrica Acta*, Suppl. XX ad vol. 24, fasc. 3, 1–30.

The following are the frequencies of births in each month of the year for the first births to 700 women in the University Hospital of Basel, Switzerland. Is there evidence that births are not spread evenly through the year?

Month	*No. of births*
January	66
February	63
March	64
April	48
May	64
June	74
July	70
August	59
September	54
October	51
November	45
December	42

100. Homing in desert ants

Duelli, P. and Wehner, R. (1973) The spectral sensitivity of polarized light orientation in *Cataglyphis bicolor* (Formicidae, Hymenoptera). *Journal of Comparative Physiology*, **86**, 37–53.

These data come from an experiment on the homing performance of a desert ant, *Cataglyphis bicolor*. Twenty ants were trained under the sun deflected by a mirror. Ten of them (controls) were observed returning to their nests under a sun of natural altitude. The other ten (experimentals) were observed returning under a sun with artificially increased altitude. The given data are the angles (in degrees) between the true home direction and the direction each ant went in. (Negative angles indicate that the ant was diverted anticlockwise from the true home direction.) Does increasing the altitude of the sun make a difference? The data are analysed in Section 6.2 of Batschelet, E. (1981) *Circular statistics in biology*, London: Academic Press.

| | *Frequencies observed* | |
Angle	*Experimental*	*Control*
−20	1	0
−10	7	3
0	2	3
+10	0	3
+20	0	1

101. Homing in pigeons

Batschelet, E. (1981) *Circular statistics in biology*, London: Academic Press, Table 6.6.1.

In an experiment carried out by F. Papi and co-workers, two groups of pigeons were taken from their loft near Siena to a location near Rome. One group, the controls, was transported in a container into which the natural air along the road was blown. The other group, the experimentals, received purified air. The left was 172 km from the release site, at a bearing of 325°. The observed directions in which the birds vanished are given below (in degrees; North is 0°). Does the type of air received affects the distribution of the direction pigeons fly in, and if so how?

Experimentals	*Controls*
107	247
109	153
186	202
121	264
171	24
4	228
110	333
82	192
131	
117	

102. Breast cancer mortality and temperature

Lea, A.J. (1965) New observations on distribution of neoplasms of female breast in certain European countries. *British Medical Journal*, **1**, 488–490.

These data were collected to investigate the relationship between mean annual temperature and the mortality rate for a type of breast cancer in women. They relate to certain regions of Great Britain, Norway and Sweden.

Mean annual temperature (°F)	Mortality index
51.3	102.5
49.9	104.5
50.0	100.4
49.2	95.9
48.5	87.0
47.8	95.0
47.3	88.6
45.1	89.2
46.3	78.9
42.1	84.6
44.2	81.7
43.5	72.2
42.3	65.1
40.2	68.1
31.8	67.3
34.0	52.5

103. Deaths from sport parachuting

Metropolitan Life Insurance Company (1979), *Statistical Bulletin*, **60**, no. 3, 4.

These data collected by an American insurance company, show the number of deaths from sport parachuting in 3 years in the 1970s. The deaths are classified by the experience of the parachutist, measured by his or her number of jumps. Does the pattern of experience change over time, and if so, how?

		Year	
Number of jumps	1973	1974	1975
1–24	14	15	14
25–74	7	4	7
75–199	8	2	10
200 or more	15	9	10
Unreported	0	2	0

104. Cotton imports in the 18th century

Mitchell, B.R. and Deane, P. (1962) *Abstract of British Historical Statistics*, Cambridge: Cambridge University Press, 177–178.

These data are a time series of raw cotton imports into the UK, by weight, for each year 1770–1800. The units are thousands of pounds (weight). How are the data best summarized? A log transformation may help.

Year	Imports	Year	Imports
1770	3612	1786	19475
1771	2547	1787	23250
1772	5307	1788	20467
1773	2906	1789	32576
1774	5707	1790	31448
1775	6694	1791	28707
1776	6216	1792	34907
1777	7037	1793	19041
1778	6569	1794	24359
1779	5861	1795	26401
1780	6877	1796	32126
1781	5199	1797	23354
1782	11828	1798	31881
1783	9736	1799	43379
1784	11482	1800	56011
1785	18400		

105. British exports in the 19th century

Mitchell, B.R. and Deane, P. (1962) *Abstract of British Historical Statistics*, Cambridge: Cambridge University Press, 282.

These data give the value (in millions of £) of British exports for each year 1820–1850.

Year	Exports	Year	Exports
1820	36.4	1836	53.3
1821	36.7	1837	42.1
1822	37.0	1838	50.1
1823	35.4	1839	53.2
1824	38.4	1840	51.4
1825	38.9	1841	51.6
1826	31.5	1842	47.4

1827	37.2	1843	52.3
1828	36.8	1844	58.6
1829	35.8	1845	60.1
1830	38.3	1846	57.8
1831	37.2	1847	58.8
1832	36.5	1848	52.8
1833	39.7	1849	63.6
1834	41.6	1850	71.4
1835	47.4		

106. Size of ships in 1907

Floud, R. (1973) *An introduction to quantitative methods for historians*, 2nd edition, London: Methuen, Table 4.1.

These data give the tonnage and crew size and type of power for 25 British merchant ships in 1907. They are taken from crew lists held by the Registrar General of Shipping. How are tonnage and crew size related, and is the relationship different for different types of power?

Tonnage	Crew size	Power
44	3	Not given
144	6	Not given
150	5	Not given
236	8	Sail
739	16	Steam
970	15	Steam
2371	23	Steam
309	5	Steam
679	13	Steam
26	4	Sail
1272	19	Steam
3246	33	Steam
1904	19	Steam
357	10	Steam
1080	16	Steam
1027	22	Steam
45	2	Not given
62	3	Not given
28	2	Sail
2507	22	Steam
138	2	Sail
502	18	Steam
1501	21	Steam
2750	24	Steam
192	9	Steam

107. Car sales in Quebec

Abraham, B. and Ledolter, J. (1983) *Statistical methods for forecasting*, New York: John Wiley & Sons, 420.

These data are a time series of monthly car sales (units) in Quebec, January 1960 to December 1968. How can the data be modelled or used for forecasting future sales? There is reasonably strong seasonality in the data.

	Jan	Feb	Mar	Apr	May	Jun
1960	6550	8728	12026	14395	14587	13791
1961	7237	9374	11837	13784	15926	13821
1962	10677	10947	15200	17010	20900	16205
1963	10862	10965	14405	20379	20128	17816
1964	12267	12470	18944	21259	22015	18581
1965	12181	12965	19990	23125	23541	21247
1966	12674	12760	20249	22135	20677	19933
1967	12225	11608	20985	19692	24081	22114
1968	13210	14251	20139	21725	26099	21084

	Jul	Aug	Sep	Oct	Nov	Dec
1960	9498	8251	7049	9545	9364	8456
1961	11143	7975	7610	10015	12759	8816
1962	12143	8997	5568	11474	12256	10583
1963	12268	8642	7962	13932	15936	12628
1964	15175	10306	10792	14752	13754	11738
1965	15189	14767	10895	17130	17697	16611
1966	15388	15113	13401	16135	17562	14720
1967	14220	13434	13598	17187	16119	13713
1968	18024	16722	14385	21342	17180	14577

108. Sales and advertising

Bass, F.M. and Clarke, D.G. (1972) Testing distributed lag models of advertising effect. *Journal of Marketing Research*, **9**, 298–308.

These data consist of 36 consecutive monthly sales and monthly advertising expenditures for a dietary weight control product. The problem is to model the relationship between the variables, given that the effect of advertising in one period carries over into subsequent periods.

Sales	Advertising	Sales	Advertising
12.0	15.0	30.5	33.0
20.5	16.0	28.0	62.0
21.0	18.0	26.0	22.0
15.5	27.0	21.5	12.0
15.3	21.0	19.7	24.0
23.5	49.0	19.0	3.0
24.5	21.0	16.0	5.0
21.3	22.0	20.7	14.0
23.5	28.0	26.5	36.0
28.0	36.0	30.6	40.0
24.0	40.0	32.3	49.0
15.5	3.0	29.5	7.0
17.3	21.0	28.3	52.0
25.3	29.0	31.3	65.0
25.0	62.0	32.2	17.0
36.5	65.0	26.4	5.0
36.5	46.0	23.4	17.0
29.6	44.0	16.4	1.0

109. Canadian lynx trappings

Elton, C. and Nicholson, M. (1942) The ten-year cycle in numbers of the lynx in Canada. *Journal of Animal Ecology*, **11**, 215–244.

These data give the annual number of Canadian lynx trapped in the Mackenzie River district of north-west Canada for each year 1821–1934, according to the records of the Hudson's Bay Company. They have been analysed very many times in the time series literature. An interesting review and re-analysis is in Section 7.2 of Tong, H. (1990) *Nonlinear time series*, Oxford: Oxford University Press. There is evidence that the standard linear Gaussian process models do not describe these data adequately.

(Read across)

269	321	585	871	1475	2821	3928	5943	4950	2577
523	98	184	279	409	2285	2685	3409	1824	409
151	45	68	213	546	1033	2129	2536	957	361
377	225	360	731	1638	2725	2871	2119	684	299
236	245	552	1623	3311	6721	4254	687	255	473
358	784	1594	1676	2251	1426	756	299	201	229
469	736	2042	2811	4431	2511	389	73	39	49
59	188	377	1292	4031	3495	587	105	153	387
758	1307	3465	6991	6313	3794	1836	345	382	808
1388	2713	3800	3091	2985	3790	374	81	80	108
229	399	1132	2432	3574	2935	1537	529	485	662
1000	1590	2657	3396						

110. Jet fighters

Stanley, W. and Miller, M. (1979) Measuring technological change in jet fighter aircraft. Report no. R-2249-AF, Rand Corporation, Santa Monica, California.

These data give the values of six variables for 22 US fighter aircraft. The variables are as follows.

FFD   first flight date, in months after January 1940
SPR   specific power, proportional to power per unit weight
RGF   flight range factor
PLF   payload as a fraction of gross weight of aircraft
SLF   sustained load factor
CAR   a binary variable which takes the value 1 if the aircraft can land on a carrier, 0 otherwise.

The aim is to model *FFD* (or some transformation of it) as a function of the other variables. The original authors model the log of *FFD*, as do Cook, R.D. and Weisberg, S. (1982) *Residuals and influence in regression*, New York: Chapman and Hall, who also analyse these data fully.

Type	*FFD*	*SPR*	*RGF*	*PLF*	*SLF*	*CAR*
FH-1	82	1.468	3.30	0.166	0.10	0
FJ-1	89	1.605	3.64	0.154	0.10	0
F-86A	101	2.168	4.87	0.177	2.90	1
F9F-2	107	2.054	4.72	0.275	1.10	0
F-94A	115	2.467	4.11	0.298	1.00	1
F3D-1	122	1.294	3.75	0.150	0.90	0
F-89A	127	2.183	3.97	0.000	2.40	1
XF10F-1	137	2.426	4.65	0.117	1.80	0
F9F-6	147	2.607	3.84	0.155	2.30	0
F-100A	166	4.567	4.92	0.138	3.20	1
F4D-1	174	4.588	3.82	0.249	3.50	0
F11F-1	175	3.618	4.32	0.143	2.80	0
F-101A	177	5.855	4.53	0.172	2.50	1
F3H-2	184	2.898	4.48	0.178	3.00	0
F-102A	187	3.880	5.39	0.101	3.00	1
F-8A	189	0.455	4.99	0.008	2.64	0
F-104B	194	8.088	4.50	0.251	2.70	1
F-105B	197	6.502	5.20	0.366	2.90	1
YF-107A	201	6.081	5.65	0.106	2.90	1
F-106A	204	7.105	5.40	0.089	3.20	1
F-4B	255	8.548	4.20	0.222	2.90	0
F-111A	328	6.321	6.45	0.187	2.00	1

111. Head size in brothers

Frets, G.P. (1921) Heredity of head form in man. *Genetica*, **3**, 193–384.

These data give two head measurements (in mm) for each of the first two adult sons in 25 families. The question of interest is the relationship (if any) between the vectors of head measurements for pairs of sons.

Head length of first son	Head breadth of first son	Head length of second son	Head breadth of second son
191	155	179	145
195	149	201	152
181	148	185	149
183	153	188	149
176	144	171	142
208	157	192	152
189	150	190	149
197	159	189	152
188	152	197	159
192	150	187	151
179	158	186	148
183	147	174	147
174	150	185	152
190	159	195	157
188	151	187	158
163	137	161	130
195	155	183	158
186	153	173	148
181	145	182	146
175	140	165	137
192	154	185	152
174	143	178	147
176	139	176	143
197	167	200	158
190	163	187	150

112. Sunspots

Box, G.E.P. and Jenkins, G.M. (1970) *Time series analysis*, San Francisco: Holden-Day, 530.

These data are a time series of the annual Wölfer sunspot numbers for each year 1770–1869. They measure the average number of sunspots on the sun during each year. The aim is to find an appropriate model for the data. Box and Jenkins suggest a second or third order autoregressive model, but numerous other suggestions have appeared in the literature.

(Read across)

101	82	66	35	31	7	20	92	154	125
85	68	38	23	10	24	83	132	131	118
90	67	60	47	41	21	16	6	4	7
14	34	45	43	48	42	28	10	8	2
0	1	5	12	14	35	46	41	30	24
16	7	4	2	8	17	36	50	62	67
71	48	28	8	13	57	122	138	103	86
63	37	24	11	15	40	62	98	124	96
66	64	54	39	21	7	4	23	55	94
96	77	59	44	47	30	16	7	37	74

113. Airline passenger numbers

Brown, R.G. (1963) *Smoothing, forecasting and prediction of discrete time series*, New Jersey: Prentice-Hall, Table C.10.

These data give monthly totals of international airline passengers for January 1949–December 1960. The aim is to provide an appropriate model that takes account of the considerable seasonal variation. The data have been analysed many times in the time series literature, notably in Chapter 9 of Box, G.E.P. and Jenkins, G.M. (1970) *Time series analysis*, San Francisco: Holden-Day. The counts are in thousands.

	Jan	Feb	Mar	Apr	May	Jun	Jul	Aug	Sep	Oct	Nov	Dec
1949	112	118	132	129	121	135	148	148	136	119	104	118
1950	115	126	141	135	125	149	170	170	158	133	114	140
1951	145	150	178	163	172	178	199	199	184	162	146	166
1952	171	180	193	181	183	218	230	242	209	191	172	194
1953	196	196	236	235	229	243	264	272	237	211	180	201
1954	204	188	235	227	234	264	302	293	259	229	203	229
1955	242	233	267	269	270	315	364	347	312	274	237	278
1956	284	277	317	313	318	374	413	405	355	306	271	306
1957	315	301	356	348	355	422	465	467	404	347	305	336
1958	340	318	362	348	363	435	491	505	404	359	310	337
1959	360	342	406	396	420	472	548	559	463	407	362	405
1960	417	391	419	461	472	535	622	606	508	461	390	432

114. Yields from a batch chemical process

Box, G.E.P. and Jenkins, G.M. (1970) *Time series analysis*, San Francisco: Holden-Day, Table 2.1.

These data are the yields from 70 consecutive runs of a batch chemical process. There is autocorrelation because of 'carry-over' effects: a high-yielding batch tended to produce tarry residues which were not entirely removed from the vessel and adversely affected the yield on the next batch. The aim is to produce a satisfactory model for the data. Box and Jenkins fit an autoregressive model of order 2.

(Read across)

47	64	23	71	38	64	55	41	59	48
71	35	57	40	58	44	80	55	37	74
51	57	50	60	45	57	50	45	25	59
50	71	56	74	50	58	45	54	36	54
48	55	45	57	50	62	44	64	43	52
38	59	55	41	53	49	34	35	54	45
68	38	50	60	39	59	40	57	54	23

115. Distances in Sheffield

Gilchrist, W. (1984) *Statistical modelling*, Chichester: John Wiley & Sons, 5.

These data give the distance by road and the straight line distance between twenty different pairs of points in Sheffield. What is the relationship between the two variables? How well can the road distance be predicted from the linear distance?

Road distance	Linear distance	Road distance	Linear distance
10.7	9.5	11.7	9.8
6.5	5.0	25.6	19.0
29.4	23.0	16.3	14.6
17.2	15.2	9.5	8.3
18.4	11.4	28.8	21.6
19.7	11.8	31.2	26.5
16.6	12.1	6.5	4.8
29.0	22.0	25.7	21.7
40.5	28.2	26.5	18.0
14.2	12.1	33.1	28.0

116. University of Iowa enrolments

Abraham, B. and Ledolter, J. (1983) *Statistical methods for forecasting*, New York: John Wiley & Sons, 116.

These data are a time series of total annual student enrolment at the University of Iowa, for each year from 1951/52 to 1970/80. How can they be used to forecast future enrolment?

(Read across)

14348	14307	15197	16715	18476	19404
20173	20645	20937	21501	22788	23579
25319	28250	32191	34584	36366	37865
39173	40119	39626	39107	39796	41567
43646	43534	44157	44551	45572	

117. Insects in dunes

Gilchrist, W. (1984) *Statistical modelling*, Chichester: John Wiley & Sons, 132.

A series of 33 insect traps was set out across sand dunes and the numbers of different insects caught in a fixed time were counted. The numbers of traps containing various numbers of insects of two different taxa are shown. If the insects of these taxa are spread randomly across the dunes, the distribution of the number of insects in a trap would be Poisson. Is this hypothesis plausible for either or both taxa?

Frequency of traps:

	Individuals in a trap						
Taxa	**0**	**1**	**2**	**3**	**4**	**5**	**6**
Staphylinoidea	10	9	5	5	1	2	1
Hemiptera	6	8	12	4	3		

118. Women's heights in Bangladesh

Huffman, S.L., Alauddin Chowdhury, A.K.M. and Mosley, W. (1979) Difference between postpartum and nutritional amenorrhea (reply to Frisch and McArthur). *Science*, **203**, 922–923.

These data come from a report on two samples of women in Bangladesh. The table below gives the heights (in cm) of the women in one of the samples in grouped form. What distribution models these data? The unequal bin widths are a complication.

Height (cm)	Frequency
<140.0	71
140.0–142.9	137
143.0–144.9	154
145.0–146.9	199
147.0–149.9	279
150.0–152.9	221
153.0–154.9	94
155.0–156.9	51
>156.9	37
Total	1243

119. Size of families in California

Burks, B.S. (1933) A statistical method for estimating the distribution of sizes of completed fraternities in a population represented by a random sampling of individuals. *Journal of the American Statistical Association*, **28**, 388–394.

These data give the birth order of each of 1800 students enrolled in elementary psychology classes at the University of California between 1924 and 1929. The aim of Burks' analysis is to estimate the distribution of family size from these data. The data are also analysed on pages 98–100 of Guttorp, P. (1991) *Statistical inference for branching processes*, New York: John Wiley & Sons.

Birth order	Frequency
1	797
2	455
3	265
4	125
5	68
6	37
7	26
8	8
9	1
10	9
11	5
12	3
13	1

120. Hutterite population structure

Guttorp, P. (1991) *Statistical inference for branching processes*, New York: John Wiley & Sons, 194.

The Hutterite Brethren are a religious group living on communal farms in parts of Canada and the USA. They migrated from Europe to North America in the 1870s. The population is essentially closed to in-migration, and practically all marriages are within the group. The data here give the frequencies of each family composition by sex for the offspring of all 1236 married Dariusleut or Lehrerleut Hutterite women born between 1879 and 1936 with at least one child. How do the distributions of male and female offspring differ?

		Number of daughters											
		0	1	2	3	4	5	6	7	8	9	10	11	12
	0		28	8	2	3	2	2	0	0	0	0	1	0
	1	21	29	21	11	5	7	6	4	1	0	0	0	1
	2	11	27	22	21	21	14	10	15	5	3	1	0	0
	3	6	16	27	20	35	29	18	12	10	2	2	0	0
	4	9	10	20	21	39	28	30	24	10	2	5	1	0
Number	5	3	7	22	22	40	17	18	23	16	7	2	0	0
of sons	6	2	9	15	16	27	26	26	17	10	4	1	0	0
	7	1	4	7	27	19	20	16	7	2	2	0	0	0
	8	0	3	12	14	12	7	10	5	3	1	0	0	0
	9	0	2	4	8	11	4	5	2	0	0	0	0	0
	10	0	1	1	2	3	2	2	1	1	0	0	0	0
	11	0	0	1	2	0	1	0	0	0	0	0	0	0
	12	0	0	1	1	0	0	0	0	0	0	0	0	0

121. Whooping cranes

Miller, R.S., Botkin, D.B. and Mendelssohn, R. (1974) The whooping crane (*Grus americana*) population of North America. *Biological Conservation*, **6**, 106–111.

The whooping crane is a very rare migratory bird which breeds in the north of Canada and winters in Texas. These data are annual counts for each year 1938–1972 of the number of whooping cranes arriving in Texas each autumn. The young birds born each year have a distinctive plumage and are recorded separately. Can the data be modelled by a discrete-time birth and death process?

Year	Young	Adult	Total	Year	Young	Adult	Total
1938	4	10	14	1956	2	22	24
1939	6	16	22	1957	4	22	26
1940	5	21	26	1958	9	23	32
1941	2	13	15	1959	2	31	33
1942	4	15	19	1960	6	30	36
1943	5	16	21	1961	5	33	38
1944	3	15	18	1962	0	32	32
1945	3	14	17	1963	7	26	33
1946	3	22	25	1964	10	32	42
1947	6	25	31	1965	8	36	44
1948	3	27	30	1966	5	38	43
1949	4	30	34	1967	9	39	48
1950	5	26	31	1968	6	44	50
1951	5	20	25	1969	8	48	56
1952	2	19	21	1970	6	51	57
1953	3	21	24	1971	5	51	56
1954	0	21	21	1972	5	46	51
1955	8	20	28				

122. Pork and cirrhosis

Nanji, A.A. and French, S.W. (1985) Relationship between pork consumption and cirrhosis. *The Lancet*, **i**, 681–683.

These data come from an investigation of whether pork consumption plays a role in the causation of cirrhosis of the liver. They give data on cirrhosis mortality, alcohol consumption and pork consumption for each of the 10 provinces of Canada in 1978. The question of interest is whether these data provide any evidence that pork consumption is related to cirrhosis mortality, in the light of the well-known causal relationship (at the individual level) between alcohol consumption and cirrhosis. The authors suggest leaving Prince Edward Island and Newfoundland out, on the grounds that "more seafood is eaten". They do not suggest leaving Quebec out, although it is clearly very different from the other provinces in pork consumption as in many other ways.

The original paper is a magnificent vehicle for demonstrating misleading and spurious statistical argument.

Province	Cirrhosis mortality (per 100,000)	Alcohol consumption (l/caput/year)	Pork consumption (kg/caput/year)
Prince Edward Island	6.5	11.00	5.8
Newfoundland	10.2	10.68	6.8
Nova Scotia	10.6	10.32	3.6
Saskatchewan	13.4	10.14	4.3

New Brunswick	14.5	9.23	4.4
Alberta	16.4	13.05	5.7
Manitoba	16.6	10.68	6.9
Ontario	18.2	11.50	7.2
Quebec	19.0	10.46	14.9
British Columbia	27.5	12.82	8.4

123. Pet birds and lung cancer

Kohlmeier, L., Arminger, G., Bartolomeycik, S., Bellach, B., Rehm, J. and Thamm, M. (1992) Pet birds as an independent risk factor for lung cancer: case-control study. *British Medical Journal*, **305**, 986–989.

These data come from a case-control study in which the cases were lung cancer patients in Berlin and the controls were members of the general population of Berlin, matched to the cases for age and sex. The aim of the study was to investigate whether keeping a pet bird is a risk factor for lung cancer, independent of other known risk factors (particularly cigarette smoking). The data give the numbers of cases and of controls who had kept a pet bird during adulthood. (Obviously on the basis of these data alone, one cannot investigate whether bird keeping is an *independent* risk factor. The authors concluded that it is, though others have disagreed.)

	Cases	*Controls*
Kept pet bird(s)	98	101
Did not keep pet bird(s)	141	328
Total	239	429

124. Water injections and whiplash injuries

Byrn, C., Olsson, I., Falkheden, L., Lindh, M., Hösterey, U., Fogelberg, M., Linder, L.-E. and Bunketorp, O. (1993) Subcutaneous sterile water injections for chronic neck and shoulder pain following whiplash injuries. *The Lancet*, **341**, 449–452.

These data come from a randomized controlled trial of subcutaneous sterile water injections in the neck and shoulders as a treatment for chronic pain resulting from so-called 'whiplash' injuries. These injuries are caused by sudden jerky head movements of people in vehicles hit by another vehicle from behind. Of 40 patients who had been suffering chronic pain for 4–6 years after their accidents, 20 were randomized to the sterile water treatment and 20 to the control. The control treatment consisted of subcutaneous injections of saline, though using these did not 'blind' either the doctor giving the injections or the patients to the treatment. (Saline injections do not hurt, while sterile water injections are painful, and this is obvious to the person giving the injection as well as the patient.)   These data report the patients' own assessments of the longer-term effectiveness of the treatments, 3 months and 8 months after treatment.   They were asked on each occasion whether their condition was unchanged, improved, or much improved.   Is there evidence that the treatment works?

Treatment	No change	Improved	Much improved
3 months			
Sterile water	1	9	10
Saline	14	3	3
8 months			
Sterile water	9	5	6
Saline	12	5	3

125. Telling cancer patients about their disease

Thomsen, O.Ø., Wulff, H.R., Martin, A. and Singer, P.A. (1993) What do gastroenterologists in Europe tell cancer patients? *The Lancet*, **341**, 473–476.

A questionnaire was sent to about 600 gastroenterologists in 27 European countries, asking what they would tell a patient with newly diagnosed cancer of the colon, and his/her spouse, about the diagnosis.   The respondent gastroenterologists were asked to read a brief case history and then to answer 7 questions with a yes/no answer (with some indication of what, if anything, they would tell the patient or spouse).   The questions were as follows.

1 Would you tell this patient that he/she has a cancer, if he/she asks no questions?
2 Would you tell the wife/husband that the patient has a cancer? [in the patient's absence]
3 Would you tell the patient ... that he/she has a cancer, if he/she directly asks you to disclose the diagnosis?
(During surgery, the surgeon notices several small metastases in the liver)
4 Would you tell the patient about [the metastases] (supposing that the patient asks to be told the result of the operation)?
5 Would you tell the patient that the condition is incurable?
6 Would you tell the wife/husband that the operation revealed metastases?
7 Would you tell the wife/husband that the condition is incurable?

These data give, for each of the countries involved, the number of yes answers and the total number of respondents for each question apart from Question 7.   (No data are presented on Question 7 in the source.)   For example, 6/7 means that 6 out of 7 respondents answered Yes and 1 answered No.   Since the sample for the survey was not random, and since individual respondent's vectors of responses are not given, detailed statistical inference is probably futile.   How should the data be presented in order to make their pattern clear?

125. Telling cancer patients about their disease

Country	No. of respondents	Questions					
		1	2	3	4	5	6
Iceland	5	5/5	5/5	4/4	5/5	5/5	5/5
Norway	7	6/7	5/6	6/6	5/5	4/4	4/5
Sweden	11	11/11	7/11	7/7	11/11	5/10	6/9
Finland	6	6/6	4/6	6/6	6/6	5/6	4/6
Denmark	14	12/13	9/13	12/12	9/12	4/11	7/13
UK	20	12/19	16/18	20/20	19/20	10/19	17/17
Ireland	3	1/1	2/3	3/3	0/2	0/2	3/3
Germany	14	14/14	13/13	13/13	12/14	2/13	13/14
Netherlands	8	8/8	8/8	8/8	7/8	5/7	7/8
Belgium	2	0/2	2/2	2/2	1/2	0/2	2/2
Switzerland	5	5/5	5/5	4/4	2/4	0/5	4/4
France	10	3/10	7/8	5/8	2/10	0/10	7/8
Spain	12	1/12	12/12	8/10	6/11	0/11	11/11
Portugal	6	1/6	6/6	4/6	3/6	0/6	6/6
Italy	15	7/15	15/15	13/14	6/15	2/15	15/15
Greece	8	1/8	8/8	5/8	1/8	0/8	7/7
Yugoslavia	15	4/15	15/15	8/15	4/15	0/15	14/14
Albania	5	2/5	3/5	2/5	2/5	3/5	3/5
Bulgaria	3	0/3	3/3	1/3	0/3	0/3	3/3
Romania	7	0/7	6/6	1/7	1/7	1/7	7/7
Hungary	5	1/5	5/5	4/5	0/5	0/5	5/5
Czechoslovakia	36	2/33	34/35	3/34	0/36	0/33	20/35
Poland	19	0/18	19/19	5/19	2/19	0/18	18/19
Russia/Ukraine	8	0/7	6/7	2/7	0/7	0/7	6/7
Lithuania	8	0/8	8/8	0/8	0/8	0/8	8/8
Latvia	5	0/5	5/5	0/5	0/5	0/5	4/4
Estonia	3	2/3	3/3	3/3	0/3	0/3	3/3

126. Diet and neural tube defects

James, N., Laurence, K.M. and Miller, M. (1980) Diet as a factor in the aetiology of neural tube malformations. *Zeitschrift für Kinderchirurgie*, **31**, 302–307.

Neural tube defects are certain congenital malformations of the backbone, spinal cord, skull and brain, that arise during foetal development.   (The best-known such malformation is spina bifida.) These data come from two studies, carried out in Wales, to investigate whether the chance of a neural tube defect in a foetus was related to the mother's diet. In the first study, which had a case-control design, mothers who had had a baby with a neural tube defect (the cases) and their sisters who had not had such a baby were interviewed about their present diet, which was then rated by nutritionists as good, fair or poor in quality. The second study had a prospective cohort design. The cohort was a group of women who had already had a baby with a neural tube defect and had become pregnant again. They had all received counselling with the aim of improving their diet. Their diet was recorded during the first three months of their pregnancy, and classified as good, fair or poor. The outcome of the pregnancy was recorded.

Case-control study

| | **Quality of diet** | | | |
	Good	**Fair**	**Poor**	**Total**
Cases	34	110	100	244
Controls (sisters)	43	48	32	123

Cohort study

| **Outcome of pregnancy** | **Quality of diet** | | | |
	Good	**Fair**	**Poor**	**Total**
Normal	68	76	27	171
neural tube defect	0	0	5	5

127. Age at marriage in Guatemala

Scholl, T.O., Odell, M.E. and Johnston, F.E. (1976) Biological correlates of modernization in a Guatemalan highland municipio. *Annals of Human Biology*, **3**, 23–32.

A number of married women from a district in Guatemala, all born during the period 1935–44, were interviewed in 1969 and asked the age at which they were married. The data are given in grouped form below. How would you model this distribution?

Age (completed years) at marriage	Frequency
9, 10	5
11, 12	11
13, 14	18
15, 16	28
17, 18	8
19, 20	7
21, 22	4
23, 24	5
25, 26	2
27, 28	0
29, 30	1
31, 32	0
33, 34	1

128. Sandflies

Christensen, H.A., Herrer, A. and Telford, S.R. (1972) Enzootic cutaneous leishmaniasis in Eastern Panama. 2. Entomological investigations. *Annals of tropical medicine and parasitology*, **66**, 55–66.

The data show the numbers of male and female sandflies caught in light traps set 3 ft and 35 ft above the ground at a site in eastern Panama. Does the proportion of males vary with height?

	Height above ground	
	3 ft	*35 ft*
Males	173	125
Females	150	73
Total	323	198

129. Tonsil size and carrier status

Krzanowski, W.J. (1988) *Principles of multivariate analysis*, Oxford: Oxford University Press, 269.

Some individuals are carriers of the bacterium *Streptococcus pyogenes*. To investigate whether there is a relationship between carrier status and tonsil size in schoolchildren, 1398 children were examined and classified according to their carrier status and tonsil size.

| | Carrier status | | |
	Carrier	Non-carrier	Total
Tonsil size Normal	19	497	516
Large	29	560	589
Very large	24	269	293
Total	72	1326	1398

130. Duration of pregnancy

Newell, D.J. (1964) Statistical aspects of the demand for maternity beds. *Journal of the Royal Statistical Society , Series A*, **127**, 1–33.

These data give the duration of pregnancy for 1669 women who gave birth in a maternity hospital in Newcastle-upon-Tyne, England, in 1954. The durations are measured in completed weeks from the beginning of the last menstrual period until delivery. The data were collected at a time when about one-third of English births took place outside hospitals, so that various criteria of suitability for admission were applied. The pregnancies are thus divided into those where an admission was booked for medical reasons, those booked for social reasons (such as poor housing), and unbooked emergency admissions. How do the durations differ between these three categories of admission?

Numbers of pregnancies by admission type and duration:

| | Admission type | | |
Duration (weeks)	**Medical**	**Emergency**	**Social**
11		1	
12			
13			
14			
15		1	
16			
17	1		
18			
19			
20		1	
21			
22	1	2	
23			
24	1	3	
25		2	1
26		1	
27	2	2	1
28	1	2	1
29	3	1	
30	3	5	1

31	4	5	2
32	10	9	2
33	6	6	2
34	12	7	10
35	23	11	4
36	26	13	19
37	54	16	30
38	68	35	72
39	159	38	115
40	197	32	155
41	111	27	128
42	55	25	64
43	29	8	16
44	4	5	3
45	3	1	6
46	1	1	1
47	1		
.			
.			
.			
56		1	
Total	775	261	633

131. 20 000 die throws

Wolf, R. (1882), *Vierteljahresschrift Naturforsch. Ges. Zürich*, **207**, 242.

Wolf tossed a die 20 000 times and recorded the number of times each of the six different faces showed. Was the die fair?

Face	1	2	3	4	5	6
Frequency	3407	3631	3176	2916	3448	3422

132. Iron in slag

Roberts, H.V. and Ling, R.F. (1982) *Conversational statistics with IDA*, New York: Scientific Press/ McGraw-Hill, 11-4.

The iron content of crushed blast-furnace slag can be determined by a chemical test at a laboratory or estimated by a cheaper, quicker magnetic test. These data were collected to investigate the extent to which the results of a chemical test of iron content can be predicted from a magnetic test of iron content, and the nature of the relationship between these quantities. The observations are given in order of the time they were made: does this have any effect?

Chemical	Magnetic	Chemical	Magnetic
24	25	20	21
16	22	20	21
24	17	25	21
18	21	27	25
18	20	22	22
10	13	20	18
14	16	24	21
16	14	24	18
18	19	23	20
20	10	29	25
21	23	27	20
20	20	23	18
21	19	19	19
15	15	25	16
16	16	15	16
15	16	16	16
17	12	27	26
19	15	27	28
16	15	30	28
15	15	29	30
15	15	26	32
13	17	25	28
24	18	25	36
22	16	32	40
21	18	28	33
24	22	25	33
15	20		

133. Lowering blood pressure during surgery

Robertson, J.D. and Armitage, P. (1959) Comparison of two hypotensive agents. *Anaesthesia*, **14**, 53–64.

It is sometimes necessary to lower a patient's blood pressure during surgery, using a hypotensive drug. Such drugs are administered continuously during the relevant phases of the operation; since the duration of this phase varies, so does the total amount of drug administered. Patients also vary in the extent to which the drugs succeed in lowering blood pressure. The sooner the blood pressure rises again to normal after the drug is discontinued, the better. These data relate to one particular hypotensive drug, and give the time in minutes before the patient's systolic blood pressure returned to 100 mm of mercury (the *recovery time*), the logarithm (base 10) of the dose of the drug in milligrams, and the average systolic blood pressure achieved while the drug was being administered. How is the recovery time related to the other two variables?

Log of dose	BP during administration	Recovery time	Log of dose	BP during administration	Recovery time
2.26	66	7	2.70	73	39
1.81	52	10	1.90	56	28
1.78	72	18	2.78	83	12
1.54	67	4	2.27	67	60
2.06	69	10	1.74	84	10
1.74	71	13	2.62	68	60
2.56	88	21	1.80	64	22
2.29	68	12	1.81	60	21
1.80	59	9	1.58	62	14
2.32	73	65	2.41	76	4
2.04	68	20	1.65	60	27
1.88	58	31	2.24	60	26
1.18	61	23	1.70	59	28
2.08	68	22	2.45	84	15
1.70	69	13	1.72	66	8
1.74	55	9	2.37	68	46
1.90	67	50	2.23	65	24
1.79	67	12	1.92	69	12
2.11	68	11	1.99	72	25
1.72	59	8	1.99	63	45
1.74	68	26	2.35	56	72
1.60	63	16	1.80	70	25
2.15	65	23	2.36	69	28
2.26	72	7	1.59	60	10
1.65	58	11	2.10	51	25
1.63	69	8	1.80	61	44
2.40	70	14			

134. Crime in the USA: 1960

Vandaele, W. (1978) Participation in illegitimate activities: Erlich revisited. In *Deterrence and incapacitation*, Blumstein, A., Cohen, J. and Nagin, D., eds., Washington. D.C.: National Academy of Sciences, 270–335.

These data are for 47 states of the USA. The aim is to investigate how the crime rate R in 1960 depends on the other variables listed. The data originally came from the *Uniform Crime Report* of the FBI and other US government sources, and relate to the calendar year 1960 except for variable Ex_1.

R Crime rate: number of offences known to the police per 1 000 000 population.

Age Age distribution: the number of males aged 14–24 per 1000 of total state population.

S Binary variable distinguishing southern states ($X_2 = 1$) from the rest.

Ed   Educational level: mean number of years of schooling $\times$ 10 of the population, 25 years old and over.

Ex_0 Police expenditure: per capita expenditure on police protection by state and local government in 1960.

Ex_1 Police expenditure: as Ex_0, but for 1959.

LF   Labour force participation rate per 1000 civilian urban males in the age group 14–24.

M The number of males per 1000 females.

N State population size in hundred thousands.

NW   The number of nonwhites per 1000.

U_1 Unemployment rate of urban males per 1000 in the age group 14–24.

U_2 Unemployment rate of urban males per 1000 in the age group 35–39.

W Wealth as measured by the median value of transferable goods and assets or family income (unit 10 dollars)

X Income inequality: the number of families per 1000 earning below one half of the median income.

134. Crime in the USA: 1960

R	Age	S	Ed	Ex_0	Ex_1	LF	M	N	NW	U_1	U_2	W	X
79.1	151	1	91	58	56	510	950	33	301	108	41	394	261
163.5	143	0	113	103	95	583	1012	13	102	96	36	557	194
57.8	142	1	89	45	44	533	969	18	219	94	33	318	250
196.9	136	0	121	149	141	577	994	157	80	102	39	673	167
123.4	141	0	121	109	101	591	985	18	30	91	20	578	174
68.2	121	0	110	118	115	547	964	25	44	84	29	689	126
96.3	127	1	111	82	79	519	982	4	139	97	38	620	168
155.5	131	1	109	115	109	542	969	50	179	79	35	472	206
85.6	157	1	90	65	62	553	955	39	286	81	28	421	239
70.5	140	0	118	71	68	632	1029	7	15	100	24	526	174
167.4	124	0	105	121	116	580	966	101	106	77	35	657	170
84.9	134	0	108	75	71	595	972	47	59	83	31	580	172
51.1	128	0	113	67	60	624	972	28	10	77	25	507	206
66.4	135	0	117	62	61	595	986	22	46	77	27	529	190
79.8	152	1	87	57	53	530	986	30	72	92	43	405	264
94.6	142	1	88	81	77	497	956	33	321	116	47	427	247
53.9	143	0	110	66	63	537	977	10	6	114	35	487	166
92.9	135	1	104	123	115	537	978	31	170	89	34	631	165
75.0	130	0	116	128	128	536	934	51	24	78	34	627	135
122.5	125	0	108	113	105	567	985	78	94	130	58	626	166
74.2	126	0	108	74	67	602	984	34	12	102	33	557	195
43.9	157	1	89	47	44	512	962	22	423	97	34	288	276
121.6	132	0	96	87	83	564	953	43	92	83	32	513	227
96.8	131	0	116	78	73	574	1038	7	36	142	42	540	176
52.3	130	0	116	63	57	641	984	14	26	70	21	486	196
199.3	131	0	121	160	143	631	1071	3	77	102	41	674	196
34.2	135	0	109	69	71	540	965	6	4	80	22	564	152
121.6	152	0	112	82	76	571	1018	10	79	103	28	537	215

154	637	36	92	89	168	938	521	157	166	107	0	119	104.3
237	396	26	72	254	46	973	521	54	58	89	1	166	69.6
200	453	40	135	20	6	1045	535	54	55	93	0	140	37.3
163	617	43	105	82	97	964	586	81	90	109	0	125	75.4
233	462	24	76	95	23	972	560	64	63	104	1	147	107.2
166	589	35	102	21	18	990	542	97	97	118	0	126	92.3
158	572	50	124	76	113	948	526	87	97	102	0	123	65.3
153	559	38	87	24	9	964	531	98	109	100	0	150	127.2
254	382	28	76	349	24	974	638	56	58	87	1	177	83.1
225	425	27	99	40	7	1024	599	47	51	104	0	133	56.6
251	395	35	86	165	36	953	515	54	61	88	1	149	82.6
228	488	31	88	126	96	981	560	74	82	104	1	145	115.1
144	590	20	84	19	9	998	601	66	72	122	0	148	88.0
170	489	37	107	2	4	968	523	54	56	109	0	141	54.2
224	496	27	73	208	40	996	522	70	75	99	1	162	82.3
162	622	37	111	36	29	1012	574	96	95	121	0	136	103.0
249	457	53	135	49	19	968	480	41	46	88	1	139	45.5
171	593	25	78	24	40	989	599	97	106	104	0	126	50.8
160	588	40	113	22	3	1049	623	91	90	121	0	130	84.9

135. Prater's gasoline yields

Prater, N.H. (1954) Estimate gasoline yields from crudes. *Petroleum Refiner*, No. 5, 35, 96.

See also Daniel, C. and Wood, F.S. (1980) *Fitting equations to data*, 2nd edition, New York: John Wiley & Sons, Table 8.2.

The following data were collected by N.H. Prater in a laboratory study of how three distillation properties (X_1, X_2 and X_3) of crude oils, together with the volatility (X_4) of the gasoline produced (measured by the ASTM end point in °F), affect the percentage yield of gasoline. The aim was to produce an equation to estimate the gasoline yield, given the end point (X_4) of the gasoline desired and the properties of the available crude. The distillation properties of the crudes were as follows.

X_1 Gravity in °API.
X_2 Vapour pressure in psi.
X_3 ASTM 10% point in °F.

Ten different crude oils were used, each one in 2–4 runs of the experiment.

Identification number	Crude oil			Gasoline	Yield, %
	X_1	X_2	X_3	X_4	
1– 1	50.8	8.6	190	205	12.2
2				275	22.3
3				345	34.7
4				407	45.7
2– 1	41.3	1.8	267	235	2.8
2				275	6.4
3				358	16.1
4				416	27.8
3– 1	40.8	3.5	210	218	8.0
2				273	13.1
3				347	26.6
4– 1	40.3	4.8	231	307	14.4
2				367	26.8
3				395	34.9
5– 1	40.0	6.1	217	212	7.4
2				272	18.2
3				340	30.4
6– 1	38.4	6.1	220	235	6.9
2				300	15.2
3				365	26.0
4				410	33.6
7– 1	38.1	1.2	274	285	5.0
2				365	17.6

	3				444	32.1
8–	1	32.2	5.2	236	267	10.0
	2				360	24.8
	3				402	31.7
9–	1	32.2	2.4	284	351	14.0
	2				424	23.2
10–	1	31.8	0.2	316	365	8.5
	2				379	14.7
	3				428	18.0

136. Strength of cables

Hald, A. (1952) *Statistical theory with engineering applications*, New York: John Wiley & Sons, 434.

For a high voltage electricity transmission network, uniform cables of great tensile strength are required. In one design, each cable was composed of 12 wires. To investigate the tensile strength of cables, a sample taken from each of the wires in 9 cables was tested. The data below show the tensile strengths in kg for each wire tested. Is there evidence that the strength of the wires varies between cables? Cables 1 to 4 were made from one lot of raw material, while the other five were made of another lot. Is there evidence that the lot makes a difference to strength?

Cable number

1	2	3	4	5	6	7	8	9
345	329	340	328	347	341	339	339	342
327	327	330	344	341	340	340	340	346
335	332	325	342	345	335	342	347	347
338	348	328	350	340	336	341	345	348
330	337	338	335	350	339	336	350	355
334	328	332	332	346	340	342	348	351
335	328	335	328	345	342	347	341	333
340	330	340	340	342	345	345	342	347
337	345	336	335	340	341	341	337	350
342	334	339	337	339	338	340	346	347
333	328	335	337	330	346	336	340	348
335	330	329	340	338	347	342	345	341

137. Control of leatherjackets

Bartlett, M.S. (1936) Some notes on insecticide tests in the laboratory and in the field. *Journal of the Royal Statistical Society, Supplement*, **3**, 185–194.

In an experiment by F.J.D. Thomas on the control of leatherjackets (cranefly larvae living in the soil), counts of surviving leatherjackets brought to the ground surface by a standard emulsion were made some days after the application of toxic emulsions. In each of six randomized blocks, there were six square plots of side 1 yard, four treated by four different toxic emulsions and two untreated as controls. Two sample counts (conducted in squares of side 1 foot) were made in each plot. Do the toxic emulsions work? A transformation of the raw data seems appropriate.

			Treatment			
Block	*Control 1*	*Control 2*	*Emulsion 1*	*Emulsion 2*	*Emulsion 3*	*Emulsion 4*
1	33	30	8	12	6	17
	59	36	11	17	10	8
2	36	23	15	6	4	3
	24	23	20	4	7	2
3	19	42	10	12	4	6
	27	39	7	10	12	3
4	71	39	17	5	5	1
	49	20	26	8	5	1
5	22	42	14	12	2	2
	27	22	11	12	6	5
6	84	23	22	16	17	6
	50	37	30	4	11	5

138. Spinning synthetic yarn

Box, G.E.P., Hunter, W.G. and Hunter, J.S. (1978) *Statistics for Experimenters*, New York: John Wiley & Sons, 251.

Synthetic yarn is produced by extruding the synthetic material from a spinneret, which contains small holes, and spinning the resulting fibres to form a yarn. These data were obtained in an experiment to determine whether the 'draw ratio' affected the yarn breaking strength of an experimental synthetic fibre. The draw ratio determines the tension applied to the yarn by the spinning machine. In this experiment, three different spinnerets were used, each supplying yarn to a different bobbin. At any given time during the experiment, the three spinnerets and their corresponding bobbins were operating at different draw ratios; the usual (denoted by *A*), 5% increase over usual (*B*), and 10% increase (*C*). When the three bobbins were full, they were removed from the machine and replaced by new ones, a process called 'doffing' — a 'doff' may be thought of as a period of filling a set of bobbins. After each doff the draw ratios applied to the spinnerets were changed round. There were 12 doffs in all. Thus there are two blocking factors (spinnerets and doffs). The three draw ratios were applied to the spinnerets and bobbins according to a plan consisting of a 3 × 3 latin square replicated four times. In each cell of the doff × spinneret table below, the treatment (*A*, *B* or *C*) and response (breaking strength) are given. The question of interest is whether, and if so how, the draw ratio affects the breaking strength of the yarn.

		Spinneret					
Doff		*I*		*II*		*III*	
1	A	19.56	B	23.16	C	29.72
2	B	22.94	C	27.51	A	23.71
3	C	25.06	A	17.70	B	22.32
4	B	23.24	C	23.54	A	18.75
5	A	16.28	B	22.29	C	28.09
6	C	18.53	A	19.89	B	20.42
7	C	23.98	A	20.46	B	19.28
8	A	15.33	B	23.02	C	24.97
9	B	24.41	C	22.44	A	19.23
10	A	16.65	B	22.69	C	24.94
11	B	18.96	C	24.19	A	21.95
12	C	21.49	A	15.78	B	24.65

139. Forearm lengths

Pearson, K. and Lee, A. (1903), On the laws of inheritance in man. I. Inheritance of physical characters. *Biometrika*, **2**, 357–462.

The data are measurements of the length of the forearm (in inches) made on 140 adult males. Is it plausible that the distribution of forearm lengths is normal?

17.3	18.4	20.9	16.8	18.7	20.5	17.9	20.4	18.3	20.5
19.0	17.5	18.1	17.1	18.8	20.0	19.1	19.1	17.9	18.3
18.2	18.9	19.4	18.9	19.4	20.8	17.3	18.5	18.3	19.4
19.0	19.0	20.5	19.7	18.5	17.7	19.4	18.3	19.6	21.4
19.0	20.5	20.4	19.7	18.6	19.9	18.3	19.8	19.6	19.0
20.4	17.3	16.1	19.2	19.6	18.8	19.3	19.1	21.0	18.6
18.3	18.3	18.7	20.6	18.5	16.4	17.2	17.5	18.0	19.5
19.9	18.4	18.8	20.1	20.0	18.5	17.5	18.5	17.9	17.4

18.7	18.6	17.3	18.8	17.8	19.0	19.6	19.3	18.1	18.5
20.9	19.8	18.1	17.1	19.8	20.6	17.6	19.1	19.5	18.4
17.7	20.2	19.9	18.6	16.6	19.2	20.0	17.4	17.1	18.3
19.1	18.5	19.6	18.0	19.4	17.1	19.9	16.3	18.9	20.7
19.7	18.5	18.4	18.7	19.3	16.3	16.9	18.2	18.5	19.3
18.1	18.0	19.5	20.3	20.1	17.2	19.5	18.8	19.2	17.7

140. Lifetime of electric lamps

Gupta, A.K. (1952), Estimation of the mean and standard deviation of a normal population from a censored sample. *Biometrika*, **39**, 260–273.

David, F.N. and Pearson, E.S. (1961) *Elementary statistical exercises*, Cambridge: Cambridge University Press, 7.

The data give (in grouped form) the lifetimes in hours of 300 electric lamps. The normal distribution fits the data remarkably well, which might make one smell a rat. Gupta was interested in censoring and gave only the lifetimes of the first 119 lamps, which failed in 1450 hours or less. David and Pearson give the rest of the data as well, but give only Gupta as their source. Gupta's paper, however, acknowledges comments from Pearson. We conclude that the data are probably real, but some doubt remains.

Lifetime (hours)	frequency
950–1000	2
1000–1050	2
1050–1100	3
1100–1150	6
1150–1200	7
1200–1250	12
1250–1300	16
1300–1350	20
1350–1400	24
1400–1450	27
1450–1500	29
1500–1550	29
1550–1600	28
1600–1650	25
1650–1700	21
1700–1750	16
1750–1800	12
1800–1850	8
1850–1900	6
1900–1950	3
1950–2000	2
2000–2050	1
2050–2100	1

141. Computer failures

Bertie, A., McConway, K. and the M345 Course Team (1987) *M345 Statistical Methods, Unit 11: Statistical computing III*, Milton Keynes: The Open University, 16.

These data relate to a DEC-20 computer which operated at the Open University in the 1980s. They give the number of times that the computer broke down in each of 128 consecutive weeks of operation, starting in late 1983. For these purposes, a breakdown was defined as occurring if the computer stopped running or was turned off for maintenance to deal with a fault. Routine preventive maintenance was not counted as a breakdown. The data seem to indicate that the breakdowns do not follow a time-homogeneous Poisson process. How might they be modelled instead? (Read across)

4	0	0	0	3	2	0	0	6	7
6	2	1	11	6	1	2	1	1	2
0	2	2	1	0	12	8	4	5	0
5	4	1	0	8	2	5	2	1	12
8	9	10	17	2	3	4	8	1	2
5	1	2	2	3	1	2	0	2	1
6	3	3	6	11	10	4	3	0	2
4	2	1	5	3	3	2	5	3	4
1	3	6	4	4	5	2	10	4	1
5	6	9	7	3	1	3	0	2	2
1	4	2	13	0	2	1	1	0	3
16	22	5	1	2	4	7	8	6	11
3	0	4	7	8	4	4	5		

142. Replacement value of books

Data collected by Kevin McConway in 1986.

Some household contents insurance policies require an estimate to be made of what it would cost to replace the existing contents. These data were collected in an attempt to use sampling methodology to estimate the total replacement cost of a collection of 1554 books on the basis of a sample of 100 of them. Replacement prices of the books sampled were found from publishers' catalogues; if a book was out of print, the cost of a similar book on a similar topic was used. The prices are in pence (100 p = £1). Since book prices vary considerably, an attempt was made to improve the accuracy of the resulting estimate by recording the value of an auxiliary variable, the width of the spine of the book (in millimetres). The total width of all 1554 books was 25 182 mm. What is the estimate of the total replacement value? Did it really help to do all that measuring of the spine thickness?

Price (p)	Width (mm)	Price (p)	Width (mm)	Price (p)	Width (mm)
995	24	395	8	100	7
1250	13	295	19	595	22
295	33	195	8	350	19
295	2	100	2	895	9
250	11	495	19	150	6
150	7	195	12	295	2
30	4	795	14	495	20
295	28	1950	39	395	8
295	10	230	13	895	56
250	12	395	13	150	6
250	4	595	13	595	8
1495	30	200	2	1495	6
295	15	495	17	175	6
175	11	195	15	495	14
150	8	350	11	195	10
1225	68	1495	7	125	6
595	13	895	33	495	30
100	3	1095	18	495	14
695	8	195	3	525	21
995	18	195	16	495	8
795	4	95	4	595	19
595	11	95	16	195	8
1295	23	250	6	1095	17
495	22	475	8	795	13
295	2	100	2	795	33
250	4	385	8	695	17
695	6	750	22	695	25
695	7	1550	5	750	34
250	13	395	13	295	23
495	10	195	8	195	17
200	2	250	10	250	13
695	6	695	29	895	16
395	20	350	29	695	4
795	27				

143. Sulphinpyrazone and heart attacks

Anturane Reinfarction Trial Research Group (1980), Sulfinpyrazone in the prevention of sudden death after myocardial infarction. *New England Journal of Medicine*, **302**, 250.

These data come from a clinical trial and show the effect of the drug sulphinpyrazone on deaths after myocardial infarction. Does this drug have an effect in reducing mortality?

	Deaths (all causes)	Survivors
Sulphinpyrazole	41	692
Placebo	60	682

144. Tibetan skulls

Morant, G.M. (1923) A first study of the Tibetan skull. *Biometrika*, **14**, 193–260.

Colonel L.A. Waddell collected 32 skulls in the south-western and eastern districts of Tibet. The collection can be divided into two groups. The first, type A, comprises 17 skulls which came from graves in Sikkim and neighbouring areas of Tibet. The remaining 15 skulls, type B, were picked up on a battlefield in the Lhasa district and were believed to be those of native soldiers from the eastern province of Khams. These skulls were of particular interest because it was thought at the time that Tibetans from Khams might be survivors of a particular fundamental human type, unrelated to the Mongolian and Indian types which surrounded them. These data consist of five measurements made on each skull, together with an indicator for the skull type. The five measurements (all in millimetres) are as follows.

X_1 Greatest length of skull
X_2 Greatest horizontal breadth of skull
X_3 Height of skull
X_4 Upper face height
X_5 Face breadth, between outermost points of cheek bones

The original source gives 45 other measurements on each skull.

Type A skulls

X_1	X_2	X_3	X_4	X_5
190.5	152.5	145.0	73.5	136.5
172.5	132.0	125.5	63.0	121.0
167.0	130.0	125.5	69.5	119.5
169.5	150.5	133.5	64.5	128.0
175.0	138.5	126.0	77.5	135.5
177.5	142.5	142.5	71.5	131.0
179.5	142.5	127.5	70.5	134.5
179.5	138.0	133.5	73.5	132.5
173.5	135.5	130.5	70.0	133.5
162.5	139.0	131.0	62.0	126.0
178.5	135.0	136.0	71.0	124.0
171.5	148.5	132.5	65.0	146.5
180.5	139.0	132.0	74.5	134.5
183.0	149.0	121.5	76.5	142.0
169.5	130.0	131.0	68.0	119.0
172.0	140.0	136.0	70.5	133.5
170.0	126.5	134.5	66.0	118.5

Type B skulls

X_1	X_2	X_3	X_4	X_5
182.5	136.0	138.5	76.0	134.0
179.5	135.0	128.5	74.0	132.0
191.0	140.5	140.5	72.5	131.5
184.5	141.5	134.5	76.5	141.5

181.0	142.0	132.5	79.0	136.5
173.5	136.5	126.0	71.5	136.5
188.5	130.0	143.0	79.5	136.0
175.0	153.0	130.0	76.5	142.0
196.0	142.5	123.5	76.0	134.0
200.0	139.5	143.5	82.5	146.0
185.0	134.5	140.0	81.5	137.0
174.5	143.5	132.5	74.0	136.5
195.5	144.0	138.5	78.5	144.0
197.0	131.5	135.0	80.5	139.0
182.5	131.0	135.0	68.5	136.0

145. East Midlands village dialects

Morgan, B.J.T. (1981) Three applications of methods of cluster analysis. *The Statistician*, **30**, 205–223.

The University of Leeds conducted a major survey of English dialects in which carefully chosen individuals from 311 rural English localities were interviewed to assess, among other things, their vocabulary. Individuals from different localities frequently use different words for the same item. The table below uses data from 25 villages in the East Midlands region of England. It records, as a measure of similarity between any pair of villages, the percentage of 60 of the items covered in the study for which the same words are used in the two villages. The data can be used to investigate, among other things, clusters of similar dialects.

The villages are as follows:

Code	Village	County
A	North Wheatley	Nottinghamshire
B	South Clifton	Nottinghamshire
C	Oxton	Nottinghamshire
D	Eastoft	Lincolnshire
E	Keelby	Lincolnshire
F	Willoughton	Lincolnshire
G	Wragby	Lincolnshire
H	Old Bolingbroke	Lincolnshire
I	Fulbeck	Lincolnshire
J	Sutterton	Lincolnshire
K	Swinstead	Lincolnshire
L	Crowland	Lincolnshire
M	Harby	Leicestershire
N	Packington	Leicestershire
O	Goadby	Leicestershire
P	Ullesthorpe	Leicestershire
Q	Empingham	Rutland
R	Warmington	Northamptonshire
S	Little Harrowden	Northamptonshire
T	Kislingbury	Northamptonshire
U	Sulgrave	Northamptonshire
V	Warboys	Huntingdonshire
W	Little Downham	Cambridgeshire
X	Tingewick	Buckinghamshire
Y	Turvey	Bedfordshire

The dissimilarities are as follows:

	A	B	C	D	E	F	G	H	I	J	K	L	M	N	O	P	Q	R	S	T	U	V	W	X
B	71																							
C	58	57																						
D	49	45	48																					
E	63	63	47	59																				
F	64	66	50	53	71																			
G	71	75	52	53	71	68																		
H	52	56	36	34	60	58	69																	
I	46	50	57	33	42	43	43	44																
J	61	49	52	40	58	61	56	61	52															
K	57	60	56	35	53	48	55	48	63	59														
L	39	46	45	30	42	40	47	44	50	53	60													
M	42	50	53	28	41	36	43	39	48	47	58	48												
N	32	34	47	20	27	29	31	23	44	39	43	39	63											
O	32	39	50	19	25	25	36	37	43	41	48	49	64	62										
P	23	27	42	14	20	22	24	28	36	27	38	35	54	57	57									
Q	41	47	56	25	38	42	46	38	48	54	54	48	72	51	61	59								
R	39	42	48	24	37	34	36	42	43	49	60	56	59	46	56	47	54							
S	32	36	43	22	22	24	34	29	47	34	45	47	38	49	51	42	44	53						
T	27	36	38	19	22	20	25	25	40	25	40	40	45	48	54	49	42	44	63					
U	28	37	37	20	25	25	31	33	42	29	41	37	46	48	49	47	43	44	58	59				
V	26	26	30	20	21	28	28	28	41	33	39	55	34	33	40	33	38	40	58	54	47			
W	30	33	32	16	25	33	33	32	41	37	37	46	47	46	49	39	46	58	42	44	42	50		
X	36	49	45	26	29	41	41	32	47	32	52	46	57	49	56	49	54	53	63	68	73	51	51	
Y	31	44	40	23	29	32	32	31	47	33	43	45	47	53	43	46	53	60	47	61	62	55	54	72

146. Lung cancer mortality

The Open University (1983) *MDST242 Statistics in Society, Unit C4: Smoking, statistics and society*, Milton Keynes: The Open University, Figures 1.1 and 1.2.

The following data were collated from various UK government publications. They give, for 1980, numbers of male deaths from lung cancer and the male population, classified into four broad age groups. These figures are given for the whole of England and Wales, and for two English counties, Cleveland and East Sussex, which have very different age structures. They can be used to illustrate age-standardization techniques.

	Age group	0–44	45–64	65–74	75 and over	Total
Cleveland	Male deaths from lung cancer	6	144	142	79	371
	Male population	190,200	60,400	19,000	8,100	277,700
East Sussex	Male deaths from lung cancer	3	94	178	146	421
	Male population	175,800	67,800	34,700	20,400	298,700
England and Wales	Male deaths from lung cancer	277	8,076	11,276	7,154	26,783
	Male population	15,673,800	5,398,700	2,008,800	903,500	23,984,800

147. Aphids

Jeffers, J.N.R. (1967) Two case studies in the application of principal components analysis, *Applied Statistics*, **16**, 225–236.

A study was carried out on 40 winged aphids caught in a light trap during 1964. For each aphid, the following 19 variables were observed.

1	Body length	13	Leg length, tibia III
2	Body width	14	Leg length, femur III
3	Fore-wing length	15	Rostrum width
4	Hind-wing length	16	Ovipositor width
5	Number of spiracles	17	Number of ovipositor spines
6–10	Length of antennal segments I–V	18	Anal fold (0 if absent, 1 if present)
11	Number of antennal spines	19	Number of hind-wing hooks
12	Leg length, tarsus III		

The object of the study was to assess whether the group of aphids was homogeneous or whether it was composed of distinct subgroups, possibly corresponding to different species of aphid. The following is the correlation matrix for the 19 variables.

	1	2	3	4	5	6	7	8	9	10	11	12	13	14	15	16	17	18
2	0.934																	
3	0.927	0.941																
4	0.909	0.944	0.933															
5	0.524	0.487	0.543	0.499														
6	0.799	0.821	0.856	0.833	0.703													
7	0.854	0.865	0.886	0.889	0.719	0.923												
8	0.789	0.834	0.846	0.885	0.253	0.699	0.751											
9	0.835	0.863	0.862	0.850	0.462	0.752	0.793	0.745										
10	0.845	0.878	0.863	0.881	0.567	0.836	0.913	0.787	0.805									
11	-0.458	-0.496	-0.522	-0.488	-0.174	-0.317	-0.383	-0.497	-0.356	-0.371								
12	0.917	0.942	0.940	0.945	0.516	0.846	0.907	0.861	0.848	0.902	-0.465							
13	0.939	0.961	0.956	0.952	0.494	0.849	0.914	0.876	0.877	0.901	-0.447	0.981						
14	0.953	0.954	0.946	0.949	0.452	0.823	0.886	0.878	0.883	0.891	-0.439	0.971	0.991					
15	0.895	0.899	0.882	0.908	0.551	0.831	0.891	0.794	0.818	0.848	-0.405	0.908	0.920	0.921				
16	0.691	0.652	0.694	0.623	0.815	0.812	0.855	0.410	0.620	0.712	-0.198	0.725	0.714	0.676	0.720			
17	0.327	0.305	0.356	0.272	0.746	0.553	0.567	0.067	0.300	0.384	-0.032	0.396	0.360	0.298	0.378	0.781		
18	-0.676	-0.712	-0.667	-0.736	-0.233	-0.504	-0.502	-0.758	-0.666	-0.629	0.492	-0.657	-0.655	-0.687	-0.633	-0.186	0.169	
19	0.702	0.729	0.746	0.777	0.285	0.499	0.592	0.793	0.671	0.668	-0.425	0.696	0.724	0.731	0.694	0.287	-0.026	-0.775

148. Pit props

Jeffers, J.N.R. (1967) Two case studies in the application of principal components analysis, *Applied Statistics*, **16**, 225–236.

A study was carried out in which thirteen variables were measured on each of 180 pit props made of Corsican pine grown in East Anglia. The aim was to investigate how these variables were related to one another (and also to investigate how these variables were related to the strength of the prop, though data on strength are not given here). The variables were as follows.

1   Top diameter of the prop in inches
2   Length of the prop in inches
3   Moisture content of the prop, as a percentage of its dry weight
4   Specific gravity of the prop at the time of strength testing
5   Oven-dry specific gravity of the timber
6   Number of annual rings at the top of the prop
7   Number of annual rings at the base of the prop
8   Maximum bow in inches
9   Distance of the maximum bow from the top of the prop in inches
10   Number of knot whorls
11   Length of clear prop from the top of the prop in inches
12   Average number of knots per whorl
13   Average diameter of the knots in inches

The correlation matrix for these variables is as follows:

	1	2	3	4	5	6	7	8	9	10	11	12
2	0.954											
3	0.364	0.297										
4	0.342	0.284	0.882									
5	-0.129	-0.118	-0.148	0.220								
6	0.313	0.291	0.153	0.381	0.364							
7	0.496	0.503	-0.029	0.174	0.296	0.813						
8	0.424	0.419	-0.054	-0.059	0.004	0.090	0.327					
9	0.592	0.648	0.125	0.137	-0.039	0.211	0.465	0.482				
10	0.545	0.569	-0.081	-0.014	0.037	0.274	0.679	0.557	0.526			
11	0.084	0.076	0.162	0.097	-0.091	-0.036	-0.113	0.061	0.085	-0.319		
12	-0.019	-0.036	0.220	0.169	-0.145	0.024	-0.232	-0.357	-0.127	-0.368	0.029	
13	0.134	0.144	0.126	0.015	-0.208	-0.329	-0.424	-0.202	-0.076	-0.291	0.007	0.184

149. Silver content of Byzantine coins

Hendy, M.F. and Charles, J.A. (1970) The production techniques, silver content and circulation history of the twelfth-century Byzantine Trachy. *Archaeometry*, **12**, 13–21.

The silver content (% Ag) of a number of Byzantine coins discovered in Cyprus was determined. Nine of the coins came from the first coinage of the reign of King Manuel I, Comnenus (1143-1180); there were seven from the second coinage minted several years later and four from the third coinage (later still); another seven were from a fourth coinage. The question arose of whether there were significant differences in the silver content of coins minted early and late in Manuel's reign.

5.9	6.9	4.9	5.3
6.8	9.0	5.5	5.6
6.4	6.6	4.6	5.5
7.0	8.1	4.5	5.1
6.6	9.3		6.2
7.7	9.2		5.8
7.2	8.6		5.8
6.9			
6.2			

150. Shoshoni American Indians

Dubois, C. (ed) (1970) *Lowie's Selected Papers in Anthropology*, California: University of California Press.

The ancient Greeks called a rectangle with a height-to-width ratio of $1 : \frac{1}{2} (\sqrt{5} + 1) = 0.618034$ a "golden rectangle", and used it often in their architecture (e.g. the Parthenon). These data are breadth-to-length ratios of beaded rectangles used by the Shoshoni American Indians to decorate their leather goods. One question is: is it reasonable to suppose they also were using golden rectangles?

0.693	0.662	0.690	0.606	0.570
0.749	0.672	0.628	0.609	0.844
0.654	0.615	0.668	0.601	0.576
0.670	0.606	0.611	0.553	0.933

151. Component lifetimes

Angus J.E. (1982) Goodness-of-fit tests for exponentiality based on a loss-of-memory type functional equation. *Journal of Statistical Planning and Inference,* **6**, 241–251.

These data list 20 operational lifetimes for components, measured in hours. The investigation was part of a test of exponentiality based on a loss-of-memory type functional equation.

6278	3113	5236	11584	12628
7725	8604	14266	6125	9350
3212	9003	3523	12888	9460
13431	17809	2812	11825	2398

152. Duckweed

Bottomley, W.B. (1914) Some accessory factors in plant growth and nutrition. *Proceedings of the Royal Society B,* **88**, 237–247.

This data set is from an experiment recorded by d'Arcy Thompson in the book "On Growth and Form". Duckweed fronds were counted weekly (weeks 0, 1, ..., 8) for plants growing in two different growth media, pure water and water with a bacterized peat auxitone. An appropriate model is sought.

(One cannot help wondering about the accuracy of the count. How was the number 19763 actually arrived at?)

Weeks	Pure water	Peat auxitone
0	20	20
1	30	38
2	52	102
3	77	326
4	135	1100
5	211	3064
6	326	6723
7	550	19763
8	1052	69350

153. Simulated polynomial regression data

Wampler, R.H. (1970) A report on the accuracy of some widely used least squares computer programs. *Journal of the American Statistical Association,* **62,** 819–841.

This is Wampler's famous simulated data set for catching out computer software at least squares polynomial fitting. There are two different response vectors, y_1 and y_2, which give, in fact, perfect quintic fits.

x	y_1	y_2
0	1	1
1	6	1.11111
2	63	1.24992
3	364	1.42753
4	1365	1.65984
5	3906	1.96875
6	9331	2.38336
7	19608	2.94117
8	37449	3.68928
9	66430	4.68559
10	111111	6
11	177156	7.71561
12	271453	9.92992
13	402234	12.75603
14	579195	16.32384
15	813616	20.78125
16	1118481	26.29536
17	1508598	33.05367
18	2000719	41.26528
19	2613660	51.16209
20	3368421	63

154. Heights of elderly females

Hand, D.J. (1992) personal communication

These are the heights in centimetres of a sample of 351 elderly women randomly selected from the community in a study of osteoporosis. These data can be used for modelling distributions, fitting normal models, and illustrating graphical techniques.

156	163	169	161	154	156	163	164	156	166	177	158
150	164	159	157	166	163	153	161	170	159	170	157
156	156	153	178	161	164	158	158	162	160	150	162
155	161	158	163	158	162	163	152	173	159	154	155
164	163	164	157	152	154	173	154	162	163	163	165
160	162	155	160	151	163	160	165	166	178	153	160
156	151	165	169	157	152	164	166	160	165	163	158
153	162	163	162	164	155	155	161	162	156	169	159

159	159	158	160	165	152	157	149	169	154	146	156
157	163	166	165	155	151	157	156	160	170	158	165
167	162	153	156	163	157	147	163	161	161	153	155
166	159	157	152	159	166	160	157	153	159	156	152
151	171	162	158	152	157	162	168	155	155	155	161
157	158	153	155	161	160	160	170	163	153	159	169
155	161	156	153	156	158	164	160	157	158	157	156
160	161	167	162	158	163	147	153	155	159	156	161
158	164	163	155	155	158	165	176	158	155	150	154
164	145	153	169	160	159	159	163	148	171	158	158
157	158	168	161	165	167	158	158	161	160	163	163
169	163	164	150	154	165	158	161	156	171	163	170
154	158	162	164	158	165	158	156	162	160	164	165
157	167	142	166	163	163	151	163	153	157	159	152
169	154	155	167	164	170	174	155	157	170	159	170
155	168	152	165	158	162	173	154	167	158	159	152
158	167	164	170	164	166	170	160	148	168	151	153
150	165	165	147	162	165	158	145	150	164	161	157
163	166	162	163	160	162	153	168	163	160	165	156
158	155	168	160	153	163	161	145	161	166	154	147
161	155	158	161	163	157	156	152	156	165	159	170
160	152	153									

155. Etruscan and Italian skulls

Barnicot, N.A. and Brothwell, D.R. (1959) The evaluation of metrical data in the comparison of ancient and modern bones. In *Medical biology and Etruscan origins*, G.E.W Wolstenholme and C.M. O'Connor, eds., Little, Brown & Co., 136.

The origins of the Etruscan empire remain something of a mystery to anthropologists. A particular question is whether Etruscans were native Italians or immigrants from elsewhere. In an anthropometric study, observations on the maximum head breadth (measured in mm) were taken on 84 skulls of Etruscan males. These data were compared with the same skull dimension for a sample of 70 modern Italian males.

Etruscan skulls

141	148	132	138	154	142	150	146	155	158	150	140
147	148	144	150	149	145	149	158	143	141	144	144
126	140	144	142	141	140	145	135	147	146	141	136
140	146	142	137	148	154	137	139	143	140	131	143
141	149	148	135	148	152	143	144	141	143	147	146
150	132	142	142	143	153	149	146	149	138	142	149
142	137	134	144	146	147	140	142	140	137	152	145

Italian skulls

133	138	130	138	134	127	128	138	136	131	126	120
124	132	132	125	139	127	133	136	121	131	125	130
129	125	136	131	132	127	129	132	116	134	125	128
139	132	130	132	128	139	135	133	128	130	130	143
144	137	140	136	135	126	139	131	133	138	133	137
140	130	137	134	130	148	135	138	135	138		

156. A random pattern screen

Laner, S., Morris, P. and Oldfield, R.C. (1957) A random pattern screen. *Quarterly Journal of Experimental Psychology*, **9**, 105–108.

In an experiment on visual perception, it was necessary to create a screen of small squares of side 1/12 inch and, at random, to colour the squares black or white. The size of the screen was 27 in by 40 in, so there were 155520 small squares to colour. A computer was used to make the decision according to the rule $P(\text{Black}) = 0.29$, $P(\text{White}) = 0.71$. After this was done, the screen was sampled to see whether the colouring algorithm had operated successfully.

A total of 1000 larger non-overlapping squares (1/3 in by 1/3 in) each containing 16 of the small squares were randomly selected, and the number of black small squares was counted in each case.

Count	Frequency
0	2
1	28
2	93
3	159
4	184
5	195
6	171
7	92
8	45
9	24
10	6
11	1
12	0
13	0
14	0
15	0
16	0

157. Rainfall in Australia

Rayner J.C.W. and Best D.J. (1989) *Smooth tests of goodness of fit*. Oxford: Oxford University Press.

Daily rainfall (in millimetres) was recorded over a 47-year period in Turramurra, Sydney, Australia. For each year the wettest day was identified (that having the greatest rainfall). The data show the rainfall recorded for the 47 annual maxima.

Surely there has *got* to be a mistake in the units: 3830 mm is more than 12 feet.

1468	909	841	475	846	452
3830	1397	556	978	1715	747
909	2002	1331	1227	2543	2649
1781	1717	2718	584	1859	1138
2675	1872	1359	1544	1372	1334
955	1849	719	1737	1389	681
1565	701	994	1188	962	1564
1800	580	1106	880	850	

158. Scottish soldiers

Stigler, Stephen M. (1986) *The History of Statistics — The Measurement of Uncertainty before 1900*. Cambridge, Massachusetts: The Belknap Press of Harvard University Press, 208.

The following data are taken from the *Edinburgh Medical and Surgical Journal* (1817). The chest circumference was measured (in inches, to the nearest inch) of 5732 Scottish soldiers. Stigler recounts the interesting history of these data. The Belgian mathematician Adolphe Quetelet (1796-1874) was interested in probability and social statistics, and amongst other things in the fitting of statistical models to data. Quetelet's summary of these data found 5738 soldiers (not 5732); his total for the number of soldiers with chest measurements between $33\frac{1}{2}$ and $34\frac{1}{2}$ inches was 18 (not 19). Stigler writes: 'Although errors have no important bearing on the explanation [of Quetelet's method for fitting a statistical model], they do exemplify Quetelet's tendency to calculate somewhat hastily, without checking his work'.

Measurement	Frequency	Measurement	Frequency
33	3	41	935
34	19	42	646
35	81	43	313
36	189	44	168
37	409	45	50
38	753	46	18
39	1062	47	3
40	1082	48	1

159. Borrowing library books

Burrell, Q.L. and Cane, V.R. (1982) The analysis of library data. *Journal of the Royal Statistical Society, Series A*, **145**, 439–471.

Large libraries often collect data on the circulation of books. One useful variable is the number of times a book is borrowed in a given time period (say, one year), counted for each book in the collection. Another variable of common interest is the number of books borrowed by a user in a year, counted for each borrower on the library register.

These data show the number of books that were borrowed k times ($k \geq 1$) from the Hillman Library of the University of Pittsburgh. The authors consider geometric and negative binomial probability fits to the data.

k	1	2	3	4	5	6	7	8
n_k	63526	25653	11855	6055	3264	1727	931	497

k	9	10	11	12	13	14	15	16
n_k	275	124	68	28	13	6	9	4

These data show the numbers of books that were borrowed k times from the long-loan collection at Sussex University over the period of a year.

k	1	2	3	4	5	6	7	8
n_k	9674	4351	2275	1250	663	355	154	72

k	9	10	11	12	13	14
n_k	37	14	6	2	0	1

160. Nerve impulse times

Cox, D.R. and Lewis, P.A.W. (1966) *The statistical analysis of series of events.* Chapman and Hall, London.

(Data collected by Dr P. Fatt & Professor B. Katz FRS, University College London.)

Time intervals between successive pulses along a nerve fibre were measured in seconds. There were 800 pulses: these are the 799 recorded waiting times.

0.21	0.03	0.05	0.11	0.59	0.06
0.18	0.55	0.37	0.09	0.14	0.19
0.02	0.14	0.09	0.05	0.15	0.23
0.15	0.08	0.24	0.16	0.06	0.11
0.15	0.09	0.03	0.21	0.02	0.14
0.24	0.29	0.16	0.07	0.07	0.04

0.02	0.15	0.12	0.26	0.15	0.33
0.06	0.51	0.11	0.28	0.36	0.14
0.55	0.28	0.04	0.01	0.94	0.73
0.05	0.07	0.11	0.38	0.21	0.49
0.38	0.38	0.01	0.06	0.13	0.06
0.01	0.16	0.05	0.10	0.16	0.06
0.06	0.06	0.06	0.11	0.44	0.05
0.09	0.04	0.27	0.50	0.25	0.25
0.08	0.01	0.70	0.04	0.08	0.16
0.38	0.08	0.32	0.39	0.58	0.56
0.74	0.15	0.07	0.26	0.25	0.01
0.17	0.64	0.61	0.15	0.26	0.03
0.05	0.34	0.07	0.10	0.09	0.02
0.30	0.07	0.12	0.01	0.16	0.14
0.49	0.07	0.11	0.35	1.21	0.17
0.01	0.35	0.45	0.07	0.93	0.04
0.96	0.14	1.38	0.15	0.01	0.05
0.23	0.31	0.05	0.05	0.29	0.01
0.74	0.30	0.09	0.02	0.19	0.47
0.01	0.51	0.12	0.12	0.43	0.32
0.09	0.20	0.03	0.05	0.13	0.15
0.05	0.08	0.04	0.09	0.10	0.10
0.26	0.07	0.68	0.15	0.01	0.27
0.05	0.03	0.40	0.04	0.21	0.29
0.24	0.08	0.23	0.10	0.19	0.20
0.26	0.06	0.40	0.51	0.15	1.10
0.16	0.78	0.04	0.27	0.35	0.71
0.15	0.29	0.04	0.01	0.28	0.21
0.09	0.17	0.09	0.17	0.15	0.62
0.50	0.07	0.39	0.28	0.20	0.34
0.16	0.65	0.04	0.67	0.10	0.51
0.26	0.07	0.71	0.11	0.47	0.02
0.38	0.04	0.43	0.11	0.23	0.14
0.08	1.12	0.50	0.25	0.18	0.12
0.02	0.15	0.12	0.08	0.38	0.22
0.16	0.04	0.58	0.05	0.07	0.28
0.27	0.24	0.07	0.02	0.27	0.27
0.16	0.05	0.34	0.10	0.02	0.04
0.10	0.22	0.24	0.04	0.28	0.10
0.23	0.03	0.34	0.21	0.41	0.15
0.05	0.17	0.53	0.30	0.15	0.19
0.07	0.83	0.04	0.04	0.14	0.34
0.10	0.15	0.05	0.04	0.05	0.65
0.16	0.32	0.87	0.07	0.17	0.10
0.03	0.17	0.38	0.28	0.14	0.07
0.14	0.03	0.21	0.40	0.04	0.11
0.44	0.90	0.10	0.49	0.09	0.01
0.08	0.06	0.08	0.01	0.15	0.50
0.36	0.08	0.34	0.02	0.21	0.32
0.22	0.51	0.12	0.16	0.52	0.21
0.05	0.46	0.44	0.04	0.05	0.04
0.14	0.08	0.21	0.02	0.63	0.35
0.01	0.38	0.43	0.03	0.39	0.04

0.17	0.23	0.78	0.14	0.08	0.11
0.07	0.45	0.46	0.20	0.19	0.50
0.09	0.22	0.29	0.01	0.19	0.06
0.39	0.08	0.03	0.28	0.09	0.17
0.45	0.40	0.07	0.30	0.16	0.24
0.81	1.35	0.01	0.02	0.03	0.06
0.12	0.31	0.64	0.08	0.15	0.06
0.06	0.15	0.68	0.30	0.02	0.04
0.02	0.81	0.09	0.19	0.14	0.12
0.36	0.02	0.11	0.04	0.08	0.17
0.04	0.05	0.14	0.07	0.39	0.13
0.56	0.12	0.31	0.05	0.10	0.13
0.05	0.01	0.09	0.03	0.27	0.17
0.03	0.05	0.26	0.23	0.20	0.76
0.05	0.02	0.01	0.20	0.21	0.02
0.04	0.16	0.32	0.43	0.20	0.13
0.10	0.20	0.08	0.81	0.11	0.09
0.26	0.15	0.36	0.18	0.10	0.34
0.56	0.09	0.15	0.14	0.15	0.22
0.33	0.04	0.07	0.09	0.18	0.08
0.07	0.07	0.68	0.27	0.21	0.11
0.07	0.44	0.13	0.04	0.39	0.14
0.10	0.08	0.02	0.57	0.35	0.17
0.21	0.14	0.77	0.06	0.34	0.15
0.29	0.08	0.72	0.31	0.20	0.10
0.01	0.24	0.07	0.22	0.49	0.03
0.18	0.47	0.37	0.17	0.42	0.02
0.22	0.12	0.01	0.34	0.41	0.27
0.07	0.30	0.09	0.27	0.28	0.15
0.26	0.01	0.06	0.35	0.03	0.26
0.05	0.18	0.46	0.12	0.23	0.32
0.08	0.26	0.82	0.10	0.69	0.15
0.01	0.39	0.04	0.13	0.34	0.13
0.13	0.30	0.29	0.23	0.01	0.38
0.04	0.08	0.15	0.10	0.62	0.83
0.11	0.71	0.08	0.61	0.18	0.05
0.20	0.12	0.10	0.03	0.11	0.20
0.16	0.10	0.03	0.23	0.12	0.01
0.12	0.17	0.14	0.10	0.02	0.13
0.06	0.21	0.50	0.04	0.42	0.29
0.08	0.01	0.30	0.45	0.06	0.25
0.02	0.06	0.02	0.17	0.10	0.28
0.21	0.28	0.30	0.02	0.02	0.28
0.09	0.71	0.06	0.12	0.29	0.05
0.27	0.25	0.10	0.16	0.08	0.52
0.44	0.19	0.72	0.12	0.30	0.14
0.45	0.42	0.09	0.07	0.62	0.51
0.50	0.47	0.28	0.04	0.66	0.08
0.11	0.03	0.32	0.16	0.11	0.26
0.05	0.07	0.04	0.22	0.08	0.08
0.01	0.06	0.05	0.05	0.16	0.05

0.13	0.42	0.21	0.36	0.05	0.01
0.44	0.14	0.14	0.14	0.08	0.51
0.18	0.02	0.51	0.06	0.22	0.01
0.09	0.22	0.59	0.03	0.71	0.14
0.02	0.51	0.03	0.41	0.17	0.37
0.39	0.82	0.81	0.24	0.52	0.40
0.24	0.06	0.73	0.27	0.18	0.01
0.17	0.02	0.11	0.26	0.13	0.68
0.13	0.08	0.71	0.04	0.11	0.13
0.17	0.34	0.23	0.08	0.26	0.03
0.21	0.45	0.40	0.03	0.16	0.06
0.29	0.43	0.03	0.10	0.10	0.31
0.27	0.27	0.33	0.14	0.09	0.27
0.14	0.09	0.08	0.06	0.16	0.02
0.07	0.19	0.11	0.10	0.17	0.24
0.01	0.13	0.21	0.03	0.39	0.01
0.27	0.19	0.02	0.21	0.04	0.10
0.06	0.48	0.12	0.15	0.12	0.52
0.48	0.29	0.57	0.22	0.01	0.44
0.05	0.49	0.10	0.19	0.44	0.02
0.72	0.09	0.04	0.02	0.02	0.06
0.22	0.53	0.18	0.10	0.10	0.03
0.08	0.15	0.05	0.13	0.02	0.10
0.51					

161. Corneal thickness of eyes

Ehlers, N. On corneal thickness and introcular pressure, II. *Acta Opthalmologica*, **48**, 1107–1112.

Eight people who each had one eye affected with glaucoma and one not had the corneal thicknesses (in microns) of both eyes measured.

Affected	Not affected
488	484
478	478
480	492
426	444
440	436
410	398
458	464
460	476

162. Memory retention

Mosteller, F., Rourke, R.E.K. and Thomas, G.B. (1970) *Probability with statistical applications,* second edition, Reading, Massachusetts: Addison-Wesley, 383, Table 11–1.

This is the psychologist Strong's famous data set on memory retention. Average percentage memory retention (p) was measured against passing time (t, minutes). The measurements were taken five times during the first hour after subjects memorized a list of disconnected items, and then at various times up to a week later.

One's first thoughts might be to attempt a model of the form $p = \exp(-\beta t)$, suggesting geometric loss of memory. In fact a scatter plot of log p against t still gives a pronounced curve. A model giving a better fit (but one less easy to explain) results from plotting p against log t.

t	p
1	0.84
5	0.71
15	0.61
30	0.56
60	0.54
120	0.47
240	0.45
480	0.38
720	0.36
1440	0.26
2880	0.20
5760	0.16
10080	0.08

163. Student's yeast cell counts

"Student" (1906) On the error of counting with a haemocytometer. *Biometrika,* **5**, 351–360.

Yeast cell counts were made on each of 400 small regions on a microscope slide. There are two main sources of error: the drop taken may not be representative of the bulk of the liquid, and the distribution of the cells or corpuscles over the area which is examined is never absolutely uniform, so that there is an "error of random sampling".

The experiment was repeated four times. This is a famous data set. Notice that no zeroes were observed in (D).

(A)

Count	Frequency
0	213
1	128
2	37
3	18
4	3
5	1

(B)

Count	Frequency
0	103
1	143
2	98
3	42
4	8
5	4
6	2

(C)

Count	Frequency
0	75
1	103
2	121
3	54
4	30
5	13
6	2
7	1
8	0
9	1

(D)

Count	Frequency
0	0
1	20
2	43
3	53
4	86
5	70
6	54
7	37
8	18
9	10
10	5
11	2
12	2

Student also included in his paper the original data from which the summary counts in (D) were calculated. The 400 squares were arranged as a 20 × 20 grid; each small square was of side 1/20 mm. The individual counts were as follows:

```
2   2   4   4   4   5   2   4   7   7   4   7   5   2   8   6   7   4   3   4
3   3   2   4   2   5   4   2   8   6   3   6   6  10   8   3   5   6   4   4
7   9   5   2   7   4   4   2   4   4   4   3   5   6   5   4   1   4   2   6
4   1   4   7   3   2   3   5   8   2   9   5   3   9   5   5   2   4   3   4
4   1   5   9   3   4   4   6   6   5   4   6   5   5   4   3   5   9   6   4
4   4   5  10   4   4   3   8   3   2   1   4   1   5   6   4   2   3   3   3
3   7   4   5   1   8   5   7   9   5   8   9   5   6   6   4   3   7   4   4
7   5   6   3   6   7   4   5   8   6   3   3   4   3   7   4   4   4   5   3
8  10   6   3   3   6   5   2   5   3  11   3   7   4   7   3   5   5   3   4
1   3   7   2   5   5   5   3   3   4   6   5   6   1   6   4   4   4   6   4
4   2   5   4   8   6   3   4   6   5   2   6   6   1   2   2   2   5   2   2
5   9   3   5   6   4   6   5   7   1   3   6   5   4   2   8   9   5   4   3
2   2  11   4   6   6   4   6   2   5   3   5   7   2   6   5   5   1   2   7
5  12   5   8   2   4   2   1   6   4   5   1   2   9   1   3   4   7   3   6
5   6   5   4   4   5   2   7   6   2   7   3   5   4   4   5   4   7   5   4
8   4   6   6   5   3   3   5   7   4   5   5   5   6  10   2   3   8   3   5
6   6   4   2   6   6   7   5   4   5   8   6   7   6   4   2   6   1   1   4
7   2   5   7   4   6   4   5   1   5  10   8   7   5   4   6   4   4   7   5
4   3   1   6   2   5   3   3   3   7   4   3   7   8   4   7   3   1   4   4
7   6   7   2   4   5   1   3  12   4   2   2   8   7   6   7   6   3   5   4
```

164. A comparison of growing conditions

Bliss, C.I. (1967) *Statistics in Biology,* Volume I, New York: McGraw-Hill.

Heights were measured (to the nearest inch) of maize plants in adjacent rows which differed only in a pollen sterility factor.

Fertile	Sterile
92	87
107	88
98	98
97	94
95	93
94	93
92	98
96	86
98	79
104	90
97	94
89	91

165. Anacapa pelican eggs

Risebrough, R.W. (1972) Effects of environmental pollutants upon animals other than man. *Proceedings of the 6th Berkeley Symposium on Mathematics and Statistics, VI.* California: University of California Press, 443–463.

For 65 Anacapa pelican eggs, the concentration in parts per million of PCB (polychlorinated biphenyl, an industrial pollutant) was measured, along with the thickness of the shell in millimetres.

Sample moments are $m_1 = 210.14$, $m_2 = 0.32$, $s_1 = 72.36$, $s_2 = 0.08$, $r = -0.25$. Contour plots based on density estimates suggest that the data are not bivariate normal.

Conc.	Thick.	Conc.	Thick.	Conc.	Thick.	Conc.	Thick.
452	0.14	184	0.19	115	0.20	315	0.20
139	0.21	177	0.22	214	0.22	356	0.22
166	0.23	246	0.23	177	0.23	289	0.23
175	0.24	296	0.25	205	0.25	324	0.26
260	0.26	188	0.26	208	0.26	109	0.27
204	0.28	89	0.28	320	0.28	265	0.29
138	0.29	198	0.29	191	0.29	193	0.29
316	0.29	122	0.30	305	0.30	203	0.30
396	0.30	250	0.30	230	0.30	214	0.30
46	0.31	256	0.31	204	0.32	150	0.34
218	0.34	261	0.34	143	0.35	229	0.35
173	0.36	132	0.36	175	0.36	236	0.37
220	0.37	212	0.37	119	0.39	144	0.39
147	0.39	171	0.40	216	0.41	232	0.41
216	0.42	164	0.42	185	0.42	87	0.44
216	0.46	199	0.46	236	0.47	237	0.49
206	0.49						

166. Steel ball bearings

Romano, A. (1977) *Applied Statistics for Science and Industry*, Boston: Allyn and Bacon.

Production lines in a large industrial corporation are set to produce a specific type of steel ball bearing with a diameter of 1 micron. At the end of a day's production, ten ball bearings were randomly picked from the production line and their diameters measured. In a second experiment ten ball bearings were selected from a different production line.

First line

1.18	1.42	0.69	0.88	1.62
1.09	1.53	1.02	1.19	1.32

Second line

1.72	1.62	1.69	0.79	1.79
0.77	1.44	1.29	1.96	0.99

167. Murder rates

Mendenhall, W., Ott, W. and Larson, R.F. (1974) *Statistics: a tool for the social sciences*, Boston: Duxbury Press.

The murder rates (per 100000) for a sample of 30 cities in southern USA were recorded for the years 1960 and 1970.

1960	1970
10.1	20.4
10.6	22.1
8.2	10.2
4.9	9.8
11.5	13.7
17.3	24.7
12.4	15.4
11.1	12.7
8.6	13.3
10.0	18.4
4.4	3.9
13.0	14.0
9.3	11.1
11.7	16.9
9.1	16.2
7.9	8.2
4.5	12.6
8.1	17.8
17.7	13.1
11.0	15.6
10.8	14.7
12.5	12.6
8.9	7.9
4.4	11.2
6.4	14.9
3.8	10.5
14.2	15.3
6.6	11.4
6.2	5.5
3.3	6.6

168. Kulasekeva model

Kulasekeva, K.B. and Tonkyn, D.W. (1992) A new discrete distribution, with applications to survival, dispersal and dispersion. *Communications in Statistics (Simulation and Computation)*, **21**, 499–518.

The authors use three data sets to illustrate the fit of a given discrete probability distribution. (Actually the topic of the paper is the function $p_k = Ck^a q^k$ defined for $k = 1, 2, ..., (0 < q < 1$, a real) and it is not at all clear from the paper how the zeroes are fitted.)

(A)   **The distribution of spiders under boards**

Count	0	1	2	3	≥4
Frequency	159	64	13	4	0

(B)   **The distribution of sowbugs under boards**

Count	0	1	2	3	4	5	6	7	8	9	10	11	12	13	14	≥15
Frequency	28	28	14	11	8	11	2	3	3	3	3	2	0	1	2	3

(C)   **Numbers of weevil eggs on beans**

Count	0	1	2	3	≥4
Frequency	5	68	88	32	0

169. Epileptic seizures

Albert, P.S. (1991) A two-state Markov mixture model for a time series of epileptic seizure counts. *Biometrics*, **47**, 1371–1381.

Thirteen patients with intractable epilepsy controlled by anti-convulsant drugs were observed for times between three months and five years. Information about the number of daily seizures was recorded.

It was demonstrated that all but one patient had seizure frequencies that deviated from a Poisson distribution. [Balish, M., Albert, P.S. and Theodore, W.H. (1991) Seizure frequency in intractable partial epilepsy: a statistical analysis. *Epilepsia*, **32**, 642–649.]

In this paper a doubly stochastic Poisson model is suggested where the mean seizure rate changes according to the states of a two-state Markov chain. Two of the 13 patients were selected to provide an illustration of this mixture model. Patient I had been observed for 422 consecutive days, Patient II for 351 days.

Patient I

Count	0	1	2	3	4	5	6		
Frequency	263	90	32	23	9	3	2		

Patient II

Count	0	1	2	3	4	5	6	7	8
Frequency	126	80	59	42	24	8	5	4	3

170. Red deer

Holgate, P. (1965) Fitting a straight line to data from a truncated population. *Biometrics*, **21**, 715–720.

Under the usual linear regression model the conditional distribution of $(Y|x)$ is normal with mean $\alpha + \beta x$ and variance σ^2. However, situations arise in practice where the distribution of Y must necessarily be truncated at a point independent of the value of x. These data about red deer were made available by Dr B. Mitchell of the Speyside Research Station of the Nature Conservancy Council.

Suppose that for all the deer in a given herd, the teeth finish growing at the same age and that thereafter the rate of wear is the same for all animals and constant in time. The randomness is due to the fact that crown weights at maturity are normally distributed with constant mean and variance. The exact age of maturity need not be known, as long as all those in the sample are above that age.

The data give the age in years and the weight of the first molar tooth (in g) for a sample of 78 stags shot on the Invermark estate in Scotland.

Age	Weight							
4.4	2.42	4.45	5.24	3.19	3.90	3.26	3.07	
4.8	4.48	3.18						
5.4	3.36	3.61	3.71	3.57	3.33	2.72	3.64	2.61
	3.89	3.30	2.62	3.10				
5.8	4.03							
6.4	3.36	3.19	3.32	2.78	3.38	3.07	3.22	3.05
	3.79	3.15	2.69					
7.4	3.92	3.07	2.54	3.82	3.10	3.56	2.60	3.56
7.8	3.80	3.49						
8.4	3.25	1.84	2.41	2.86	2.88	2.35	2.94	2.99
	2.76	2.40	2.67	2.97	2.61			
9.4	1.89	1.80	2.62	1.92	3.75	4.60	2.31	2.26
	3.48	2.86	2.38					
9.8	2.82							
10.4	1.09	2.69	2.48	2.72				
11.4	2.10							
12.4	2.73							
12.8	1.71							
13.4	2.14	2.76						
14.4	1.57							

171. The incidence of albinism

Kocherlakota, S. and Kocherlakota, K. (1990) Tests of hypotheses for the weighted binomial distribution. *Biometrics*, **46**, 645–656.

In this paper weighted discrete probability distributions of the kind $p_x = w_x q_x / \Sigma w_x q_x$ where $w_x > 0$, $\Sigma q_x = 1$ are described. The authors consider the special case of a weighted binomial distribution with $w_x = x^a$, $a > 0$ (so $p_0 = 0$). The model is fitted to data on the number of albino children in families of differing size (from 4 to 7) and where there is at least one albino child.

(ML estimation can give negative estimates for α.)

		Family size			
		4	5	6	7
	1	22	25	18	16
	2	21	23	13	10
	3	7	10	18	14
Albinos	4	0	1	3	5
	5		1	0	1
	6			1	0
	7				0

172. Females in queues

Jinkinson, R.A. and Slater, M. (1981) Critical discussion of a graphical method for identifying discrete distributions. *The Statistician*, **30**, 239–248.

This paper is about identifying an underlying discrete probability distribution given a frequency distribution of observed counts. For instance, denoting by $f(i)$ the frequency of the count i then a plot of the points $(i, if(i)/f(i-1))$ should yield an approximate straight line with negative slope if the underlying distribution was binomial. The plot should give scatter around a horizontal if the data were Poisson.

But as is illustrated in the paper the method easily fails. The authors elicit this moral: the fact that a method is graphical does not exempt it from the usual process of investigation and verification.

The number of females was counted in each of 100 queues of length ten, at a London underground station.

Count	0	1	2	3	4	5	6	7	8	9	10
Frequency	1	3	4	23	25	19	18	5	1	1	0

173. Student absenteeism

Ishii, G. and Hayakawa, R. (1960) On the compound binomial distribution. *Annals of the Institute of Statistics and Mathematics, Tokyo,* **12**, 69–80.

The absences of each of 113 students from a lecture course were recorded over the 24 lectures for which the course ran.   There were eleven lectures for the first semester and thirteen for the second.   It is unlikely that each student has the same probability for missing a lecture — some students are simply more committed than others — and a binomial model does not give a good fit. Permitting the absence probability to vary from student to student according to some distribution could lead to a more realistic, and a better, model. The authors discuss at some length the fitting of a beta-binomial distribution.

	Second semester								
First semester	0	1	2	3	4	5	6	...	
0	15	10	4	4	2	0	0	...	35
1	6	11	9	4	2	0	0	...	32
2	5	7	6	5	0	0	0	...	23
3	1	3	2	4	3	1	0	...	14
4	1	0	2	0	1	0	0	...	4
5	0	0	0	0	0	1	1	...	2
6	0	0	0	0	0	0	2	...	2
7	0	0	0	0	0	0	0	...	0
8	0	0	0	0	0	0	0	...	0
9	1	0	0	0	0	0	0	...	1
	.	.	.	.	.	.	.	...	.
	29	31	23	17	8	2	3	...	113

The number of lectures each student missed altogether can be found by summing across diagonals, giving frequencies as follows.

Absences	0	1	2	3	4	5	6	7	8	9	10	11	12	13	...	24
Frequency	15	16	20	21	16	9	6	3	2	1	1	1	2	0	...	0

174. Leading digits

The Open University (1984) *M245 Probability and Statistics, Unit 15: Excursions in Probability*, Milton Keynes: The Open University.

These data were collected by one of the authors. Sometimes it is necessary to use random digits, for instance when performing a simulation. If tables of random digits are not available, or if a computer generator is not available, then one might try "thinking" of a random digit. It has been shown empirically that sequences of digits obtained in this way demonstrate too great a regularity. One might try getting them from some sort of list instead. Possible sources include the telephone directory, an atlas or an encyclopaedia. An atlas includes data, based on a recent census, on the population of towns. The numbers seem very specific (it does not matter whether they are accurate). In the nature of things the first digit is never a zero, so this will not do to generate a random sequence of digits 0, 1, ..., 9, but the last digit might.

Data were collected, nevertheless, on the first digit for all the 305 towns mentioned on a randomly selected page of the gazetteer. The distribution of these digits is nothing like uniform: more than half (in particular, more than 2/9) begin with a 1 or a 2. On theoretical grounds [see, say, Feller W. (1971) *An introduction to probability theory and its applications,* Vol II, London: John Wiley & Sons, 62] a good probability model for the first significant digit in lists of numbers in an almanac can be shown to have the probability distribution

$$P(N \le n) = \log_{10}(n+1), \quad n = 1, 2, ..., 9.$$

Digit	1	2	3	4	5	6	7	8	9
Frequency	107	55	39	22	13	18	13	23	15

175. Spores of the fungus *Sordaria*

Ingold, C.T. and Hadland, S.A. (1959) *New Phytologist,* **58**, 46–57.

Spores of the fungus *Sordaria* are generated in chains of eight. Any of the seven links in the chain may break, independently of one another and with the same probability from link to link and from chain to chain. Ultimately the spores escape as projectiles comprising from one to eight spores. In one experiment a total of 907 projectiles was collected, involving altogether 7251 spores. The frequencies of the different observed lengths (singlets, doublets, triplets, ..., octuplets) were recorded.

The probability distribution of projectile lengths is actually a complicated matter. One can imagine a "new" octuplet flung to the floor, where its links may or may not break: repeated experiments will give rise to a probability distribution for the length of resulting projectiles. Alternatively, and more sensibly in nature, one might consider link breakages occurring as a Poisson process in time. Then one has to consider the definition and age-structure of the population from which one is sampling.

Length	1	2	3	4	5	6	7	8
Frequency	490	343	265	199	200	134	72	272

176. Litters of pigs

Parkes, A.S. (1923) Studies on the sex-ratio and related phenomena. *Biometrika,* **15**, 373–381.

Data were collected on a total of 2020 different litters of Duroc-Jersey pigs. In each case the size of the litter (none less than 2, none more than 14) was noted, and the number of males in each litter was counted. The results for the 1961 litters of sizes varying between four and twelve are given.

Number of Males	Litter size								

Number of Males	4	5	6	7	8	9	10	11	12
0	1	2	3	0	1	0	0	0	0
1	14	20	16	21	8	2	7	1	0
2	23	41	53	63	37	23	8	3	1
3	14	35	78	117	81	72	19	15	8
4	1	14	53	104	162	101	79	15	4
5		4	18	46	77	83	82	33	9
6			0	21	30	46	48	13	18
7				2	5	12	24	12	11
8					1	7	10	8	15
9						0	0	1	4
10							0	1	0
11								0	0
12									0
Total	53	116	221	374	402	346	277	102	70

177. Shoots of *Armeria maritima*

Thomas, M. (1949) A generalisation of Poisson's binomial limit for use in ecology. *Biometrika,* **36**, 18–25.

The author discusses a generalization of the Poisson distribution for application to plant counts. (What she calls the "double Poisson" is in fact a Poisson sum of (Poisson + 1)'s.) The model is fitted to two data sets.

A region of land was split into 100 square quadrats of equal area, and the number of shoots of *Armeria maritima* was counted in each region:

Count	0	1	2	3	4	5	6	7	8	9	10
Frequency	57	6	12	5	5	5	7	1	0	1	1

Similarly, shoots of *Plantago maritima* were counted over a region:

Count	0	1	2	3	4	5	6	7	8	9	10
Frequency	12	8	9	13	6	8	11	7	8	7	3

Count	11	12	13	14	15	16	17	18	19
Frequency	4	1	1	0	0	1	0	0	1

178. American N.F.L. matches

Csörgö, S. and Welsh, A.S. (1989) Testing for exponential and Marshall-Olkin distributions. *Journal of Statistical Planning and Inference*, **23**, 287–300.

The bivariate Marshall-Olkin distribution has survivor function

$$P(X_1 > x_1, X_2 > x_2)$$

$$= \exp -[\alpha_1 x_1 + \alpha_2 x_2 + \alpha_{12} \max(x_1, x_2)]$$

defined for $x_1, x_2 > 0$, $\alpha_1, \alpha_2 > 0$ and $\alpha_{12} \geq 0$. This distribution has exponential marginals and a singular component along the diagonal $x_1 = x_2$.

The data are from American National Football League matches played on three consecutive weekends in 1986. The variable X_1 is the game-time to the first points scored by kicking the ball between the end-posts, and X_2 is the game-time to the first points scored by moving the ball into the end-zone. Times are given in minutes and seconds. If $X_1 < X_2$ the first score is a field goal; if $X_1 = X_2$ the first score is a converted touchdown; if $X_1 > X_2$ the first score is an unconverted touchdown. Since X_1 and X_2 are waiting times, the constant marginal hazard assumption might be reasonable; and since ties occur, it is reasonable at least to conjecture a Marshall-Olkin model. Exponential tests on the marginals showed that X_2 was arguably exponential, X_1 less arguably so.

X_1	X_2	X_1	X_2	X_1	X_2	X_1	X_2
2.03	3.59	9.03	9.03	0.51	0.51	3.26	3.26
7.47	7.47	10.34	14.17	7.03	7.03	2.35	2.35
7.14	9.41	6.51	34.35	32.27	42.21	8.32	14.34
31.08	49.53	14.35	20.34	5.47	25.59	13.48	49.45
7.15	7.15	4.15	4.15	1.39	1.39	6.25	15.05
4.13	9.29	15.32	15.32	2.54	2.54	7.01	7.01
6.25	6.25	8.59	8.59	10.09	10.09	8.52	8.52
10.24	14.15	2.59	2.59	3.53	6.26	0.45	0.45
11.38	17.22	1.23	1.23	10.21	10.21	12.08	12.08
14.35	14.35	11.49	11.49	5.31	11.16	19.39	10.42
17.50	17.50	10.51	38.04				

179. Survival times in leukaemia

Bryson, M.C. and Siddiqui M.M. (1969) Survival times: some criteria for aging. *Journal of the American Statistical Association*, **64**, 1472–1483.

The data give the survival times of patients suffering from chronic granulocytic leukaemia, measured in days from the time of diagnosis.

7	47	58	74	177	232	273	285	317	429
440	445	455	468	495	497	532	571	579	581
650	702	715	779	881	900	930	968	1077	1109
1314	1334	1367	1534	1712	1784	1877	1886	2045	2056
2260	2429	2509							

180. Tensile strength

Quesenberry, C.P. and Hales, C. (1980) Concentration bands for uniformity plots. *Journal of Statistical Computation and Simulation*, **11**, 41–53.

Measurements were taken on the tensile strength of polyester fibres to see if they were consistent with the lognormal distribution. These data follow from a preliminary transformation. If the lognormal hypothesis was correct, these data should have been uniformly distributed.

0.023	0.032	0.054	0.069	0.081	0.094
0.105	0.127	0.148	0.169	0.188	0.216
0.255	0.277	0.311	0.361	0.376	0.395
0.432	0.463	0.481	0.519	0.529	0.567
0.642	0.674	0.752	0.823	0.887	0.926

181. Fatigue-life failures

Singpurwalla, N.D. (1988) An interactive PC-based procedure for reliability assessment incorporating expert opinion and survival data. *Journal of the American Statistical Association*, **83**, 43–51.

The author quotes data [from Lieblein, J. and Zelen, M. (1956) Statistical investigation of the fatigue-life of deep-groove ball-bearings, *Journal of Research, National Bureau of Standards*, **57**, 273–316] on the fatigue-life failures of ball-bearings. The data give the number of cycles to failure. Lieblein and Zelen argued that the Weibull distribution is a suitable model for describing these data.

17.88	28.92	33.00	41.52	42.12
45.60	48.48	51.84	51.96	54.12
55.56	67.80	68.64	68.88	84.12
93.12	98.64	105.12	105.84	127.92
128.04	173.40			

182. Linked transmission failures

Tapan Kumar Nayak (1988) Testing equality of conditionally independent exponential distributions. *Communications in Statistics (Theory and Methods)*, **17**, 807–820.

The paper describes the following model: the two random variables $(X|V=v)$ and $(Y|V=v)$ are conditionally independent and exponentially distributed with respective parameters $\alpha_1 v$ and $\alpha_2 v$. The paper considers the problem of testing the hypothesis $\alpha_1 = \alpha_2$ when the distribution of V is unknown and no data are available on its value.

The data considered as an example for the methods developed in the paper give the failure times of transmission (X) and of transmission pumps (Y) on 15 caterpillar tractors [from Barlow, R.E and Proschan, F. (1977) Techniques for analysing multivariate failure data. *Theory and Applications of Reliability*, **1**, 373–396].

	X	*Y*
1	1641	850
2	5556	1607
3	5421	2225
4	3168	3223
5	1534	3379
6	6367	3832
7	9460	3871
8	6679	4142
9	6142	4300
10	5995	4789
11	3953	6310
12	6922	6310
13	4210	6378
14	5161	6449
15	4732	6949

183. Published research papers in biology

Tripathi, R.C. and Gupta, R.C. (1988) Another generalisation of the logarithmic series and the geometric distribution. *Communications in Statistics (Theory and Methods)*, **17**, 1541–1547.

The authors consider a new generalization of the logarithmic series distribution based on a generalized negative binomial distribution obtained from a generalized Poisson distribution compounded with a truncated gamma distribution. The model can usefully be applied to long-tailed data.

The authors illustrate their models with data on 1534 biologists according to the number of research papers to their credit in the review of *Applied Entomology*, Volume 24 (1936). [See also Jain, G.C. and Gupta, R.P. (1973) A logarithmic series type distribution. *Trab. Estadistica*, **24**, 99–105.]

Number of papers	Frequency
1	1062
2	263
3	120
4	50
5	22
6	7
7	6
8	2
9	0
10	1
11	1

184. Counting cockroaches

Routledge, R.D. (1989) The removal method for estimating natural populations: incorporating auxiliary information. *Biometrics*, **45**, 111–121.

The author describes the removal method for estimating the size of a population. The technique involves monitoring the decline in the success rate in catching animals as the population is itself reduced through animal collection. The method is subject to bias that is difficult to quantify for instance, the remaining animals might be better at avoiding capture). Auxiliary information in the form of other signs of activity can also provide indicators to the size of the population.

An example using observations of a population of German cockroaches (*Blattella germanica* Linn.) was used to illustrate the methods. A number of adult roaches was confined in a tank. A baited live trap in the tank was cleared each morning and trapped roaches were removed from the population. Counts of faecal pellets on filter papers were used as signs of activity. The pellets were counted and the papers removed and replaced at the same time that the trap was cleared. The first paper was positioned 24 hours before the trap was put into operation.

The data give the number of cockroaches removed at day i (R_i) and numbers of faecal pellets removed (Y_i).

(The initial number of cockroaches was known: it was 228. The author's method of analysis leads to the maximum likelihood estimate 216.)

i	R_i	Y_i
0	—	117
1	40	97
2	19	95
3	10	78

185. Survival times of green sunfish

Matis, J.H. and Wehrly, T.E. (1979) Stochastic models of compartmental systems. *Biometrics*, **35**, 199–220.

The objective of this study was to investigate the resistance of green sunfish *Lepomis cyanellus* to various levels of thermal pollution. Twenty fish were introduced into heated water and the numbers of survivors recorded at pre-specified times. This followed a five-day acclimatization at 35°C.

Water at 39.5°C		Water at 39.7°C	
t, s	Survivors	*t*, s	Survivors
136	20	45	20
141	18	50	19
146	18	55	19
151	18	60	18
156	17	65	17
161	15	70	14
166	15	75	13
171	15	80	12
176	12	85	11
181	11	90	8
186	8	95	7
191	8	100	6
196	7	105	6
201	7	110	4
206	7	115	3
211	6	120	3
216	6	125	1
221	6	130	1
226	4	135	0
231	4		
236	3		
241	2		
246	1		
251	1		
256	1		
261	0		

186. The great plague of 1665

Defoe D. (1722) *A Journal of the Plague Year*. Penguin English Library, 1966 ed., Britain.

In his *Journal of the Plague Year*, describing the outbreak of the bubonic plague that in London in 1665, Defoe uses published bills of mortality as indicators of the progress of the epidemic.

It was about the beginning of September, 1664, that I, among the rest of my neighbours, heard in ordinary discourse that the plague was returned again in Holland; for it had been very violent there, and particularly at Amsterdam and Rotterdam, in the year 1663 ... several councils were held about ways to prevent its coming over; but all was kept very quiet. Hence it was that this rumour died off again, and people began to forget it as a thing we were very little concerned in, and that we hoped was not true; till the latter end of November or the beginning of December 1664 when two men, said to be Frenchmen, died of the plague in Long Acre, or rather at the upper end of Drury Lane ...

... It was observed with great uneasiness by the people that the weekly bills in general increased very much during these weeks, although it was at a time of year when usually the bills are very moderate. The usual number of burials within the bills of mortality for a week was from about 240 or thereabouts to 300. The last was esteemed a pretty high bill; but after this we found the bills successively increasing as follows:

	Buried
December the 20th to the 27th	*291*
December the 27th to the 3rd January	*349*
January the 3rd to the 10th	*394*
January the 10th to the 17th	*415*
January 17th to the 24th	*474*

This last bill was really frightful, being a higher number than had been known to have been buried in one week since the preceding visitation of 1656.

These are data for the numbers of deaths published in bills of mortality for nine weeks from late summer into autumn:

		Of all Diseases	Of the Plague
8 Aug	– 15 Aug	5319	3880
15 Aug	– 22 Aug	5568	4237
22 Aug	– 29 Aug	7496	6102
29 Aug	– 5 Sep	8252	6988
5 Sep	– 12 Sep	7690	6544
12 Sep	– 19 Sep	8297	7165
19 Sep	– 26 Sep	6460	5533
26 Sep	– 3 Oct	5720	4929
3 Oct	– 10 Oct	5068	4327
		59870	49705

187. Log-series data

Tripathi, R.C. and Gupta, R.C. (1985) A generalisation of the log-series distribution. *Communications in Statistics (Theory and Methods)*, **14**, 1779–1799.

In this paper a two-parameter version of the generalized logarithmic series distribution is proposed; the model is sufficiently flexible to describe short-tailed as well as long-tailed data. The distribution (which takes values on the positive integers 1, 2, ...) is most easily specified by the rule

$$p_1 = 1 / \sum_0^\infty j! \beta^j / [(\Theta+1)(\Theta+2)...(\Theta+j)]$$

$$p_{j+1} = \beta j p_j / (\Theta+j), \quad j = 1, 2, ...$$

for constants $0 < \beta < 1$ and $\Theta > 0$. The authors refer to two data sets in their model description.

[These data are from Haight, F.A. (1970) Group size distributions, with applications to vehicle occupancy. *Random counts in scientific works, Vol 3*, Pennsylvania State University Press, 95–106.] They list the numbers of occupants in 1469 cars (including the driver).

Occupants	Frequency
1	902
2	403
3	106
4	38
5	16
≥6	4

[These data are from Kempton, R.A. (1975) A generalized form of Fisher's logarithmic series. *Biometrika*, **62**, 29–38.] They list the species frequency distribution of insect catches from three traps at Rothamsted.

(A)   *Moths per species*
Light trap in Geescroft Wilderness, 1970

1	37
2+	42
4+	30
8+	21
16+	22
32+	21
64+	17
128+	15
256+	6

(B)   *Moths per species*
Light trap on old allotments site, 1970

1	33
2+	24
4+	22
8+	9
16+	5
32+	4
64+	0
128+	5

(C)   *Aphids per species*
Suction trap, Rothamsted Tower, 1968

1	17
2+	11
4+	11
8+	8
16+	6
32+	8
64+	6
128+	3
1024+	3

188. Eye and hair colour in children

Goodman, L.A. (1981) Association models and canonical correlation in the analysis of cross-classifications having ordered categories. *Journal of the American Statistical Association*, **76**, 320–334.

These are data for the eye and hair colour of 5387 children in Caithness, Scotland.

	Hair				
Eyes	**Fair**	**Red**	**Medium**	**Dark**	**Black**
Blue	326	38	241	110	3
Light	688	116	584	188	4
Medium	343	84	909	412	26
Dark	98	48	403	681	85

These are data for the eye and hair colour of 22361 children in Aberdeen, Scotland.

	Hair				
Eyes	**Fair**	**Red**	**Medium**	**Dark**	**Black**
Blue	1368	170	1041	398	1
Light	2577	474	2703	932	11
Medium	1390	420	3826	1842	33
Dark	454	255	1848	2506	112

189. Suicide figures in prisons

The Guardian, Friday January 4th, 1991.

The figures below are for prisons in England and Wales. The first figures record the number of deaths, the second those in which suicide verdicts were returned (some inquests for 1990 were still pending at the time this report appeared).

1985	29	23
1986	21	17
1987	46	42
1988	37	30
1989	48	33
1990	50	—

190. Counting beetles

Beall, G. (1938) Methods of estimating the population of insects in a field. *Biometrika,* **30**, 422–439.

The adult Colorado potato beetle (*Leptinotarsa decemlineata* Say.) is easily counted since it is both seen and collected rapidly. A count was conducted in a potato field near Chatham, Ontario on 14th August 1935. The plot chosen for examination was 124 ft wide and 96 ft long. The field was split into a 4 × 4 Latin square design and four men (A, B, C and D) collected beetles in their allocated squares. The number of beetles they managed to collect in each square was recorded:

1127	D	1331	B	628	A	430	C
658	C	635	A	969	D	758	B
869	B	794	D	560	C	411	A
523	A	490	C	213	B	517	D

191. The number of words in a sentence

Williams, C.B. (1939) A note on the statistical analysis of sentence-length as a criterion of literary style. *Biometrika,* **31**, 356–361.

The data below come from 600 sentences selected at random from G.K. Chesterton's *Shorter History of England.*

Number of words	Frequency	Number of words	Frequency
1 – 5	3	31 – 35	68
6 – 10	27	36 – 40	41
11 – 15	71	41 – 45	28
16 – 20	113	46 – 50	18
21 – 25	107	51 – 60	12
26 – 30	109	61 – 100	3

192. The size of gangs

Thrasher, F.M. (1927) *The Gang*, Chicago: University Press of Chicago.

A study was made of 1313 gangs in Chicago. (*The gang is an interstitial group originally formed spontaneously, and then integrated through conflict ... The lone wolf and the dual group are not gangs.* Thrasher, 1927) Thrasher gives information on the approximate numbers of members of 895 of the gangs studied:

Size of group	Frequency	Size of group	Frequency
3 – 5	37	41 – 50	51
6 – 10	198	51 – 75	26
11 – 15	191	76 – 100	25
16 – 20	149	101 – 200	25
21 – 25	79	201 – 500	11
26 – 30	46	501 – 1000	2
31 – 40	55		

193. Skin graft survival times

Woolson, R.F. and Lachenbruch P.A. (1980) Rank tests for censored matched pairs. *Biometrika*, **67**, 597–606.

The authors re-present data of Holt, J.D. and Prentice, R.L. (1974) Survival analysis in twin studies and matched pairs experiments. *Biometrika*, **61**, 17–30. These data consist of the survival times, in days, of closely and poorly matched skin grafts on the same burn patient.

Patient	Close match	Poor match
1	37	29
2	19	13
3	57*	15
4	93	26

5	16	11
6	22	17
7	20	26
8	18	21
9	63	43
10	29	15
11	60 *	40

* *censored*

194. Angles of spiders' webs

Gadsden, R.J. and Kanji, G.K. (1981) Sequential analysis for angular data. *The Statistician,* **30**, 119–129.

Spiders webs' angles made with the vertical to the Earth's surface have a von Mises circular distribution with known mean direction μ and concentration parameter k as follows (these parameter values are based on many past observations).

Isoxya cicatricosa	μ = 28.12°	$k = 38.17$
Araneus rufipalpus	μ = 15.66°	$k = 37.94$

The question arose of which species had constructed the webs when the following angles for 10 webs were observed:

25°   12°   31°   26°   17°   15°   24°   10°   16°   12°

195. Finger length

Macdonell, W.R. (1902) On criminal anthopometry and the identification of criminals. *Biometrika,* **1**, 177–227.

A total of 3000 criminals had their left middle finger measured (in cm). The table gives the central values for grouped classifications.

Length, cm	Frequency	Length, cm	Frequency
9.5	1	11.6	691
9.8	4	11.9	509
10.1	24	12.2	306
10.4	67	12.5	131
10.7	193	12.8	63
11.0	417	13.1	16
11.3	575	13.4	3

196. The Charlier model

Medhi, J. and Borah, M. (1986) On [the] generalized four-parameter Charlier distribution. *Journal of Statistical Planning and Inference*, **14**, 69–77.

The authors consider various discrete distributions as special cases of a four-parameter Charlier distribution. Methods for fitting a three-parameter generalized Charlier distribution are shown, and the distribution is fitted with success to two data sets. The first (Lespedeza) is from Beall, G. and Rescia, R. (1953) A generalisation of Neyman's contagious distribution. *Biometrics*, **9**, 354–386. The second (Leptinotarsa) is from Katti, S.K. and Gurland, J. (1961) The Poisson-Pascal distribution. *Biometrics*, **17**, 527–538. In each case the Charlier fit is better than the one originally attempted.

Count	*Lespedeza capitate*	*Leptinotarsa decemlineata*
0	7178	33
1	286	12
2	93	5
3	40	6
4	24	5
5	7	0
6	5	2
7	1	2
8	2	2
9	1	0
10	2	1
11+	1	2
	7640	70

197. Disease clusters

Tango, T. (1984) The detection of disease clustering in time. *Biometrics*, **40**, 15–26.

Tango's index for disease clustering C is given by the matrix relation $C = \mathbf{r'Ar}$, where $n\mathbf{r'} = (n_1, n_2, ..., n_m)$ $(n = n_1 + n_2 + ... n_m)$ denotes a vector of observed frequencies of an event in m successive time intervals, and $\mathbf{A}$ is a symmetric matrix whose elements a_{ij} are arbitrary measures of closeness between the ith and jth intervals. Tango's data set illustrating the use of C list the frequency of trisomy among karyotyped spontaneous abortions of pregnancies by calendar month of the last menstrual period, for the months July 1975 to June 1977 in three New York hospitals.

Month	7	8	9	10	11	12	1	2	3	4	5	6
Frequency	0	4	1	2	1	3	1	3	2	2	3	4

Month	7	8	9	10	11	12	1	2	3	4	5	6
Frequency	1	1	1	2	4	7	7	2	2	6	1	2

198. Multiple sclerosis

Joseph, L., Wolfson, C. and Wolfson, D.B. (1990) Is multiple sclerosis an infectious disease? Inference about an input process based on the output. *Biometrics*, **46**, 337–349.

This paper considers a controversial issue in the study of multiple sclerosis (MS) — whether the disease is infectious. [This problem is also considered by Kurtzke, J.K. and Hyllested, K. (1986) Multiple sclerosis in the Faroe Islands II. Clinical update, transmission and the nature of multiple sclerosis. *Neurology*, **36**, 307–328.]

The possibility that the disease is infectious resides in the apparently sudden occurrence of MS in the Faroe Islands after British troops arrived there in 1941. There were two groups of patients: Those in Group A had not been off the islands before onset, while those in Group B had been off the islands for less than two years before onset.

The time origin is 1941.

Patient	Sex	Age at onset	Onset time	Group
1	M	30	2	A
2	M	15	3	A
3	M	24	3	A
4	M	48	3	A
5	F	19	4	A
6	F	44	4	B
7	F	39	4	B
8	M	24	5	A
9	M	38	5	A
10	F	26	5	A
11	M	32	6	A
12	F	19	6	A
13	F	16	7	A
14	M	25	7	B
15	F	32	8	A
16	M	20	8	A
17	F	42	11	A
18	F	18	12	A
19	F	14	13	A
20	M	17	14	A
21	F	19	15	B
22	M	19	16	A
23	M	37	17	A
24	F	19	18	A
25	M	29	18	A
26	M	21	19	A
27	M	40	20	A
28	F	27	24	B
29	F	21	27	B
30	F	17	28	A
31	F	20	29	A
32	F	33	32	B

One moderately interesting item is that for the method of analysis pursued in the paper it was necessary to avoid ties in the time of onset of the disease: ties were eliminated by distributing those observations tied at year j independently and uniformly over the interval $(j-\frac{1}{2}, j+\frac{1}{2})$.

199. The number of pigs in a litter

Brande, R., Clarke P.M. and Mitchell K.G. (1955) *Journal of Agricultural Science,* **45,** 19.

The number of pigs born alive in 378 different litters was counted.

Number	1	2	3	4	5	6	7	8	9	10	11
Frequency	1	2	2	6	7	8	16	18	18	34	55

Number	12	13	14	15	16	17	18	19	20	21
Frequency	45	52	35	37	20	10	5	2	4	1

200. The lengths of scallops

Jorgensen, M.A. (1990) Inference-based diagnostics for finite mixture models. *Biometrics,* **46,** 1047–1058.

In this paper a three-component mixture of normal distributions is fitted to data on the lengths of 222 scallops caught during a dredge survey of Mercury Bay, Whitianga, New Zealand. The lengths recorded range from 62 mm to 126 mm.

Length	61	62	63	64	65	66	67	68	69	70
Frequency	0	2	3	2	5	5	10	13	10	14

Length	71	72	73	74	75	76	77	78	79	80
Frequency	25	21	13	9	7	12	5	1	4	4

Length	81	82	83	84	85	86	87	88	89	90
Frequency	2	3	0	0	0	3	2	3	3	1

Length	91	92	93	94	95	96	97	98	99	100
Frequency	3	5	1	1	4	0	3	1	1	2

Length	101	102	103	104	105	106	107	108	109	110
Frequency	2	1	1	1	2	1	1	2	0	0

Length	111	112	113	114	115	116	117	118	119	120
Frequency	1	2	1	0	0	1	0	2	0	0

Length	121	122	123	124	125	126	127	128	129	130
Frequency	0	0	0	0	0	1	0	0	0	0

201. Smoking habits in children

Kalbfleish, J.D. and Lawless, J.F. (1985) The analysis of panel data under a Markov assumption. *Journal of the American Statistical Association*, **80**, 863–871.

A study was conducted on smoking behaviour in children from two Ontario counties (Waterloo and Oxford). At times $t_0 = 0$, $t_1 = 0.15$, $t_2 = 0.75$, $t_3 = 1.10$, $t_4 = 1.90$ (measured in years) the children were classified into one of three states as follows.

State 0   *child has never smoked*
State 2   *child is currently a smoker*
State 3   *child has smoked, but has now quit*

The children were split into a control group and a group that received educational material about smoking during the first two months after the start of the trial.

The first table gives the transitions for the 125 children in the Oxford "treatment" group. The second table gives the transitions for a group of 88 children classified as "high risk" — for example, their parents smoked.

State	0	1	2	0	1	2	0	1	2	0	1	2
0	93	3	2	89	2	2	83	3	3	76	3	4
1	0	8	10	0	7	5	0	9	5	0	6	8
2	0	1	8	0	5	15	0	2	20	0	0	28
0	61	1	2	59	1	1	56	2	1	51	2	3
1	0	8	8	0	7	3	0	8	3	0	6	6
2	0	1	7	0	3	14	0	2	16	0	0	20

202. The genesis of the t-test

"Student" (1908) The probable error of a mean. *Biometrika*, **6**, 1–25.

In this seminal paper (in which the t-test was developed) and writing under the pseudonym "Student", William Sealy Gosset uses several illustrative data sets. This one is taken from a table by A.R. Cushny and A.R. Peebles in the *Journal of Physiology* (1904), showing the different effects of the optical isomers of *hyoscyamine hydrobromide* in producing sleep.

The sleep of ten patients was measured without hypnotic, and also after treatment (a) with *D. hyoscyamine hydrobromide* and (b) with *L. hyoscyamine hydrobromide*. The average number of hours' sleep gained by the use of the drug was tabulated.

The table below is as printed in *Biometrika*. It is a pity that in this fundamental paper there should have been a typographical error: this has not been corrected here (but it is not difficult to isolate).

Patient	(a)	(b)	(b)-(a)
1	+0.7	+1.9	+1.2
2	-1.6	+0.8	+2.4
3	-0.2	+1.1	+1.3
4	-1.2	+0.1	+1.3
5	-1.0	-0.1	0
6	+3.4	+4.4	+1.0
7	+3.7	+5.5	+1.8
8	+0.8	+1.6	+0.8
9	0	+4.6	+4.6
10	+2.0	+3.4	+1.4
Mean	0.75	2.33	1.58
SD	1.70	1.90	1.17

203. Dimensions of cuckoos' eggs

Latter, O.H. (1901-02) The egg of Cuculus Canorus. *Biometrika*, **1**, 164–176.

The length and breadth of 243 cuckoos' eggs were measured (in mm, to the nearest half mm). The table below gives the frequencies. Lengths shown are the midpoint of the class interval.

Length	Frequency	Breadth	Frequency
19	1	14	1
$19\frac{1}{2}$	1	$14\frac{1}{2}$	1
20	7	15	5
$20\frac{1}{2}$	3	$15\frac{1}{2}$	9
21	29	16	73
$21\frac{1}{2}$	13	$16\frac{1}{2}$	51
22	54	17	80
$22\frac{1}{2}$	38	$17\frac{1}{2}$	15
23	47	18	7
$23\frac{1}{2}$	22	$18\frac{1}{2}$	0
24	21	19	1
$24\frac{1}{2}$	5		
25	2		

204. Time intervals between coal mining disasters

Jarrett, R.G. (1979) A note on the intervals between coal mining disasters. *Biometrika*, **66**, 191–193.

Original data [Maguire, B.A., Pearson, E.S. and Wynn, A.H.A. (1952) The time intervals between industrial accidents. *Biometrika*, **39**, 168–180] on the time intervals between successive coal mining disasters involving ten or more men killed contain errors. In this paper, errors in the original paper are corrected, and the data set is extended to cover altogether 191 disasters from 1851 to 1962.

These data give the time intervals in days between explosions in coal mines from 15th March 1851 to 22nd March 1962 inclusive. (The data are to be read across rows.)

This time interval covers 40550 days. There were 191 explosions altogether, including ones on each of the two bound days. So the data involve 190 numbers whose sum is 40549. The 0 occurs because there were two accidents on 6th December 1875.

157	123	2	124	12	4	10	216	80	12
33	66	232	826	40	12	29	190	97	65
186	23	92	197	431	16	154	95	25	19
78	202	36	110	276	16	88	225	53	17
538	187	34	101	41	139	42	1	250	80
3	324	56	31	96	70	41	93	24	91
143	16	27	144	45	6	208	29	112	43
193	134	420	95	125	34	127	218	2	0
378	36	15	31	215	11	137	4	15	72
96	124	50	120	203	176	55	93	59	315
59	61	1	13	189	345	20	81	286	114
108	188	233	28	22	61	78	99	326	275
54	217	113	32	388	151	361	312	354	307
275	78	17	1205	644	467	871	48	123	456
498	49	131	182	255	194	224	566	462	228
806	517	1643	54	326	1312	348	745	217	120
275	20	66	292	4	368	307	336	19	329
330	312	536	145	75	364	37	19	156	47
129	1630	29	217	7	18	1358	2366	952	632

The author also splits the data into disasters (191 of them again) per day of week, and per month of year:

Mon	Tue	Wed	Thu	Fri	Sat	Sun
19	34	33	36	35	29	5

J	F	M	A	M	J	J	A	S	O	N	D
14	20	20	13	14	10	18	15	11	16	16	24

205. Ticks on sheep

Fisher, R.A. (1941) The negative binomial distribution, *Annals of Eugenics* OR
Fisher, R.A. (1950) *Contributions to Mathematical Statistics*. Wiley. **Paper 38**.

The number of ticks was counted on each of 82 sheep:

No. of ticks	0	1	2	3	4	5	6	7	8	9
Frequency	4	5	11	10	9	11	3	5	3	2
	10	11	12	13	14	15	16	17	18	19
	2	5	0	2	2	1	1	0	0	1
	20	21	22	23	24	25				
	0	1	1	1	0	2				

206. Brownlee's stack loss data

Brownlee, K.A. (1965) *Statistical theory and methodology in science and engineering*,
2nd edition, London: John Wiley & Sons, 454.

These classic data are observations from 21 days' operation of a plant for the
oxidation of ammonia as a stage in the production of nitric acid. The carrier variables
are:

X_1 = air flow
X_2 = cooling water inlet temperature (°C)
X_3 = acid concentration (%)

and the response variable is:

Y = stack loss

Stack loss is the percentage of the ingoing ammonia that escapes unabsorbed:

	Stack loss			
	X_1	X_2	X_3	Y
1	80	27	58.9	4.2
2	80	27	58.8	3.7
3	75	25	59.0	3.7
4	62	24	58.7	2.8
5	62	22	58.7	1.8
6	62	23	58.7	1.8
7	62	24	59.3	1.9
8	62	24	59.3	2.0

9	58	23	58.7	1.5
10	58	18	58.0	1.4
11	58	18	58.9	1.4
12	58	17	58.8	1.3
13	58	18	58.2	1.1
14	58	19	59.3	1.2
15	50	18	58.9	0.8
16	50	18	58.6	0.7
17	50	19	57.2	0.8
18	50	19	57.9	0.8
19	50	20	58.0	0.9
20	56	20	58.2	1.5
21	70	20	59.1	1.5

207. Heine-Euler extensions

Kemp, A.W. (1992) Heine-Euler extensions of the Poisson distribution. *Communications in Statistics (Theory and Methods)*, **21**, 571–580

In this paper, relations between the Heine, Euler, pseudo-Euler, Poisson and geometric distributions are explored.

The first data set is from David, F.N. (1971) *A first course in statistics*, 2nd edition, Griffin. In fact, the aim here was to estimate the maximum number of fish that could be caught in one trap.

Fish catch data

Fish per trap	0	1	2	3	4	5	6	7	8	9
Frequency	1	2	11	20	29	23	10	3	1	0

The second data set is from Hasselblad, V. (1969) Estimation of finite mixtures of distributions from the exponential family. *Journal of the American Statistical Association*, **64**, 1459–1471. It lists the number of death notices of women aged 80 or over appearing in *The Times* on each day for three consecutive years, 1910–1912.

Death notices in *The Times*

Notices per day	0	1	2	3	4	5	6	7	8	≥9
Frequency	162	267	271	185	111	61	27	8	3	1

The third data set is from Jeffers, J.N.R. (1978) *An introduction to systems analysis with ecological applications*, London: Edward Arnold. Here a Poisson model would provide a very poor fit for the variation observed: there are more pairs caught together than a Poisson model would suggest.

Catches of the leech Helobdella in water samples

Leeches	0	1	2	3	4	5	6	7	8	≥9
Frequency	58	25	13	2	2	1	1	0	1	0

208. Compound normal distributions

Luceno, A (1992) A new family of probability distributions with applications to data analysis. *Communications in Statistics (Theory and Methods)*, **21**, 391–409.

Compound normal distributions are obtained by ascribing a probability distribution to the normal mean. The family of distribution functions generated can be used effectively for skewed and long-tailed data.

(A)   Measurements of the coefficient of friction for a metal

< 0.020	10
0.020 - 0.025	30
0.025 - 0.030	44
0.030 - 0.035	58
0.035 - 0.040	45
0.040 - 0 045	29
0.045 - 0 050	17
0.050 - 0 055	9
> 0.055	8

(B)   Heights of fragmentation bomb bases (in inches)

< 0.8215	9
0.8215 - 0.8245	5
0.8245 - 0.8275	14
0.8275 - 0.8305	21
0.8305 - 0.8335	55
0.8335 - 0.8365	23
0.8365 - 0.8395	7
0.8395 - 0.8425	6
> 0.8425	5

209. Accident-repeatedness among children

Mellinger, C.D., Gaffey, W.R., Sylwester, D.L. and Manheimer, D.I. (1965) A mathematical model with applications to a study of accident repeatedness among children. *Journal of the American Statistical Association*, **60**, 1046–1059.

Injuries were counted for children aged between 4 and 11, the counts taking place over two time intervals (T_1, aged 4 to 7; T_2, aged 8 to 11).

		Injuries in T_1								
		0	1	2	3	4	5	6	≥7	
	0	101	76	35	15	7	3	3	0	240
	1	67	61	32	14	12	4	1	1	192
	2	24	36	22	15	6	1	2	1	107
Injuries in T_2	3	10	19	10	5	2	4	0	2	52
	4	1	7	3	4	2	0	0	0	17
	5	2	1	4	2	0	0	0	0	9
	6	1	1	1	1	0	0	0	0	4
		206	201	107	56	29	12	6	4	621

210. Volume of black cherry trees

Ryan, T.A. Jr., Joiner, B.L. and Ryan, B.F. (1985) *The Minitab Student Handbook*, Boston: Duxbury Press, 328–329. See also Atkinson, A.C. (1982) Regression diagnostics, transformations and constructed variables (with discussion). *Journal of the Royal Statistical Society, Series B*, **44**, 1–36.

These data give the volume (cubic feet), height (feet) and diameter (inches) (at 54 inches above ground) for a sample of 31 black cherry trees in the Allegheny National Forest, Pennsylvania. The data were collected in order to find an estimate for the volume of a tree (and therefore for the timber yield), given its height and diameter.

Diameter	Height	Volume
8.3	70	10.3
8.6	65	10.3
8.8	63	10.2
10.5	72	16.4
10.7	81	18.8
10.8	83	19.7
11.0	66	15.6
11.0	75	18.2
11.1	80	22.6
11.2	75	19.9
11.3	79	24.2
11.4	76	21.0
11.4	76	21.4
11.7	69	21.3
12.0	75	19.1
12.9	74	22.2
12.9	85	33.8
13.3	86	27.4

13.7	71	25.7
13.8	64	24.9
14.0	78	34.5
14.2	80	31.7
14.5	74	36.3
16.0	72	38.3
16.3	77	42.6
17.3	81	55.4
17.5	82	55.7
17.9	80	58.3
18.0	80	51.5
18.0	80	51.0
20.6	87	77.0

211. Digit counts in pi and e

Using the computer algebra program *Mathematica*, the number π was found to be equal to

$$\pi = 3 \cdot (141592 \ldots 5678)5667 \ldots$$

The 10 000 digits in brackets were tallied. Similarly,

$$e = 2 \cdot (718281 \ldots 6788)5674 \ldots$$

and the 10 000 digits in brackets were again tallied.

These are the frequencies of the digits 0, 1, ..., 9 in the first 10 000 places of the decimal expansions of pi and e:

	0	1	2	3	4	5	6	7	8	9
π	968	1026	1021	974	1012	1046	1021	970	948	1014
e	974	989	1004	1008	982	992	1079	1008	996	968

212. Drug content of rat livers

Cook, R.D. and Weisberg, S. (1982) *Residuals and influence in regression*. London: Chapman and Hall.

In an experiment to investigate the amount of a drug retained in the liver of a rat, 19 rats were weighed and dosed. The dose was approximately 40 mg per 1 kg of body weight, since liver weight is known to be strongly correlated with body weight. After a fixed length of time the rat was sacrificed, the liver weighed, and the percentage of the dose in the liver was determined.

Body wt (gm)	Liver wt (gm)	Dose (gm × 10⁻⁴)	Dose in liver (gm × 10⁻⁴)
176	6.5	0.88	0.42
176	9.5	0.88	0.25
190	9.0	1.00	0.56
176	8.9	0.88	0.23
200	7.2	1.00	0.23
167	8.9	0.83	0.32
188	8.0	0.94	0.37
195	10.0	0.98	0.41
176	8.0	0.88	0.33
165	7.9	0.84	0.38
158	6.9	0.80	0.27
148	7.3	0.74	0.36
149	5.2	0.75	0.21
163	8.4	0.81	0.28
170	7.2	0.85	0.34
186	6.8	0.94	0.28
146	7.3	0.73	0.30
181	9.0	0.90	0.37
149	6.4	0.75	0.46

213. Envelope usage

Data provided by Dr W.J. Sutherland, School of Biological Sciences, University of East Anglia.

Dr Sutherland works in a large organization in which internal notes and memoranda are sent in re-usable envelopes. Each envelope has twelve spaces (windows) for the names of recipients; new users cross out their own name, and write in the next window the name of the person they wish to contact. The contributor kept a count of how many names, including his, were written on 311 of the envelopes he received, thus obtaining some idea of the age structure of envelopes in circulation.

The original article [Sutherland, W. (1990) The great pigeonhole in the sky. *New Scientist*, **9th June 1990**, 73–74] gave only a histogram of the data. We obtained from Dr Sutherland the original frequencies, which are reproduced here, with thanks.

No. of windows	Frequency	No. of windows	Frequency
1	91	7	11
2	64	8	8
3	39	9	7
4	30	10	9
5	27	11	4
6	20	12	1

214. American Psychological Association election data

Diaconis, P. (1989) A generalization of spectral analysis with application to ranked data. *Annals of Statistics*, **17**, 949–979.

In a paper about measures of concordance, data are included on results in the ballot for the election of officers in the American Psychological Association. Five candidates were ranked in order of precedence by 5738 voters. The 60 different rankings were recorded as follows:

54321	29	43521	91	32541	41	21543	36
54312	67	43512	84	32514	64	21534	42
54231	37	43251	30	32451	34	21453	24
54213	24	43215	35	32415	75	21435	26
54132	43	43152	38	32154	82	21354	30
54123	28	43125	35	32145	74	21345	40
53421	57	42531	58	31542	30	15432	40
53412	49	42513	66	31524	34	15423	35
53241	22	42351	24	31452	40	15342	36
53214	22	42315	51	31425	42	15324	17
53142	34	42153	52	31254	30	15243	70
53124	26	42135	40	31245	34	15234	50
52431	54	41532	50	25431	35	14532	52
52413	44	41523	45	25413	34	14523	48
52341	26	41352	31	25341	40	14352	51
52314	24	41325	23	25314	21	14325	24
52143	35	41253	22	25143	106	14253	70
52134	50	41235	16	25134	79	14235	45
51432	50	35421	71	24531	63	13542	35
51423	46	35412	61	24513	53	13524	28
51342	25	35241	41	24351	44	13452	37
51324	19	35214	27	24315	28	13425	35
51243	11	35142	45	24153	162	13254	95
51234	29	35124	36	24135	96	13245	102
45321	31	34521	107	23541	45	12543	34
45312	54	34512	133	23514	52	12534	35
45231	34	34251	62	23451	53	12453	29
45213	24	34215	28	23415	52	12435	27
45132	38	34152	87	23154	186	12354	28
45123	30	34125	35	23145	172	12345	30

215. Distribution of seabirds

Solow, A.R. and Smith, W. (1991) Detecting cluster in a heterogeneous community sampled by quadrats. *Biometrics*, **47**, 311–317.

Counts were made on four species of seabird in ten square quadrats in the Anadyr Strait off the coast of Alaska during the summer of 1988. The quadrat side was $\frac{1}{4}$ km. The species were murra (1), crested auklet (2), least auklet (3) and puffin (4).

		Species		
Quadrat	1	2	3	4
1	0	0	1	1
2	0	0	2	0
3	0	0	0	1
4	1	2	0	1
5	1	3	0	0
6	0	1	0	0
7	0	5	1	3
8	1	0	3	1
9	1	1	2	1
10	3	5	3	0

216. Spermarche

Jorgensen, M., Keiding, N. and Skakkebaek, N.E. (1991) Estimation of spermarche from longitudinal spermaturia data. *Biometrics*, **47**, 177–193.

This paper explores the statistical modelling required to extract evidence on the age at onset of sperm emission (spermarche) in urine for a group of boys followed from an age that was certainly below spermarche and through puberty. Every three months a urine sample was collected from each of 40 boys and analysed for the presence of spermatozoa. Of the 40 boys, seven did not show spermaturia during the observation period.

The table gives the age at which the first sample was taken, the age at first evidence of spermaturia and the age at which the last sample was taken. The +/- sequence indicates positive and negative samples. The total number of samples taken is given.

Age at first sample	Age evidence	Age last sample		
10.3	13.4	16.7	---------++----+++--	21
10.0	12.1	17.0	--------+--++-+--+-+-----++	27
9.8	12.1	16.4	--------+-++-+++++++--++-+	25
10.6	13.5	17.7	-----------++---+----	21
9.3	12.5	16.3	-----------++----+--------	27
9.2	13.9	16.2	-----------------+-------	25
9.6	15.1	16.7	------------------+---+	24
9.2	-	12.2	------------	12
9.7	-	12.1	---------	9
9.6	12.7	16.4	-----------+-+++++--++-+	25
9.6	12.5	16.7	----------+--+-+--+++	21
9.3	15.7	16.0	----------------------++	24
9.6	-	12.0	---------	9
9.4	12.6	13.1	----------++++	14
10.5	12.6	17.5	-------+-+++++++++--+--++	24
10.5	13.5	14.1	----------+--	13
9.9	14.3	16.8	---------------+-----+-+	24
9.3	15.3	16.2	--------------------+++	24
10.4	13.5	17.3	--------++-+-++-+-+++	21
9.8	12.9	16.7	----------++++-++++-++-+--	27
10.8	14.2	17.3	-----------+--+++-+	20
10.9	13.3	17.8	--------++++-+++++-++--	23
10.6	-	13.8	-----------	11
10.6	14.3	16.3	-------------+---+---	21
10.5	12.9	17.4	--------+-++++---++--++++	25
11.0	-	12.4	------	6
8.7	-	12.3	--------------	14
10.9	-	14.5	-------------	13
11.0	14.6	17.5	-----------+++++++++++-+	24
10.8	14.1	17.6	-----------++--+------	22
11.3	14.4	18.2	-----------++-++--+-----	24
11.4	13.8	18.3	-------+---+---+++--+-+	23
11.3	13.7	17.8	-------+++-+---+++-++	21
11.2	13.5	15.7	---------+--------	18
11.3	14.5	16.3	-----------+-++---	18
11.2	14.3	17.2	-----------+--+-++++++-	23
11.6	13.9	14.7	-----+---	9
11.8	14.1	17.9	----+-+-+-++++----	18
11.4	13.3	18.2	----+++-+-----+++++--	21
11.5	14.0	17.9	-------++-------++-+-	21

217. Capture experiments

Huggins, R.M. (1991) Some practical aspects of a conditional likelihood approach to capture experiments. *Biometrics*, **47**, 725–732.

The author describes procedures for estimating the size of a closed population when the capture probabilities are heterogeneous, by modelling the capture probabilities in terms of observable covariates.

These data are captures of *peromyscus maniculatus* collected by V. Reid at East Stuart Gully, Colorado. The columns represent the sex (M or F); the age (Y young, S semi-adult, A adult), the weights (in g) and the capture histories of 36 individuals over six trapping occasions (1 trapped, 0 not).

(The author writes of "38" individuals: either this is a misprint; or two rows of the data were lost.)

M	Y	12	1	1	1	1	1	1
F	Y	15	1	0	0	1	1	1
M	Y	15	1	1	0	0	1	1
M	Y	15	1	1	0	1	1	1
M	Y	13	1	1	1	1	1	1
M	Y	13	0	1	1	0	1	0
F	Y	5	0	1	0	1	0	1
F	A	20	0	1	0	0	0	1
M	Y	12	0	1	0	0	1	1
F	Y	6	0	0	1	0	0	0
M	A	21	1	1	0	1	1	1
M	Y	11	1	1	1	1	1	0
M	S	15	1	1	1	0	0	1
M	Y	14	1	1	1	1	1	1
M	Y	13	1	1	0	1	1	1
F	A	22	1	1	1	0	1	1
M	Y	14	1	1	1	1	1	1
M	Y	11	1	0	1	1	1	0
F	Y	10	1	0	0	1	0	0
F	A	23	0	1	0	0	1	0
F	Y	7	0	1	1	0	0	1
M	Y	8	0	1	0	0	0	1
M	A	19	0	1	0	1	0	1
F	A	22	0	0	1	1	1	1
F	Y	10	0	0	1	0	1	1
F	Y	14	0	0	1	1	1	1
F	A	19	0	0	1	0	0	0
F	A	20	0	0	0	1	0	0
M	S	16	0	0	0	1	1	1
F	Y	11	0	0	0	1	1	0
M	Y	14	0	0	0	0	1	0
F	Y	11	0	0	0	0	1	0
M	A	24	0	0	0	0	1	0
M	Y	9	0	0	0	0	0	1
M	S	16	0	0	0	0	0	1
F	A	19	0	0	0	0	0	1

218. Unaided distance vision

Stuart, A. (1955) A test for homogeneity of the marginal distributions in a two-way classification. *Biometrika*, **42**, 412–416.

The unaided distance vision of 7477 women aged 30-39 employed in Royal Ordnance factories was measured. These data have been analysed by many statisticians using various statistical models and methods.

		Left eye grade (1=high)			
		1	**2**	**3**	**4**
Right	**1**	1520	266	124	66
eye	**2**	234	1512	432	78
grade	**3**	117	362	1772	205
	4	36	82	179	492

219. Dental records

Potthoff, R.F. and Roy, S.N. (1964) A generalized multivariate analysis of variance model useful especially for growth curve problems. *Biometrika*, **51**, 313–326.

Data were collected on 11 girls and 16 boys by investigators at the University of North Carolina Dental School. Each measurement is the distance, in mm, from the centre of the pituitary to the pteryomaxillary. (The reason why this distance occasionally decreases with age is that the distance represents the relative position of two moving points.)

Girls

	Age			
	8	**10**	**12**	**14**
1	21	20	$21\frac{1}{2}$	23
2	21	$21\frac{1}{2}$	24	$25\frac{1}{2}$
3	$20\frac{1}{2}$	24	$24\frac{1}{2}$	26
4	$23\frac{1}{2}$	$24\frac{1}{2}$	25	$26\frac{1}{2}$
5	$21\frac{1}{2}$	23	$22\frac{1}{2}$	$25\frac{1}{2}$
6	20	21	21	$22\frac{1}{2}$
7	$21\frac{1}{2}$	$22\frac{1}{2}$	23	25
8	23	23	$23\frac{1}{2}$	24
9	20	21	22	$21\frac{1}{2}$
10	$16\frac{1}{2}$	19	19	$19\frac{1}{2}$
11	$24\frac{1}{2}$	25	28	28

Boys

	Age			
	8	10	12	14
1	26	25	29	31
2	$21\frac{1}{2}$	$22\frac{1}{2}$	23	$26\frac{1}{2}$
3	23	$22\frac{1}{2}$	24	$27\frac{1}{2}$
4	$25\frac{1}{2}$	$27\frac{1}{2}$	$26\frac{1}{2}$	27
5	20	$23\frac{1}{2}$	$22\frac{1}{2}$	26
6	$24\frac{1}{2}$	$25\frac{1}{2}$	27	$28\frac{1}{2}$
7	22	22	$24\frac{1}{2}$	$26\frac{1}{2}$
8	24	$21\frac{1}{2}$	$24\frac{1}{2}$	$25\frac{1}{2}$
9	23	$20\frac{1}{2}$	31	26
10	$27\frac{1}{2}$	28	31	$31\frac{1}{2}$
11	23	23	$23\frac{1}{2}$	25
12	$21\frac{1}{2}$	$23\frac{1}{2}$	24	28
13	17	$24\frac{1}{2}$	26	$29\frac{1}{2}$
14	$22\frac{1}{2}$	$25\frac{1}{2}$	$25\frac{1}{2}$	26
15	23	$24\frac{1}{2}$	26	30
16	22	$21\frac{1}{2}$	$23\frac{1}{2}$	25

220. Nigerian smallpox data

Becker, N.G. (1983) Analysis of data from a single epidemic. *Australian Journal of Statistics*, **25**, 191–197.

These are data from a smallpox outbreak in the community of Abakaliki in southeastern Nigeria. There were 30 cases in a population of 120 individuals at risk. The data give the 29 inter-removal times (in days). A zero corresponds to cases appearing on the same day.

13	7	2	3	0	0	1	4	5	3	2	0	2	0	5
3	1	4	0	1	1	1	2	0	1	5	0	5	5	

221. Disease lesions in wheat

Shaw, M.W. (1990) A test of spatial randomness on small scales, combining information from mapped locations within several quadrats. *Biometrics*, **46**, 447–458.

Sixteen square quadrats with side 1 m were laid out to sample from within a plot of winter wheat cv. Longbow on 25 September 1984. When the disease caused by the fungus *Septoria tritici* was first found in the crop, the leaves of each plant within each quadrat were searched for lesions caused by the disease. The data give the lesion co-ordinates (in cm from one corner).

221. Disease lesions in wheat

1	40,11	45,22	54,13	87,31	90,70	62,95	30,18									
2	22,05	27,15	67,02	55,12	20,12	54,25	43,52	42,80								
3	15,14	22,08	43,30	34,39	44,50	89,09	15,67	13,42								
4	15,45	25,05	39,00	69,08	78,17	72,35	53,52	61,73	65,89	43,82	10,18	17,45				
5	73,92	51,80	43,50	78,08							98,02					
6	75,10	73,19	09,06	89,56	70,60	60,60	57,68	57,73	11,45	20,40						
7	95,65	65,57	55,79	92,91	06,05	04,60	14,20	98,60								
8	19,24	40,37	70,82	85,80	16,17											
9	87,16	91,11	70,60	86,94	50,70	56,21	48,97	12,62	02,30	65,02						
10	65,82	50,16	51,33	52,77	41,97	21,19	06,06	09,89	75,32							
11	77,20	85,61	98,92	62,02	45,20	34,02	27,25	28,60	21,70	90,15						
12	82,46	90,84	61,32	52,55	42,50	25,84	09,35	06,16	89,08							
13	30,84	12,40	09,14	66,32												
14	72,56	62,23	58,32	63,50	58,91	54,90	48,50	70,41	35,35	08,83	92,55					
15	01,10	37,11	41,10	38,45	48,36	42,87	48,80	86,55	78,75	80,59	76,08	89,33	91,02	99,19	72,84	
16	48,32	15,22	20,45	12,40	11,44	03,35	29,78	30,70	09,60	10,92	77,40					05,03

222. Measuring drug diffusion

Nievergelt, Y. (1990) Fitting density functions and diffusion tensors to three-dimensional drug transport within brain tissue. *Biometrics*, **46**, 1111–1121.

The paper deals with a study into drug transport within brain tissue. A common method for determining the extent and shape of the spatial distribution of drugs in brain tissue involves removing cylindrical punches from the tissue and counting by scintillation the total amount of drug in each punch. The data are the scintillation counts (each 4.753×10^{13} mole) in a $5 \times 5 \times 4$ spatial array. From a visual inspection of the injection site, it was concluded that the location of the injection coincided with the cylinder having the highest scintillation count, indicated by the number 28353. Interest lay in the diffusion of the drug away from the injection site.

Rear

157	162	132	177	153
170	182	161	143	174
144	183	358	231	154
155	221	287	217	142
160	192	240	166	177

Next

155	155	197	158	187
169	218	532	406	154
201	935	4435	1099	212
236	1392	2102	1986	260
160	224	581	1161	319

Next

164	442	1320	414	188
480	7022	14411	5158	352
2091	23027	28353	13138	681
789	21260	20921	11731	727
213	1303	3765	1715	453

Front

163	324	432	243	166
712	4055	6098	1048	232
2137	15531	19742	4785	330
444	11431	14960	3182	301
294	2061	1036	258	188

223. Judgement concordance

Darroch, J.N. and McCloud, P.I. (1986) Category distinguishability and observer agreement. *Australian Journal of Statistics*, **28**, 371–388.

The authors discuss data for judgement concordance.

(1)   Two neurologists independently classified 149 patients into four classes: (A) certainly suffering from multiple sclerosis; (B) probable MS; (C) possible MS ("50:50"); and (D) doubtful, unlikely and definitely not MS.

	A	B	C	D	
A	38	5	0	1	44
B	33	11	3	0	47
C	10	14	5	6	35
D	3	7	3	10	23
	84	37	11	17	149

(2)   Two observers classified the health of 992 trees and shrubs from (1) poor to (4) good.

	(1)	(2)	(3)	(4)	
(1)	239	18	9	11	277
(2)	24	38	41	11	114
(3)	15	49	113	94	271
(4)	6	22	109	193	330
	284	127	272	309	992

(3)   One observer classified the health of 46 plants one week apart. It can be assumed that in that week the actual health of the plant did not alter.

			Late		
	(1)	(2)	(3)	(4)	
(1)	6	0	0	0	6
(2)	1	4	1	0	6
(3)	0	1	3	5	9
(4)	0	0	4	21	25
	7	5	8	26	46

Data set (1) is from Westlund, K.B. and Kurland, L.T. (1953) Studies on multiple sclerosis in Winnipeg, Manitoba and New Orleans, Loiuisiana. *American Journal of Hygiene*, **57**, 380–396. Data sets (2) and (3) are from Lay, B.G. and Meissner, A.P. (1985) An objective method for assessing the performance of amenity plantings. *Journal of Adelaide Botanical Gardens*, **7**, 159–166.

224. Bird colouring

Anderson, J.A. and Pemberton, J.D. (1985) The grouped continuous model for multivariate ordered categorical variables and covariate adjustment. *Biometrics*, **41**, 875–885.

For each of 90 birds, the colour of the lower mandible (LM), upper mandible (UM) and orbital ring (OR) were recorded as ordered categorical variables 1-3. The question arose: are the colours of the three features related?

		OR 1 LM			OR 2 LM			OR 3 LM		
		1	2	3	1	2	3	1	2	3
UM	1	40	1	1	19	6	2	0	0	0
	2	0	1	0	0	2	1	0	1	1
	3	0	0	0	1	1	6	0	0	7

225. Incidence and mortality for lung cancer

O'Neill, T.J., Tallis, G.M. and Leppard, P. (1985) The epidemiology of a disease using hazard functions. *Australian Journal of Statistics*, **27**, 283–297.

The paper reports incidence and mortality data for lung cancer in South Australia in 1981.

Age interval	Estimated population size		New cases of cancer		Deaths	
	Male	Female	Male	Female	Male	Female
0-4	47589	45273	0	0	0	0
5-9	53814	50672	0	0	0	0
10-14	58561	55645	0	0	0	0
15-19	59408	57756	0	0	0	0
20-24	58443	57249	0	0	0	0
25-29	54341	53376	0	0	1	0
30-34	53456	52978	1	0	1	0
35-39	42113	41988	0	2	0	0
40-44	35648	35547	2	5	3	3
45-49	32911	31799	8	2	10	2
50-54	36485	35333	38	8	26	8

55-59	35192	35555	61	18	43	8
60-64	28131	30868	67	16	57	15
65-69	24419	27390	88	15	69	17
70-74	16613	21402	60	21	61	21
75-79	9958	14546	46	10	46	9
80-84	4852	9749	24	6	23	4
85+	2790	7477	7	2	8	3

226. Bump-hunting

Good, I.J. and Gaskins, R.A. (1980) Density estimation and bump-hunting by the penalized likelihood method exemplified by scattering and meteorite data. *Journal of the American Statistical Association*, **75**, 42–56.

A total of 25752 scattering 'events' were counted from a scattering reaction, over 172 bins each of width 10 MeV. There are peaks of activity ("bumps") and the aim is to identify where these peaks occur (read across rows).

5	11	17	21	15	17	23	25	30	22
36	29	33	43	54	55	59	44	58	66
59	55	67	75	82	98	94	85	92	102
113	122	153	155	193	197	207	258	305	332
318	378	57	540	592	646	773	787	783	695
774	759	692	559	557	499	431	421	353	315
343	306	262	265	254	225	246	225	196	150
118	114	99	121	106	112	122	120	126	126
141	122	122	115	119	166	135	154	120	162
156	175	193	162	178	201	214	230	216	229
214	197	170	181	183	144	114	120	132	109
108	97	102	89	71	92	58	65	55	5
40	42	46	47	37	49	38	29	34	42
45	42	40	59	42	35	41	35	48	4
47	49	37	40	33	33	37	29	26	38
22	27	27	13	18	25	24	21	16	24
14	23	21	17	17	21	10	14	18	16
21	6								

227. 'Bliss' data sets

Bliss, C.I. (1953) Fitting the negative binomial distribution to biological data. *Biometrics*, **9**, 176–200. (And including *A Note on the efficient fitting of the negative binomial* by R.A. Fisher.)

(1)   Twenty-five leaves were selected at random from each of six similar apple trees. The number of adult female European red mites on each was counted.

Mites	0	1	2	3	4	5	6	7
Frequency	70	38	17	10	9	3	2	1

(2)   Bacterial clumps per field were counted in a milk film. (A microscope slide was split into 400 regions of equal area: each region is called a 'field'.)

0	1	2	3	4	5	6	7	8	9
56	104	80	62	42	27	9	9	5	3

10	11	12	13	14	15	16	17	18	19
2	0	0	0	0	0	0	0	0	1

(3)   Yeast cells per square in a haemocytometer were counted.

Cells	0	1	2	3	4	5
Frequency	213	128	37	18	3	1

Original source: Student

(4)   The number of accidents experienced by 414 machinists was counted. (Time period not specified.)

Accidents	0	1	2	3	4	5	6	7	8
Frequency	296	74	26	8	4	4	1	0	1

Original source: Greenwood, M. and Yule, G.U. (1920) An inquiry into the nature of frequency distributions representative of multiple happenings with particular reference to multiple attacks of disease or of repeated accidents. *Journal of the Royal Statistical Society*, **83**, 255–279.

(5)   Soil organisms were counted.

Colonies per field

0	1	2	3	4	5	≥6
11	37	64	55	37	24	12

Bacteria per colony

1	2	3	4	5	6	≥7
359	146	57	41	26	17	27

Bacteria per field

0	1	2	3	4	5	6
11	17	31	24	29	18	19

7	8	9	10	11	≥12
16	13	17	6	8	31

Original source: Jones, P.C.T., Mollison, J.E. and Quenouille, M.H. (1948) A technique for the quantitative estimation of soil micro-organisms. *Journal Gen. Microbiology*, **2**, 54–69.

(6)   *Primula auricula* in a grassland association.

109 quadrats were sampled. The number of plants in each was counted.

Plants	0	1	2	3	4	5	6	7	8	9
Frequency	26	21	23	14	11	4	5	4	0	1

Original source: Blackman, G.E. (1935) A study by statistical methods of the distribution of species in grassland associations. *Ann. Bot.*, **49**, 749–777.

(7)   *Microcalanus nauplii* in 150 samples of marine plankton

0	1	2	3	4	5	6	7	8	9
0	2	4	3	5	8	16	13	12	13

10	11	12	13	14	15	16	17	18	19
15	15	9	9	7	4	4	6	2	0

20	21
2	1

Original source: Barnes, H. and Marshall, S.M. (1951) On the variability of replicate plankton samples and some applications of "contagious" series to the statistical distribution of catches over restricted periods. *Journal of Marine Biol. Association UK*, **30**, 233–263.

(8)   *Tanytarsus* in 189 Ekman hauls. (An Ekman haul is a collection of fish by weight.)

0	1	2	3	4	5	6	7	8	9
32	28	25	34	13	14	17	5	6	1

10	11	12	13	14	15	16	17	18
9	0	0	2	1	0	0	1	1

Oligochaetes in 164 Petersen hauls. (A Petersen haul is a collection of fish by volume.)

0	1	2	3	4	5	6	7	8	9
39	24	18	21	15	15	6	8	6	2

10	11	12	13	14	15
1	2	3	3	0	1

(9)   Isopods under 122 boards

0	1	2	3	4	5	6	7	8	9
28	28	14	11	8	11	2	3	3	3

10	11	12	13	14	15	16	17
3	2	0	1	2	1	0	2

Original source: Cole L.C. (1946) A theory for analysing contagiously distributed populations. *Ecology*, **27**, 329–341.

(10)   The mite *Liponysus bacoti* counted on the backs of 227 Savannah rats.

0	1	2	3	4	5	6	7	8	9
160	19	11	6	5	4	4	3	2	2

10	11	12	13	14	15	16	17	18	19
0	0	1	0	0	1	0	0	1	1

20
6

Original source: Cole, L.C. (1949) The measurement of interspecific association. *Ecology*, **30**, 411–424.

(11)   The distribution of the weed *Chenopodium album* on arable land. (Plants per quadrat.)

0	1	2	3	4	5	6	7	8	9	10
19	5	6	9	5	20	14	8	4	3	2

Original source: Singh, B.N. and Chalam, G.V. (1937) A quantitative analysis of the weed flora on arable land. *Journal of Ecology*, **25**, 213–221.

(12)   Lice per head of Hindu male prisoners in Cannamore, South India, 1937–1939.

0	612
1	106
2	50
3	29
4	33
5	20
6	14
7	12
8	18
9	11
10	11
12	10
13	8
14	6
15	3
16	6
17	7
18	4
19	7
20	7
21	3
22-23	8
24-25	7
26-27	10
28-29	6
30-32	2
33-35	7
36-38	9
39-41	5
42-45	5
46-49	7
≥ 50	27

Original source: Williams, C.B. (1944) Some applications of the logarithmic series and the index of diversity to ecological problems. *Journal of Ecology*, **32**, 1–44.

228. Bird survival

Pollock, K.H., Winterstein, S.R. and Conroy, M.J. (1989) Estimation and analysis of survival distributions for radio-tagged animals. *Biometrics*, **45**, 99–109.

Fifty female black ducks from two locations in New Jersey were captured and fitted with radios. The birds included 31 hatch-year birds (born during the previous breeding season) and 19 after-hatch-year birds (all at least one year old). Only for some birds was death observed, shown in the column headed Indicator.

Bird	Survival time	Indicator	Age	Weight	Length
1	2	1	1	1160	277
2	6	0	0	1140	266
3	6	0	1	1260	280
4	7	1	0	1160	264
5	13	1	1	1080	267
6	14	0	0	1120	262
7	16	0	1	1140	277
8	16	1	1	1200	283
9	17	0	1	1100	264
10	17	1	1	1420	270
11	20	0	1	1120	272
12	21	1	1	1110	271
13	22	1	0	1070	268
14	26	1	0	940	252
15	26	1	0	1240	271
16	27	1	0	1120	265
17	28	0	1	1340	275
18	29	1	0	1010	272
19	32	1	0	1040	270
20	32	0	1	1250	276
21	34	1	0	1200	276
22	34	1	0	1280	270
23	37	1	0	1250	272
24	40	1	0	1090	275
25	41	1	1	1050	275
26	44	1	0	1040	255
27	49	0	0	1130	268
28	54	0	1	1320	285
29	56	0	0	1180	259
30	56	0	0	1070	267
31	57	0	1	1260	269
32	57	0	0	1270	276
33	58	0	0	1080	260
34	63	0	1	1110	270
35	63	0	0	1150	271
36	63	0	0	1030	265
37	63	0	0	1160	275

38	63	0	0	1180	263
39	63	0	0	1050	271
40	63	0	1	1280	281
41	63	0	0	1050	275
42	63	0	0	1160	266
43	63	0	0	1150	263
44	63	0	1	1270	270
45	63	0	1	1370	275
46	63	0	1	1220	265
47	63	0	0	1220	268
48	63	0	0	1140	262
49	63	0	0	1140	270
50	63	0	0	1120	274

229. Trailing digits in data

Preece, D.A. (1981) Distributions of final digits in data. *The Statistician*, **30**, 31–60.

The author gives several examples of the distribution of the final digit in data.

(1)   Frequencies of final digits in thermometer readings:

	1899-1902		1922-1924	
	max	min	max	min
0	314	422	194	277
1	30	12	148	149
2	144	84	108	113
3	59	39	72	70
4	39	19	29	35
5	109	192	51	50
6	51	49	32	27
7	8	11	54	68
8	114	111	102	98
9	132	61	210	113
	1000	**1000**	**1000**	**1000**

(2)   Frequencies of final digits in some motor-car stopping distances (feet):

0	1	2	3	4	5	6	7	8	9
11	0	8	1	10	1	13	1	5	0

(3)   Frequencies of final digits in the quoted length of 141 North American rivers (miles):

0	1	2	3	4	5	6	7	8	9
74	8	5	4	5	22	8	5	6	4

(4)   Frequencies of final digits in measurements of the antero-posterior curves of 100 "ordinary" prisoners (as opposed to "lunatic"):

0	1	2	3	4	5	6	7	8	9
59	0	1	0	3	35	0	0	2	0

(5)   Frequencies of final digits of the chest circumference of 125 one-year-old children.

0	1	2	3	4	5	6	7	8	9
60	0	9	0	0	41	3	0	12	0

230. Silica in meteors

Good, I.J. and Gaskins, R.A. (1980) Density estimation and bump-hunting by the penalized likelihood method exemplified by scattering and meteorite data. *Journal of the American Statistical Association*, **75**, 42–56.

The percentage of silica was calculated in each of 22 chondrites meteors:

20.77   22.56   22.71   22.99   26.39   27.08   27.32   27.33   27.57   27.81   28.69   29.36
30.25   31.89   32.88   33.23   33.28   33.40   33.52   33.83   33.95   34.82

231. Husbands and wives

Marsh, C. (1988) *Exploring Data*. Cambridge, UK: Polity Press, 315.

These data are taken from the OPCS study of the heights and weights of the adult population of Great Britain in 1980. They represent a random sample of two hundred married men and their wives. The five variables are husband's age (years), husband's height (mm), wife's age (years), wife's height (mm) and husband's age at the time of the marriage. A * indicates a missing observation.

The data are interesting for at least two reasons. A look at the trailing digits suggests that husbands' heights were measured with greater precision than wives' heights. And is there an indication that if the wife's age was unrecorded (refused?) then her height was measured more accurately?

Husband		Wife		Husband age at marriage
Age	Height	Age	Height	
49	1809	43	1590	25
25	1841	28	1560	19
40	1659	30	1620	38
52	1779	57	1540	26
58	1616	52	1420	30
32	1695	27	1660	23
43	1730	52	1610	33
42	1753	*	1635	30
47	1740	43	1580	26
31	1685	23	1610	26
26	1735	25	1590	23
40	1713	39	1610	23
35	1736	32	1700	31
45	1715	*	1522	41
35	1799	35	1680	19
35	1785	33	1680	24
47	1758	43	1630	24
38	1729	35	1570	27
33	1720	32	1720	28
32	1810	30	1740	22
38	1725	40	1600	31
45	1764	*	1689	24
29	1683	29	1600	25
59	1585	55	1550	23
26	1684	25	1540	18
50	1674	45	1640	25
49	1724	44	1640	27
42	1630	40	1630	28
33	1855	31	1560	22
31	1796	*	1652	25
27	1700	25	1580	21
57	1765	51	1570	32
34	1700	31	1590	28
28	1721	25	1650	23
46	1823	*	1591	*
37	1829	35	1670	22
56	1710	55	1600	44
27	1745	23	1610	25
36	1698	35	1610	22
31	1853	28	1670	20
57	1610	52	1510	25
55	1680	53	1520	21
47	1809	43	1620	25
64	1580	61	1530	21
60	1600	*	1451	26
31	1585	23	1570	28
35	1705	35	1580	25
36	1675	35	1590	22

40	1735	39	1670	23
30	1686	24	1630	27
32	1768	29	1510	21
27	1721	*	1560	26
20	1754	21	1660	19
45	1739	39	1610	25
59	1699	52	1440	27
43	1825	52	1570	25
29	1740	26	1670	24
48	1704	*	1635	27
39	1719	*	1670	25
47	1731	48	1730	21
54	1679	53	1560	*
43	1755	42	1590	20
54	1713	50	1600	23
61	1723	64	1490	26
27	1783	26	1660	20
51	1585	*	1504	50
27	1749	32	1580	24
32	1710	31	1500	31
54	1724	53	1640	20
37	1620	39	1650	21
55	1764	45	1620	29
36	1791	33	1550	30
32	1795	32	1640	25
57	1738	55	1560	24
51	1639	*	1552	25
62	1734	*	1600	33
57	1695	*	1545	22
51	1666	52	1570	24
50	1745	50	1550	22
32	1775	32	1600	20
54	1669	54	1660	20
34	1700	32	1640	22
45	1804	41	1670	27
64	1700	61	1560	24
55	1664	43	1760	31
27	1753	28	1640	23
55	1788	51	1600	26
27	1765	*	1571	*
41	1680	41	1550	22
44	1715	41	1570	24
22	1755	21	1590	21
30	1764	28	1650	29
53	1793	47	1690	31
42	1731	37	1580	23
31	1713	28	1590	28
36	1725	35	1510	26
56	1828	55	1600	30
46	1735	45	1660	22
34	1760	34	1700	23

55	1685	51	1530	34
44	1685	39	1490	27
45	1559	35	1580	34
48	1705	45	1500	28
44	1723	44	1600	41
59	1700	47	1570	39
64	1660	57	1620	32
34	1681	33	1410	22
37	1803	38	1560	23
54	1866	59	1590	49
49	1884	46	1710	25
63	1705	60	1580	27
48	1780	47	1690	22
64	1801	55	1610	37
33	1795	45	1660	17
52	1669	47	1610	23
27	1708	24	1590	26
33	1691	32	1530	21
46	1825	47	1690	23
54	1760	57	1600	23
27	1949	*	1693	25
50	1685	*	1580	21
42	1806	*	1636	22
54	1905	46	1670	32
49	1739	42	1600	28
62	1736	63	1570	22
34	1845	32	1700	24
23	1868	24	1740	19
36	1765	32	1540	27
53	1736	*	1555	30
32	1741	*	1614	22
59	1720	56	1530	24
53	1871	50	1690	25
55	1720	55	1590	21
62	1629	58	1610	23
42	1624	38	1670	22
50	1653	44	1690	35
37	1786	35	1550	21
51	1620	44	1650	30
25	1695	25	1540	19
54	1674	43	1660	35
34	1864	31	1620	23
43	1643	35	1630	29
43	1705	41	1610	22
58	1736	50	1540	32
28	1691	23	1610	23
45	1753	43	1630	21
47	1680	49	1530	20
57	1724	59	1520	24
27	1710	*	1544	20

34	1638	38	1570	33
57	1725	42	1580	52
27	1725	21	1550	24
54	1630	*	1570	34
24	1810	*	1521	16
48	1774	42	1580	30
37	1771	35	1630	28
25	1815	26	1650	20
57	1575	57	1640	20
40	1729	34	1650	26
61	1749	63	1520	21
25	1705	23	1620	24
32	1875	*	1744	22
37	1784	*	1647	22
45	1584	*	1615	29
24	1774	23	1680	22
47	1658	46	1670	24
44	1790	40	1620	24
52	1798	53	1570	25
45	1824	40	1660	23
20	1796	22	1550	19
60	1725	60	1590	21
36	1685	32	1620	25
25	1769	24	1560	18
25	1749	28	1670	21
35	1716	40	1650	17
35	1664	*	1539	22
49	1773	48	1470	21
33	1760	33	1580	20
50	1725	49	1670	23
63	1645	64	1520	28
57	1694	55	1620	24
41	1851	41	1710	23
38	1691	38	1530	20
30	1880	31	1630	22
52	1835	52	1720	30
51	1730	43	1570	22
46	1644	51	1560	27
50	1723	47	1650	25
32	1758	*	1635	24
52	1718	32	1590	25
30	1723	33	1590	22
33	1708	*	1566	21
20	1786	18	1590	19
32	1764	*	1662	*
51	1675	45	1550	25
64	1641	64	1570	30
44	1743	43	1560	25
40	1823	39	1630	23
59	1720	56	1530	24

232. Presurgical stress

Hoaglin, D.C., Mosteller, F. and Tukey, J.W. (1985) *Exploring data tables, trends and shapes.* New York: John Wiley & Sons, 420.

People have higher levels of beta-endorphin in the blood under conditions of emotional stress. For 19 patients scheduled to undergo surgery, blood samples were taken (a) 12-14 hours before surgery and (b) 10 minutes before surgery. Beta-endorphin levels were measured in fmol/ml (a femto-mole is 10^{-15} grams times the molecular weight of the substance).

(a)	10.0	6.5	8.0	12.0	5.0	11.5	5.0	3.5	7.5	5.8
	4.7	8.0	7.0	17.0	8.8	17.0	15.0	4.4	2.0	

(b)	6.5	14.0	13.5	18.0	14.5	9.0	18.0	42.0	7.5	6.0
	25.0	12.0	52.0	20.0	16.0	15.0	11.5	2.5	2.0	

233. Bladder cancer study

Davis, C.S. and Wei, L.J. (1988) Nonparametric methods for analysing incomplete nondecreasing repeated measurements. *Biometrics*, **44**, 1005–1018.

A bladder cancer study was conducted by the Veterans Administration Cooperative Urological Research Group. All patients had superficial bladder tumours when they entered the trial. These tumours were removed and patients were assigned to one of three treatment groups: placebo, thiotepa and pyridoxine. At subsequent visits, new tumours were removed and the treatment continued. Visits were theoretically quarterly, but there were many missed appointments (indicated in the tables by the symbol -).

The data show for each patient in Groups I and II the accumulated number of tumours found.

Group I (Placebo)

Patient	Visit (months)											
	3	6	9	12	15	18	21	24	27	30	33	36
1	0	-	-	-	-	-	-	-	-	-	-	-
2	-	0	-	-	-	-	-	-	-	-	-	-
3	0	-	0	-	-	-	-	-	-	-	-	-
4	0	1	1	-	-	-	-	-	-	-	-	-
5	0	-	0	-	0	-	-	-	-	-	-	-
6	0	-	0	2	5	5	-	-	-	-	-	-

7	0	-	-	-	0	0	-	-	-	-	-	-
8	-	2	2	2	-	2	-	-	-	-	-	-
9	0	0	6	-	9	9	-	9	-	-	-	-
10	8	-	8	8	16	16	-	24	-	-	-	-
11	1	1	2	2	2	2	10	10	-	-	-	-
12	0	0	0	0	0	0	-	0	-	-	-	-
13	0	8	15	15	20	-	-	27	-	-	-	-
14	1	1	1	1	2	2	2	5	-	-	-	-
15	8	8	-	8	8	8	8	8	8	-	-	-
16	4	4	-	-	-	-	-	-	12	-	-	-
17	0	0	-	0	0	-	0	0	0	-	-	-
18	-	0	-	-	-	-	-	3	3	-	-	-
19	-	0	0	-	-	0	-	-	-	0	-	-
20	0	0	-	-	0	-	-	-	-	0	-	-
21	0	0	0	0	-	0	-	0	-	0	-	-
22	4	-	4	-	-	6	10	10	10	10	-	-
23	1	4	7	10	10	10	10	10	13	13	-	-
24	0	0	0	0	-	-	0	-	2	3	-	-
25	0	0	0	2	5	5	-	6	6	6	-	-
26	0	0	0	0	0	0	0	-	0	-	0	-
27	0	-	0	-	0	-	0	-	-	0	0	-
28	0	0	0	0	0	0	0	0	-	0	-	0
29	-	-	0	-	-	-	0	-	-	8	-	8
30	0	-	0	-	-	0	-	0	-	0	-	0
31	0	-	8	-	8	10	15	16	16	16	-	16
32	0	0	0	0	0	9	-	10	10	12	13	13
33	0	0	0	0	0	0	0	0	0	0	-	0
34	3	3	3	-	3	-	3	-	3	3	3	3
35	0	1	1	1	1	1	1	1	-	-	1	1
36	5	8	12	-	12	12	12	12	12	12	12	12
37	0	0	1	4	-	4	5	5	9	12	12	12
38	0	0	0	0	0	1	1	1	1	1	1	1
39	-	0	0	0	-	0	-	0	-	0	-	0
40	0	0	-	0	-	0	-	0	-	0	-	1
41	0	-	0	0	0	1	1	1	1	1	1	1
42	7	-	-	7	9	-	-	-	9	9	-	9
43	0	0	0	0	0	0	0	-	0	-	-	0
44	1	1	1	1	4	4	-	8	8	11	15	15
45	0	3	3	3	7	7	9	9	14	14	14	14
46	1	1	4	10	-	14	15	15	15	15	16	16

Group II (Thiotepa)

Patient	Visit (months)											
	3	6	9	12	15	18	21	24	27	30	33	36
1	-	8	-	-	-	-	-	-	-	-	-	-
2	0	0	0	-	-	-	-	-	-	-	-	-
3	-	0	0	-	-	-	-	-	-	-	-	-
4	0	-	-	0	-	-	-	-	-	-	-	-
5	1	1	1	1	1	-	-	-	-	-	-	-
6	7	14	16	16	16	16	-	-	-	-	-	-
7	-	0	-	0	-	2	-	-	-	-	-	-

8	-	-	-	-	-	0	-	-	-	-	-	-
9	2	-	-	-	-	2	-	-	-	-	-	-
10	0	0	0	0	0	2	2	-	-	-	-	-
11	0	0	0	-	0	0	0	-	-	-	-	-
12	0	-	0	0	-	0	0	0	-	-	-	-
13	-	-	0	0	-	0	-	0	-	-	-	-
14	0	0	0	0	0	0	0	0	-	-	-	-
15	0	2	2	5	-	-	6	6	6	-	-	-
16	0	1	1	1	1	-	1	-	1	-	-	-
17	2	-	-	-	-	-	-	-	-	2	-	-
18	0	0	0	-	-	-	0	0	3	3	3	6
19	0	0	0	0	0	0	0	0	0	-	-	-
20	0	1	1	1	4	4	4	7	10	10	18	27
21	0	0	0	0	0	0	2	3	5	-	8	-
22	-	0	-	-	0	0	-	3	5	6	-	-
23	0	0	0	0	0	0	0	0	0	0	0	0
24	0	0	0	0	0	0	0	0	0	0	0	0
25	1	1	-	1	-	1	1	1	2	2	2	2
26	-	-	0	-	-	-	-	0	-	-	-	-
27	2	2	2	2	2	2	3	5	6	6	-	-
28	0	0	0	0	0	-	0	0	0	-	0	-
29	1	1	1	1	1	1	1	1	1	-	1	-
30	0	0	0	0	0	0	0	0	0	0	0	0
31	-	-	-	0	0	-	-	0	-	0	-	-
32	1	-	1	1	-	-	1	2	-	2	2	2
33	0	-	-	-	-	-	0	0	-	-	0	-
34	0	0	0	0	0	0	0	0	0	0	0	-
35	0	0	0	0	0	0	0	0	0	0	0	0
36	0	0	0	0	0	0	0	0	0	0	0	0

234. Interspike waiting times

Zeger, S.L. and Bahjat Qaqish (1988) Markov regression models for time series: a quasi-likelihood approach. *Biometrics*, **44**, 1019–1031.

Motor cortex neuron interspike intervals were measured (in ms) for an unstimulated monkey. (Times to be read across rows.) The objects of the analysis discussed in this paper were to estimate the firing rate prior to stimulation and to characterize the time dependence.

68	41	82	66	101	66	57	41	27	78	59	73	6	44	72	66	59	60
39	52	50	29	30	56	76	55	73	104	104	52	25	33	20	60	47	6
47	22	35	30	29	58	24	34	36	34	6	19	28	16	36	33	12	26
36	39	24	14	28	13	2	30	18	17	28	9	28	20	17	12	19	18
14	23	18	22	18	19	26	27	23	24	35	22	29	28	17	30	34	17
20	49	29	35	49	25	55	42	29	16								

235. Traffic flow

Miller, A.J. (1961) A queueing model for road traffic flow. *Journal of the Royal Statistical Society, Series B*, **23**, 64–75.

Data were supplied by the Swedish State Roads Institute on a straight section of road between Stockholm and Uppsala. The time intervals were measured (in s) between consecutive vehicles in an analysis of the nature of traffic flow on this stretch of road.

0-1	1-2	2-3	3-4	4-5	5-6	6-8	8-10	10-16	16-25	25-50	>50
26	86	40	19	20	18	22	13	34	27	32	12

236. Waiting times for planned pregnancies

Weinberg, C.R. and Gladen, B.C. (1986) The beta-geometric distribution applied to comparative fecundability studies. *Biometrics*, **42**, 547–560.

Women who were pregnant with planned pregnancies were asked how many cycles it took to get them pregnant. (Data from Baird and Wilcox (1985).) A total of 678 women with planned pregnancies were interviewed: 654 became pregnant within 12 cycles after discontinuing contraception. Women were classified as smokers if they reported smoking at least an average of one cigarette a day during at least the first cycle they were trying to get pregnant.

	Observed cycles to pregnancy	
	Smokers	**Non-smokers**
1	29	198
2	16	107
3	17	55
4	4	38
5	3	18
6	9	22
7	4	7
8	5	9
9	1	5
10	1	3
11	1	6
12	3	6
>12	7	12

237. Iowa soil samples

Snapinn, S.M. and Small, R.D. (1986) Tests of significance using regression models for ordered categorical data. *Biometrics*, **42**, 583–592.

In this paper a probit model approach to a data analysis is adopted where multiple regression could also have been used. Chemical determinations of inorganic phosphorus (X_1) and a component of organic phosphorus (X_2) in the soil are the explanatory variables, and plant-available phosphorus (Y) of corn grown in the soil is the response. The units are parts per million; there were 18 soil samples taken at 20°C.

Sample	X_1	X_2	Y
1	0.4	53	64
2	0.4	23	60
3	3.1	19	71
4	0.6	34	61
5	4.7	24	54
6	1.7	65	77
7	9.4	44	81
8	10.1	31	93
9	11.6	29	93
10	12.6	58	51
11	10.9	37	76
12	23.1	46	96
13	23.1	50	77
14	21.6	44	93
15	23.1	56	95
16	1.9	36	54
17	26.8	58	168
18	29.9	51	99

238. Greenland turbot

Kimura, D.K. and Chikuni, S. (1987) Mixtures of empirical distributions: an iterative application of the age-length key. *Biometrics*, **43**, 23–35.

The data give the estimated catch of Greenland turbot (in numbers-at-length) by small Japanese freezer trawlers in the Eastern Bering Sea, 1978–1983.

Catch in thousands per year

Length (cm)	1978	1979	1980	1981	1982	1983
32	48	27	43	141	27	8
33	68	83	75	310	41	12
34	167	131	174	573	85	24
35	229	292	305	801	150	47
36	287	483	448	1140	252	77
37	397	732	604	1360	401	106
38	548	997	674	1709	552	143
39	664	1095	794	1898	796	191
40	808	1180	943	2019	1041	239
41	851	1158	977	2140	1270	303
42	984	1238	1110	2173	1370	359
43	1045	1303	1154	2106	1454	433
44	981	1286	1037	1973	1374	495
45	964	1236	978	1816	1327	571
46	935	1161	794	1520	1180	620
47	839	1039	728	1292	1074	679
48	714	897	605	1233	895	690
49	622	817	488	1056	814	680
50	638	741	448	800	701	668
51	580	636	405	691	638	610
52	483	587	326	586	498	547
53	462	539	268	521	468	477
54	428	450	248	444	411	428
55	395	377	203	387	332	397
56	392	326	223	332	307	361
57	335	275	192	279	296	353
58	327	234	190	263	278	346
59	295	235	189	247	283	378
60	277	211	162	228	344	386
61	324	238	179	235	297	409
62	312	242	217	244	294	406
63	374	232	229	256	299	387
64	360	274	256	233	282	362
65	396	273	257	238	260	326
66	328	257	269	208	251	278
67	326	232	241	219	227	235
68	262	200	234	212	195	197
69	196	177	191	187	173	168
70	147	121	165	161	152	145
71	118	95	148	135	144	135
72	102	86	113	101	119	113
73	103	61	111	95	117	112
74	135	66	110	141	116	113
75	114	67	118	100	125	128
76	158	82	138	105	120	140
77	171	86	130	139	159	158
78	215	109	183	142	156	179
79	215	102	221	166	172	204
80	220	121	190	155	192	226
81	218	108	224	167	197	234
82	192	90	204	171	190	249

83	212	82	198	154	191	240
84	172	70	156	150	187	224
85	144	69	132	131	163	213
86	114	49	135	143	139	178
87	93	49	97	80	112	152
88	95	36	81	58	82	124
89	57	23	49	48	78	100
90	55	25	45	42	69	75
91	38	16	33	33	37	54
92	25	10	22	23	27	41
93	27	11	19	16	21	31
94	19	8	12	16	16	25
95	15	8	16	10	12	18
96	10	7	10	12	9	11
97	6	5	6	7	5	7
98	7	4	10	3	2	5
99	6	4	3	4	3	4
Total	21844	23561	19937	34778	24049	17034

239. Insect data

Lindsey, J.C., Herzberg, A.M. and Watts, D.G. (1987) A method for cluster analysis based on projections and quantile-quantile plots. *Biometrics*, **43**, 327–341.

Three variables were measured on each of ten insects for each of three species of a type of insect, Chaetocnema. The variable X_1 (microns) is the width of the first joint of the first tarsus; X_2 (microns) is the width of the first joint of the second tarsus; X_3 (microns) is the maximal width of the aedegus.

In a real situation one would not know which insect belonged to which species, and the object would be to group similar insects.

Species I			Species II			Species III		
X_1	X_2	X_3	X_1	X_2	X_3	X_1	X_2	$X3$
191	131	53	186	107	49	158	141	58
185	134	50	211	122	49	146	119	51
200	137	52	201	144	47	151	130	51
173	127	50	242	131	54	122	113	45
171	128	49	184	108	43	138	121	53
160	118	47	211	118	51	132	115	49
188	134	54	217	122	49	131	127	51
186	129	51	223	127	51	135	123	50
174	131	52	208	125	50	125	119	51
163	115	47	199	124	46	130	120	48

The authors go on to say that the methods developed in the paper were applied to six new insects, to determine which species they belonged to:

X_1	X_2	X_3
190	143	52
174	131	50
211	129	49
218	126	49
130	131	51
138	127	52

240. Skin resistance

Berry, D.A. (1987) Logarithmic transformations in ANOVA. *Biometrics*, **43**, 439–456.

Five different types of electrodes were applied to the arms of 16 subjects and the resistance measured (in kilohms). The experiment was designed to see whether all electrode types performed similarly. After obtaining the results it was decided that the reason for the two extreme readings on Subject 15 was that the subject had very hairy arms.

The author goes on "If one uses $p < 0.05$ to indicate statistical significance (a deplorable but widespread practice!) then some tests show significant differences among electrodes while others do not. Which to use? ..."

Subject	Electrode type				
	1	2	3	4	5
1	500	400	98	200	250
2	660	600	600	75	310
3	250	370	220	250	220
4	72	140	240	33	54
5	135	300	450	430	70
6	27	84	135	190	180
7	100	50	82	73	78
8	105	180	32	58	32
9	90	180	220	34	64
10	200	290	320	280	135
11	15	45	75	88	80
12	160	200	300	300	220
13	250	400	50	50	92
14	170	310	230	20	150
15	66	1000	1050	280	220
16	107	48	26	45	51

241. Energy maintenance in sheep

Wallach, D. and Goffinet, B. (1987) Mean square error of prediction in models for studying ecological and agronomic systems. *Biometrics*, **43**, 561–573.

The energy maintenance requirements were assessed for a sample of grazing Merino wether sheep in Australia. There were 64 sheep in the sample. For each sheep its weight (x, kg) was measured and its daily energy requirement (Y, Mcal/day) assessed. The main interest arises because energy requirements play an important role in predicting meat production in grazing sheep systems.

x	Y	x	Y	x	Y	x	Y	x	Y
22.1	1.31	25.1	1.46	25.1	1.00	25.7	1.20	25.9	1.36
26.2	1.27	27.0	1.21	30.0	1.23	30.2	1.01	30.2	1.12
33.2	1.25	33.2	1.32	33.2	1.47	33.9	1.03	33.8	1.46
34.3	1.14	34.9	1.00	42.6	1.81	43.7	1.73	44.9	1.93
49.0	1.78	49.2	2.53	51.8	1.87	51.8	1.92	52.5	1.65
52.6	1.70	53.3	2.66	23.9	1.37	25.1	1.39	26.7	1.26
27.6	1.39	28.4	1.27	28.9	1.74	29.3	1.54	29.7	1.44
31.0	1.47	31.0	1.50	31.8	1.60	32.0	1.67	32.1	1.80
32.6	1.75	33.1	1.82	34.1	1.36	34.2	1.59	44.4	2.33
44.6	2.25	52.1	2.67	52.4	2.28	52.7	3.15	53.1	2.73
52.6	3.73	46.7	2.21	37.1	2.11	31.8	1.39	36.1	1.79
28.6	2.13	29.2	1.80	26.2	1.05	45.9	2.36	36.8	2.31
34.4	1.85	34.4	1.63	26.4	1.27	27.5	0.94		

242. Lengths of lives of rats

Berger, R.L., Boos, D.D. and Guess, F.M. (1988) Tests and confidence sets for comparing two mean residual life functions. *Biometrics*, **44**, 103–115.

The effects on the total lifelength of rats of a restricted diet versus an *ad libitum* diet (i.e. free eating) were studied. Research indicates that diet restriction promotes longevity. Treatments were begun after an initial weaning period.

242. Lengths of lives of rats

Restricted diet

105	193	211	236	302	363	389	390	391	403	530	604	605	630
716	718	727	731	749	769	770	789	804	810	811	833	868	871
875	893	897	901	906	907	919	923	931	940	957	958	961	962
974	979	982	1001	1008	1010	1011	1012	1014	1017	1032	1039	1045	1046
1047	1057	1063	1070	1073	1076	1085	1090	1094	1099	1107	1119	1120	1128
1129	1131	1133	1136	1138	1144	1149	1160	1166	1170	1173	1181	1183	1188
1190	1203	1206	1209	1218	1220	1221	1228	1230	1231	1233	1239	1244	1258
1268	1294	1316	1327	1328	1369	1393	1435						

Ad libitum diet

89	104	387	465	479	494	496	514	532	536	545	547	548	582
606	609	619	620	621	630	635	639	648	652	653	654	660	665
667	668	670	675	677	678	678	681	684	688	694	695	697	698
702	704	710	711	712	715	716	717	720	721	730	731	732	733
735	736	738	739	741	743	746	749	751	753	764	765	768	770
773	777	779	780	788	791	794	796	799	801	806	807	815	836
838	850	859	894	963									

243. Crying babies

Coffey, M. (1988) A random effects model for binary data from dependent samples. *Biometrics*, **44**, 787–801.

An investigation was pursued to explore the possible beneficial effects of rocking on babies' crying. On each of 18 days, babies not crying at a specified time in a hospital ward were the subjects. These groups varied in size from six to ten. One of the subjects was chosen at random and rocked; the remainder (the controls) were not rocked. At the end of a specified time it was noted whether babies were crying or not.

Day	Controls	Controls crying	Rocked crying
1	8	5	0
2	6	4	0
3	5	4	0
4	6	5	1
5	5	1	0
6	9	5	0
7	8	3	0
8	8	4	0
9	5	2	0
10	9	1	1
11	6	1	0
12	9	1	0
13	8	3	0
14	5	1	0
15	6	2	0
16	8	1	0
17	6	2	1
18	8	3	0

244. Plant competition data

Gleeson, A.C. and McGilchrist, C.A. (1980) Bilateral processes on a rectangular lattice. *Australian Journal of Statistics*, **22**, 197–206.

Measurements (g) were taken on individual plant weights in an investigation into plant competition. The data are taken from an experiment whose three replicates included a monoculture plot in which 49 individual plants of Morocco 18952, a genotype of tall fescue, were sown on a square grid with 9 cm spacings.

1st	5.2	3.2	1.9	4.7	4.5	2.6	4.1
	3.0	2.3	2.0	1.1	2.8	2.2	0.5
	3.0	3.1	1.6	0.4	0.7	4.0	3.0
	2.7	2.2	1.6	0.6	3.8	3.4	2.4
	1.7	2.5	1.7	1.5	2.5	2.6	2.8
	7.2	1.5	3.7	2.1	1.0	3.9	0.8
	3.8	1.9	1.5	1.6	4.0	0.5	1.9
2nd	5.8	2.8	6.7	3.4	3.8	4.9	6.2
	5.8	2.6	3.7	3.2	1.1	2.0	2.4
	4.8	1.4	3.2	1.8	2.3	2.8	1.5
	6.4	2.9	3.1	2.5	0.8	2.9	2.8
	1.8	2.0	4.0	2.1	2.4	1.9	0.9
	3.9	2.4	0.9	1.7	3.3	1.6	2.2
	3.7	3.1	2.3	0.8	3.4	0.1	2.1
3rd	2.1	2.9	2.7	1.8	1.1	3.1	3.6
	2.8	2.7	2.4	4.5	2.6	0.9	3.2
	3.3	2.6	2.8	3.0	2.6	3.9	3.2
	1.8	2.9	1.8	2.5	2.0	4.8	2.7
	4.0	1.6	2.6	2.6	2.6	3.4	4.1
	3.8	1.9	3.1	2.6	3.6	1.8	1.1
	2.5	5.9	2.5	2.5	2.5	4.3	5.3

245. Product sales

Nicholls, D.F. (1979) The analysis of time series — the time domain approach. *Australian Journal of Statistics*, **21**, 93–120.

These data give the cumulative weekly sales of a plastic container used for the packaging of drugs in the United States. Time progresses along rows.

592	1208	1864	2508	3160	3792	4419	5023	5626	6250
6903	7564	8223	8900	9569	10226	10823	11304	11785	12252
12724	13339	13940	14514	15102	15648	16212	16725	17217	17839
18463	19094	19718	20342	20966	21563	22167	22802	23431	24044
24562	25131	25676	26205	26718	27215	27767	28389	28974	29502
30022	30542	31044	31556	32092	32636	33196	33772	34347	34915
35502	36078	36643	37219	37804	38381	38974	39513	40044	40634
41237	41853	42430	42977	43529	44081	44657	45277	45917	46517
47100	47692	48241	48802	49399	49992	50601	51214	51855	52499
53127	53763	54387	54979	55591	56219	56843	57491	58127	58767
59264	59777	60298	60779	61348	61942	62521	63041	63576	64082
64550	65074	65594	66168	66664	67204	67711	68272	68837	69414
70012	70617	71211	71701	72214	72740	73283	73889	74509	75115
75743	76351	76971	77585	78169	78745	79309	79918	80563	81184
81761	82318	82855	83384	83921	84476	85028	85647	86275	86889
87493	88097	88737	89361	90017	90668	91313	91938	92575	93156

93669	94182	94695	95201	95679	96183	96783	97397	97989	98603
99223	99839	100478	101095	101724	102333	102902	103516	104157	104784
105411	106026	106626	107214	107806	108418	109048	109688	110321	110961
111586	112211	112828	113437	114038	114663	115288	115904	116519	117121
117688	118280	118916	119548	120130	120647	121210	121773	122350	122947
123540	124102	124705	125255	125769	126325	126893	127501	128133	128769
129402	129994	130491	130996	131565	132142	132719	133280	133863	134435
134997	135587	136187	136812	137424	138024	138648	139272	139909	140534
141171	141768	142329	142878	143423	144000	144585	145168	145738	146294
146832	147364	147920	148520	149128	149744	150342	150902	151407	151895
152396	152885	153390	153939	154520	155129	155742	156374	157006	157648
158295	158951	159655	160339	161039	161723	162410	163050	163698	164366
165031	165680	166321	166934	167559	168193	168882	169580	170277	170954
171602	172220	172852	173500	174136	174764	175364	175956	176560	177197

246. Timber data

Brown, B.M. and Maritz, J.S. (1982) Distribution-free methods in regression. *Australian Journal of Statistics*, **24**, 318–331.

These data are for specimens of 50 varieties of timber, for modulus of rigidity (y), modulus of elasticity (z) and air dried density (x), arranged in increasing order of magnitude of x.

y	z	x	y	z	x
1000	99	25.3	1897	240	50.3
1112	173	28.2	1822	248	51.3
1033	188	28.6	2129	261	51.7
1087	133	29.1	2053	245	52.8
1069	146	30.7	1676	186	53.8
925	91	31.4	1621	188	53.9
1306	188	32.5	1990	252	54.9
1306	194	36.8	1764	222	55.1
1323	195	37.1	1909	244	55.2
1379	177	38.3	2086	274	55.3
1332	182	39.0	1916	276	56.9
1254	110	39.6	1889	254	57.3
1587	203	40.1	1870	238	58.3
1145	193	40.3	2036	264	58.6
1438	167	40.3	2570	189	58.7
1281	188	40.6	1474	223	59.5
1595	238	42.3	2116	245	60.8
1129	130	42.4	2054	272	61.3
1492	189	42.5	1994	264	61.5
1605	213	43.0	1746	196	63.2
1647	165	43.0	2604	268	63.3
1539	210	46.7	1767	205	68.1
1706	224	49.0	2649	346	68.9
1728	228	50.2	2159	246	68.9
1703	209	50.3	2078	237.5	70.8

247. Parasite data

Feigin, P.D., Tweedie, R.L. and Belyea, C. (1983) Weighted area techniques for explicit parameter estimation in multi-stage models. *Australian Journal of Statistics*, **25**, 1–16.

The table gives two sets of data on hatching times of eggs of the cattle parasite Ostertagia circumcinta. The eggs were collected in different ways, (i) using faeces deposited in bags, (ii) using faeces collected directly from the rectum of the animals concerned. Incubation took place at 20°C. Both data sets indicate a delay before hatching begins. In the paper, delayed second-order gamma densities are fitted to both data sets. A * indicates a missing observation.

(i)				(ii)		
Time	Eggs	Larvae		Time	Eggs	Larvae
1	264	0		1	326	1
2	282	0		2	291	0
3	244	1		3	261	0
4	223	1		4	359	0
5	225	1		5	270	0
6	173	1		6	231	0
7	217	4		7	280	1
8	194	11		8	289	1
9	221	20		9	344	7
10	150	29		10	286	8
11	184	62		11	255	21
12	167	103		12	222	58
13	165	131		13	235	112
14	111	176		14	182	143
15	110	180		15	127	137
16	70	217		16	128	179
17	57	227		17	98	225
18	49	224		18	71	240
19	42	229		19	66	259
20	32	231		20	52	277
21	28	271		21	70	246
22	23	298		22	*	*
23	17	259		23	42	292
24	22	244		24	39	305
25	14	230		25	23	289
26	18	323		26	28	298
27	12	305		27	28	388
28	14	244		28	14	277
29	6	243		29	17	279
30	11	272		30	18	274
31	9	252		31	17	283
32	13	307		32	13	263
33	19	247		33	31	302
34	7	267		34	8	238
35	4	218		35	8	273

248. Azimuth data

Till, R. (1974) *Statistical methods for the earth scientist.* London: Macmillan, 39.

Measurements were taken (in degrees East of North) of palaeocurrent azimuths from the Jura Quartzite, Islay. One aim of the research was to compare a von Mises fit to the data with a wrapped normal fit.

12	353	359	332	341	299	30	24	53
284	99	72	28	93	125	318	3	45

249. Perpendicular distance models

Buckland, S.T. (1985) Perpendicular distance models for line transect sampling. *Biometrics*, **41**, 177–195.

An observer walks along a transect line, recording the perpendicular distance from the line for each object of a particular type sighted. The paper raises interesting problems to do with (for instance) the probability of detection, given that there is an object present. Altogether 23 different experiments are summarized. These are the results for just three of them. (Capercaillie are a species of bird.)

(1) Capercaillie: Monaughty Forest, Scotland:

0-10	10-20	20-30	30-40	40-50	50-60	60-70	70-80	80-100
24	39	29	22	26	9	7	5	3

(2) Capercaillie: Culbin Forest, Scotland:

0-10	10-20	20-30	30-40	40-50	50-60	60-90
12	15	12	11	7	5	6

(3) Pine cones: Tarradale, Scotland:

0-1	1-2	2-3	3-4	4-5	5-6	6-8	8-10	10-16.5
10	8	9	10	8	10	7	6	6

250. Japanese black pines

Ogata Y. and Tanemura M. (1985) Estimation of interaction potentials of marked spatial point patterns through the maximum likelihood method. *Biometrics*, **41**, 421–433.

Data were collected on the position, height (cm) and age (years) of the natural stands of the seedlings and saplings of 204 Japanese black pines in a 10 m × 10 m region. The original paper includes an aerial sketch, permitting a spatial analysis. Here, only the data for height and age are given.

80,9	8,3	52,5	63,7	28,5	95,9	51,5	6,2	13,3
60,6	8,3	3,1	3,1	65,8	5,3	42,8	85,8	66,9
15,3	12,3	150,20	67,6	30,4	14,3	56,5	15,3	8,3
12,3	12,3	8,3	11,3	8,3	15,4	12,3	15,8	60,6
14,3	11,3	45,6	55,5	58,8	13,3	20,3	32,3	7,3
150,20	46,7	76,8	53,6	77,6	95,6	15,3	22,4	33,5
14,3	9,3	9,3	11,3	12,3	40,6	9,3	9,3	12,3
75,6	87,6	10,3	9,3	9,3	16,3	16,3	18,3	11,2
11,3	8,3	10,3	15,3	10,3	150,20	30,4	130,8	37,5
123,8	10,3	13,3	14,3	7,3	7,3	16,3	19,3	33,9
124,8	45,6	70,6	10,9	10,2	10,2	4,1	30,4	46,6
10,2	8,2	13,3	10,2	11,2	9,2	6,1	10,2	6,1
9,2	86,9	9,3	5,2	7,2	7,3	8,3	4,2	45,9
65,10	65,10	30,5	62,6	54,8	11,2	8,2	24,5	15,3
8,2	46,8	39,5	49,6	46,5	50,5	85,8	9,3	14,3
37,5	15,2	8,2	30,6	9,2	34,6	150,20	7,2	13,3
17,3	56,10	8,2	55,11	16,3	37,6	27,4	21,4	13,3
59,7	7,3	30,6	7,3	10,4	41,8	21,5	5,2	40,6
19,5	68,8	14,5	38,7	30,5	150,20	45,6	8,2	13,4
28,4	59,7	29,5	10,3	9,4	11,5	11,6	42,6	6,2
9,2	21,5	9,2	48,6	15,4	16,5	8,2	59,8	19,3
150,20	25,11	15,4	9,3	56,7	65,7	15,4	53,6	14,4
7,5	10,3	9,3	10,5	10,3	26,7			

251. Treatments for locally unresectable gastric cancer

Stablein, D.M. and Koutrouvelis, I.A. (1985) A two-sample test sensitive to crossing hazards in uncensored and singly censored data. *Biometrics*, **41**, 643–652.

In 1982 the Gastrointestinal Tumor Study Group (USA) reported on the results of a trial comparing chemotherapy with combined chemotherapy and radiotherapy in the treatment of locally unresectable gastric cancer. The data give the survival times (in days) for the 45 patients on each treatment. A + indicates censored data.

(1) Chemotherapy

1	63	105	129	182	216	250	262	301
301	342	354	356	358	380	383	383	388
394	408	460	489	499	523	524	535	562
569	675	676	748	778	786	797	955	968
1000	1245	1271	1420	1551	1694	2363	2754+	2950+

(2) Chemotherapy with radiotherapy

17	42	44	48	60	72	74	95	103
108	122	144	167	170	183	185	193	195
197	208	234	235	254	307	315	401	445
464	484	528	542	567	577	580	795	855
1366	1577	2060	2412+	2486+	2796+	2802+	2934+	2988+

252. Strength of beams

Draper, N.R. and Stoneman, D.M. (1966) Testing for the inclusion of variables in linear regression by a randomisation technique. *Technometrics*, **8**, 695–699.

Data were collected on the specific gravity (x_1), moisture content (x_2) and strength (y) of ten wood beams. Strength is the response variable. (Units not known.)

y	x_1	x_2
11.14	0.499	11.1
12.74	0.558	8.9
13.13	0.604	8.8
11.51	0.441	8.9
12.38	0.550	8.8
12.60	0.528	9.9
11.13	0.418	10.7
11.70	0.480	10.5
11.02	0.406	10.5
11.41	0.467	10.7

253. Salinity values

Till, R. (1974) *Statistical methods for the earth scientist*. London: Macmillan, 104.

The data give sets of salinity values (parts per thousand) for the three separate water masses in the Bimini Lagoon, Bahamas. Several samples were drawn at each site.

I	II	III
37.54	40.17	39.04
37.01	40.80	39.21
36.71	39.76	39.05
37.03	39.70	38.24
37.32	40.79	38.53
37.01	40.44	38.71
37.03	39.79	38.89
37.70	39.38	38.66
37.36		38.51
36.75		40.08
37.45		
38.85		

254. Epileptic seizures and chemotherapy

Thall, P.F. and Vail, S.C. (1990) Some covariance models for longitudinal count data with overdispersion. *Biometrics*, **46**, 657–671.

In a clinical trial including 59 people with epilepsy, patients suffering from simple or partial seizures were randomized to groups receiving either the anti-epileptic drug progabide, or a placebo, as an adjuvant to standard chemotherapy. The number of seizures was counted over four two-week periods. There was also a baseline seizure rate recorded for each patient, based on the 8-week prerandomization seizure count. The seizure counts exhibit considerable extra-Poisson variation; there is heteroscedasticity and dependence within patients. In the table the variables Y_i, $i = 1$, 2, 3, 4, are the seizure counts; the treatment X_1 is 0 (placebo) or 1 (progabide); X_2 is the baseline rate; and X_3 is the age of the patient. In their analysis, the authors omitted Patient 49 as an outlier.

	Y_1	Y_2	Y_3	Y_4	X_1	X_2	X_3
1	5	3	3	3	0	11	31
2	3	5	3	3	0	11	30
3	2	4	0	5	0	6	25
4	4	4	1	4	0	8	36
5	7	18	9	21	0	66	22
6	5	2	8	7	0	27	29
7	6	4	0	2	0	12	31
8	40	20	23	12	0	52	42
9	5	6	6	5	0	23	37

10	14	13	6	0	0	10	28
11	26	12	6	22	0	52	36
12	12	6	8	4	0	33	24
13	4	4	6	2	0	18	23
14	7	9	12	14	0	42	36
15	16	24	10	9	0	87	26
16	11	0	0	5	0	50	26
17	0	0	3	3	0	18	28
18	37	29	28	29	0	111	31
19	3	5	2	5	0	18	32
20	3	0	6	7	0	20	21
21	3	4	3	4	0	12	29
22	3	4	3	4	0	9	21
23	2	3	3	5	0	17	32
24	8	12	2	8	0	28	25
25	18	24	76	25	0	55	30
26	2	1	2	1	0	9	40
27	3	1	4	2	0	10	19
28	13	15	13	12	0	47	22
29	11	14	9	8	1	76	18
30	8	7	9	4	1	38	32
31	0	4	3	0	1	19	20
32	3	6	1	3	1	10	30
33	2	6	7	4	1	19	18
34	4	3	1	3	1	24	24
35	22	17	19	16	1	31	30
36	5	4	7	4	1	14	35
37	2	4	0	4	1	11	27
38	3	7	7	7	1	67	20
39	4	18	2	5	1	41	22
40	2	1	1	0	1	7	28
41	0	2	4	0	1	22	23
42	5	4	0	3	1	13	40
43	11	14	25	15	1	46	33
44	10	5	3	8	1	36	21
45	19	7	6	7	1	38	35
46	1	1	2	3	1	7	25
47	6	10	8	8	1	36	26
48	2	1	0	0	1	11	25
49	102	65	72	63	1	151	22
50	4	3	2	4	1	22	32
51	8	6	5	7	1	41	25
52	1	3	1	5	1	32	35
53	18	11	28	13	1	56	21
54	6	3	4	0	1	24	41
55	3	5	4	3	1	16	32
56	1	23	19	8	1	22	26
57	2	3	0	1	1	25	21
58	0	0	0	0	1	13	36
59	1	4	3	2	1	12	37

255. Period between earthquakes

The Open University (1981) S237: *The Earth: Structure, Composition and Evolution.*

The table gives the time in days between successive serious earthquakes world-wide. An earthquake is included if its magnitude was at least 7.5 on the Richter scale, or if over 1000 people were killed. Recording starts on 16th December 1902 and ends on 4th March 1977. There were 63 earthquakes recorded altogether, and so 62 recorded waiting times. If earthquakes occur at random, an exponential model for these data should provide a reasonable fit.

840	157	145	44	33	121	150	280	434	736	584	887	263
1901	695	294	562	721	76	710	46	402	194	759	319	460
40	1336	335	1354	454	36	667	40	556	99	304	375	567
139	780	203	436	30	384	129	9	209	599	83	832	328
246	1617	638	937	735	38	365	92	82	220			

256. Alveolar-bronchiolar adenomas

Tamura, R.N. and Young, S.S. (1987) A stabilized moment estimator for the beta-binomial distribution. *Biometrics*, **43**, 813–824.

The authors supply data on the proportions of mice in 23 independent groups having alveolar-bronchiolar adenomas. A binomial model does not fit: the authors fit a beta-binomial model.

0/12	0/20	1/20	1/15	3/20	4/47	0/12	0/20	1/19	2/25
3/20	6/54	0/10	0/19	1/19	2/22	3/18	8/49	1/10	0/17
1/17	2/20	4/20							

257. Woodlice

Lloyd, M. (1967) Mean crowding. *Journal of Animal Ecology*, **36**, 1.

The numbers of the large woodlouse Philoscia muscorum were counted in 37 contiguous hexagonal quadrats of beech litter at Wytham Woods near Oxford.

```
          1       2       1       2
       0     1     5     1     3
     2     1     2     0     0     2
  0     0     2     0     0     3     4
     2     2     0     0     2     3
       5     0     0     3     0
          4       0       0       0
```

The numbers of the small woodlouse Trichoniscus pusillus provisorius were counted in 24 contiguous quadrats of beech litter. The quadrats were of different shapes but had equal areas.

```
                 6
            9        27
                 9
      3      23       13      10
                23
         49             15
      26       13    49      12
         25             18
      12       14    16      8
         12             20
                24
```

258. Drilling times

Penner, R. and Watts, D.G. (1991) Mining information. *American Statistician*, **45**, 4–9.

An investigation took place to see whether drilling holes in rock is faster using "dry" or "wet" drilling. (In dry drilling, compressed air is used to flush the cuttings, and in wet drilling water is used.) In an area about 10 feet by 20 feet, six holes, three dry and three wet, were drilled to a depth of 400 feet. The time (in 1/100 minutes) taken to drill each five-foot length was recorded. Data points with the symbol < are known to be too great; * indicates a missing observation.

Depth	Dry			Wet		
	1	2	3	1	2	3
5	761	868	861	725	707	490
10	816	813	771	855	662	507
15	711	721	702	743	704	677
20	947	826	830	860	682	665
25	991	924	907	845	719	699
30	827	1022	975	795	634	741
35	987	939	847	798	588	607
40	965	904	750	780	699	704
45	902	918	804	762	665	549
50	989	814	727	772	599	603
55	959	913	785	835	663	619
60	934	850	861	758	699	643
65	726	849	891	807	728	627
70	888	882	889	810	691	603
75	<1172	855	830	842	674	634
80	867	861	896	813	579	557
85	987	792	889	762	670	570
90	876	833	903	655	560	623
95	1004	891	853	700	610	660
100	869	936	860	698	560	584
105	860	844	802	658	577	517
110	834	789	706	652	604	603
115	760	891	784	682	632	684
120	901	943	780	680	747	658
125	901	849	788	713	690	703
130	1015	918	829	837	693	689
135	886	901	902	838	807	727
140	695	812	924	760	739	692
145	1117	718	902	665	642	715
150	954	762	871	742	780	725
155	842	801	841	627	797	705
160	740	797	901	820	792	695
165	914	802	671	760	553	676
170	900	918	875	753	688	597
175	834	894	915	743	762	717
180	801	802	856	842	<934	711
185	915	851	794	668	689	707
190	871	835	827	778	790	717
195	929	890	938	737	972	691
200	850	903	1047	<1042	1076	615
205	997	871	1019	767	1024	619
210	819	880	895	832	922	629

215	805	710	902	858	866	558
220	1012	880	1006	857	<928	722
225	1096	839	1096	1047	<928	762
230	1072	868	869	1053	<12	828
235	1000	846	901	1005	<1029	759
240	1112	757	1000	925	<1189	642
245	900	783	1125	975	<1045	712
250	912	713	786	965	<1521	662
255	742	682	668	832	<1068	484
260	736	681	970	808	<667	597
265	829	944	1363	470	<1074	777
270	913	1071	1105	582	<1119	896
275	917	1037	<1168	942	<917	873
280	1176	997	990	863	<652	829
285	900	1016	1088	853	<795	805
290	1121	963	<1263	872	<921	1079
295	1319	1099	1184	1067	<1167	929
300	1320	1025	1173	1113	<1249	962
305	1094	844	1121	1077	*	891
310	1177	968	826	972	*	973
315	1224	1124	1172	<1540	*	834
320	1311	1068	1118	<1875	*	675
325	1170	<1084	1114	<1378	*	686
330	1234	953	905	1072	*	854
335	1252	963	1054	1063	*	860
340	1266	1016	1181	1023	*	841
345	1215	<1144	1262	1227	*	881
350	1200	974	1452	1430	*	646
355	1224	1014	1166	1297	*	1179
360	1239	932	1093	830	*	1024
365	<1355	909	1099	707	*	993
370	<1228	1043	1165	1010	*	1122
375	1072	862	977	713	*	1212
380	884	<1146	1225	898	*	1153
385	1324	963	<1243	1202	*	1209
390	1050	832	1250	1062	*	1274
395	861	804	1270	910	*	1155
400	1027	1035	1250	1222	*	1102

259. Failure data

Jorgenson, D.W. (1961) Multiple regression analysis of a Poisson process. *Journal of the American Statistical Association*, **56**, 235–245.

The data consist of failures of a piece of electronic equipment operating in two modes. Over nine operating periods ($i = 1, 2, ..., 9$) $T1_i$ is the time spent operating in one mode, $T2_i$ is the time spent operating in the other. The number n_i is the total number of failures recorded. The aim is to estimate the failure rate over time in each of the two operating modes.

$T1_i$	33.3	52.2	64.7	137.0	125.9	116.3	131.7	85.0	91.9
$T2_i$	25.3	14.4	32.5	20.5	97.6	53.6	56.6	87.3	47.8
n_i	15	9	14	24	27	27	23	18	22

260. Spatial presence-absence data

Strauss, D. (1992) The many faces of logistic regression. *American Statistician*, **46**, 321–327.

The paper reports and analyses data of Bartlett on the presence or absence of the plant Carex arenaria over a spatial region divided into a 24 × 24 lattice.

```
0 1 1 1 0 1 1 1 1 1 0 0 0 1 0 0 1 0 0 1 1 0 1 0
0 1 0 0 1 1 1 0 0 1 0 0 0 0 0 0 0 1 1 1 0 0 0 1
1 1 1 1 0 1 1 0 0 0 0 0 0 0 1 1 1 1 0 0 0 0 0 0
0 0 1 1 1 1 1 0 1 0 1 0 0 0 0 0 1 1 0 0 0 0 0 0
0 1 1 0 1 1 1 1 1 0 0 0 0 1 1 1 0 1 1 0 0 0 0 0
0 1 0 0 0 1 0 1 0 1 1 1 1 1 0 0 0 0 1 1 0 0 0 0
0 1 0 1 1 0 1 0 1 0 0 0 1 0 0 1 1 0 0 1 0 0 0 0
0 0 0 0 0 0 1 1 0 1 0 0 0 0 0 0 1 0 0 0 0 0 0 0
0 0 0 0 0 1 0 1 1 0 0 0 0 0 0 0 1 0 0 0 1 0 0 0
0 0 0 0 0 0 1 0 1 0 0 0 0 0 0 0 1 0 0 0 1 0 0 0
0 0 0 0 0 1 0 0 0 1 1 0 0 0 0 1 1 0 0 0 0 0 0 0
0 0 0 0 1 0 0 0 0 0 1 1 0 0 0 0 1 0 0 0 0 0 0 0
0 0 0 1 0 0 0 0 0 0 0 0 1 0 0 1 0 0 0 0 0 0 1 0
0 0 1 0 0 0 1 1 0 1 0 1 1 1 1 1 1 1 0 0 0 0 1 1
0 1 0 0 1 1 0 0 0 0 0 0 0 1 1 0 0 1 0 1 0 1 0 1
0 0 0 0 0 0 0 0 0 0 0 1 0 0 1 1 0 0 0 1 1 0 0 1
1 0 0 0 0 0 0 1 0 0 1 0 1 0 1 0 0 0 0 0 0 0 0 0
0 0 0 0 0 0 1 0 0 0 0 0 0 1 1 1 0 0 1 1 0 1 0 1
0 1 0 0 0 0 1 0 0 0 0 0 0 1 1 0 0 1 1 1 1 1 0 1
1 0 0 0 0 0 0 0 0 0 0 0 0 1 0 0 0 0 1 0 0 1 0 1
0 0 0 1 0 1 0 0 0 0 0 0 1 0 0 0 0 1 0 0 0 0 1
1 0 0 0 0 1 1 0 1 1 0 0 1 0 0 0 0 0 0 0 0 0 0
0 0 1 0 0 0 0 0 0 0 0 1 1 0 1 0 0 0 1 0 0 0 0 1
0 0 0 0 1 1 0 0 1 0 0 0 0 0 0 0 0 0 1 0 0 0 0 0
```

261. Neuralgia data

Piegorsch, W.W. (1992) Complementary log regression for generalized linear models. *American Statistician*, **46**, 94–99.

The analgesic effect of iontophoretic treatment with the nerve conduction-inhibiting chemical vincristine was studied on elderly patients complaining of post-herpetic neuralgia. Eighteen patients were interviewed six weeks after undergoing treatment to see whether any improvement had occurred. In the following table, the response Y takes the value 1 if pain cessation was recorded. The variable X_1 indicates treatment; X_2 is the patient's age; X_3 is the patient's sex; X_4 is the pretreatment duration of symptoms (in months). Patient 13 died during the six-week observation period: Y was set to 1 since in this case improvement was noticed throughout and after treatment.

	Y	X_1	X_2	X_3	X_4
1	1	1	76	M	36
2	1	1	52	M	22
3	0	0	80	F	33
4	0	1	77	M	33
5	0	1	73	F	17
6	0	0	82	F	84
7	0	1	71	M	24
8	0	0	78	F	96
9	1	1	83	F	61
10	1	1	75	F	60
11	0	0	62	M	8
12	0	0	74	F	35
13	1	1	78	F	3
14	1	1	70	F	27
15	0	0	72	M	60
16	1	1	71	F	8
17	0	0	74	F	5
18	0	0	81	F	26

262. Temperatures in America

Peixoto, J.L. (1990) A property of well-formulated polynomial regression models. *American Statistician*, **44**, 26–30.

The table gives the normal average January minimum temperature (Y, °F) with latitude (x_1) and longitude (x_2) for 56 cities in the United States. (Average minimum temperature for January is found by adding together the daily minimum temperatures and dividing by 31. For this table the January average minima for the years 1931 to 1960 were averaged over the 30 years.) The authors report a study in which a linear relationship is assumed between temperature and latitude; then, after adjusting for latitude, a cubic polynomial in longitude accurately predicts temperature.

	Y	x_1	x_2
Mobile, AL	44	31.2	88.5
Montgomery, AL	38	32.9	86.8
Phoenix, AZ	35	33.6	112.5
Little Rock, AR	31	35.4	92.8
Los Angeles, CA	47	34.3	118.7
San Francisco, CA	42	38.4	123.0
Denver, CO	15	40.7	105.3
New Haven, CT	22	41.7	73.4
Wilmington, DE	26	40.5	76.3
Washington, DC	30	39.7	77.5
Jacksonville, FL	45	31.0	82.3
Key West, FL	65	25.0	82.0
Miami, FL	58	26.3	80.7
Atlanta, GA	37	33.9	85.0
Boise, ID	22	43.7	117.1
Chicago, IL	19	42.3	88.0
Indianapolis, IN	21	39.8	86.9
des Moines, IA	11	41.8	93.6
Wichita, KS	22	38.1	97.6
Louisville, KY	27	39.0	86.5
New Orleans, LA	45	30.8	90.2
Portland, ME	12	44.2	70.5
Baltimore, MD	25	39.7	77.3
Boston, MA	23	42.7	71.4
Detroit, MI	21	43.1	83.9
Minneapolis, MN	2	45.9	93.9
St Louis, MO	24	39.3	90.5
Helena, MT	8	47.1	112.4
Omaha, NB	13	41.9	96.1
Concord, NH	11	43.5	71.9
Atlantic City, NJ	27	39.8	75.3
Albuqurque, NM	24	35.1	106.7
Albany, NY	14	42.6	73.7
New York, NY	27	40.8	74.6
Charlotte, NC	34	35.9	81.5
Raleigh, NC	31	36.4	78.9
Bismarck, ND	0	47.1	101.0
Cincinnati, OH	26	39.2	85.0
Cleveland, OH	21	42.3	82.5

Oklahoma City, OK	28	35.9	97.5
Portland, OR	33	45.6	123.2
Harrisburg, PA	24	40.9	77.8
Philadelphia, PA	24	40.9	75.5
Charleston, SC	38	33.3	80.8
Nashville, TN	31	36.7	87.6
Amarillo, TX	24	35.6	101.9
Galveston, TX	49	29.4	95.5
Houston, TX	44	30.1	95.9
Salt Lake City, UT	18	41.1	112.3
Burlington, VT	7	45.0	73.9
Norfolk, VA	32	37.0	76.6
Seattle, WA	33	48.1	122.5
Spokane, WA	19	48.1	117.9
Madison, WI	9	43.4	90.2
Milwaukee, WI	13	43.3	88.1
Cheyenne, WY	14	41.2	104.9

263. Weldon's dice data — (currently) the last word

Kemp, A.W. and Kemp, C.D. (1991) Weldon's dice data revisited. *American Statistician*, **45**, 216–222.

In a letter to Galton dated 2nd February 1894, Weldon reported the outcomes of 26306 rolls of 12 dice. There are three well-known subsets of these data reported in the paper.

I   In this experiment 12 dice were rolled 26306 times; a '5' or a '6' was counted a success.

0	1	2	3	4	5	6	7	8	9	10	11	12
185	1149	3265	5475	6114	5194	3067	1331	403	105	14	4	0

II   7006 of the 26306 experiments were performed by a clerk, deemed by Galton "reliable and accurate".

0	1	2	3	4	5	6	7	8	9	10	11	12
45	327	886	1475	1571	1404	787	367	112	29	2	1	0

III   In this subset of the data, 4096 of the rolls were scrutinized, with a '6' counted a success.

0	1	2	3	4	5	6	7	8	9	10	11	12
447	1145	1181	796	380	115	24	8	0	0	0	0	0

IV   4096 rolls, with a '4' or a '5' or a '6' counted a success.

0	1	2	3	4	5	6	7	8	9	10	11	12
0	7	60	198	430	731	948	847	536	257	71	11	0

264. Putting computers to the test

Longley, J.W. (1967) An appraisal of least squares programs for the electronic computer from the point of view of the user. *Journal of the American Statistical Association*, **62**, 819–841.

The data are used to compare the capabilities of a number of statistical computer packages for regression. The regressor variables are major USA economic indicators:

X_1 GNP implicit price deflator (1954 = 100.0)
X_2 GNP
X_3 Unemployment
X_4 Size of armed forces
X_5 Non-institutional population aged 14 and over

and X_6 is the year. The regressands are numbers in employment in different sectors (assumed to be additive). Measured annually (1947-1962) these are:

Y Total derived employment
Y_1 Census: agricultural employment
Y_2 Census: self-employed
Y_3 Census: unpaid family workers
Y_4 Census: domestics
Y_5 BLS: non-agricultural
Y_6 BLS: federal government
Y_7 BLS: state/local government

Amongst other interesting features: the means of the vectors of X and of Y have round numbers; the elements of the product moments matrix end in 5 or 0 within nine decimals. Hence the exact amount of rounding by the computer is known.

X_1	X_2	X_3	X_4	X_5	X_6
83.0	234289	2356	1590	107608	1947
88.5	259426	2325	1456	108632	1948
88.2	258054	3682	1616	109773	1949
89.5	284599	3351	1650	110929	1950
96.2	328975	2099	3099	112075	1951
98.1	346999	1932	3594	113270	1952
99.0	365385	1870	3547	115094	1953
100.0	363112	3578	3350	116219	1954
101.2	397469	2904	3048	117388	1955
104.6	419180	2822	2857	118734	1956
108.4	442769	2936	2798	120445	1957
110.8	444546	4681	2637	121950	1958
112.6	482704	3813	2552	123366	1959
114.2	502601	3931	2514	125368	1960
115.7	518173	4806	2572	127852	1961
116.9	554894	4007	2827	130081	1962

Y	Y_1	Y_2	Y_3	Y_4	Y_5	Y_6	Y_7
60323	8256	6045	427	1714	38407	1892	3582
61122	7960	6139	401	1731	39241	1863	3787
60171	8017	6208	396	1772	37922	1908	3948
61187	7497	6069	404	1995	39196	1928	4098
63221	7048	5869	400	2055	41460	2302	4087
63639	6792	5670	431	1922	42216	2420	4188
64989	6555	5794	423	1985	43587	2305	4340
63761	6495	5880	445	1919	42271	2188	4563
66019	6718	5886	524	2216	43761	2187	4727
67857	6572	5936	581	2359	45131	2209	5069
68169	6222	6089	626	2328	45278	2217	5409
66513	5844	6185	605	2456	43530	2191	5702
68655	5836	6298	597	2520	45214	2233	5957
69564	5723	6367	615	2489	45850	2270	6250
69331	5463	6388	662	2594	45397	2279	6548
70551	5190	6271	623	2626	46652	2340	6849

265. Assessing the effects of toxicity

Shirley, E. (1977) A non-parametric equivalent of Williams' test for contrasting increasing dose levels of a treatment. *Biometrics*, **33**, 386–389.

The toxicity of substances is often assessed at some stage by animal experiments. Increasing doses of the substance are given to different groups, and the effects noted. There is usually also a control group (zero dose, or a placebo). The experimenter wishes to know whether there is evidence of toxicity, and at what levels the evidence is apparent. In this experiment there were four levels of dose (0 – 3) and the reaction times of mice to stimuli to their tails were measured (seconds). There is reason to believe that the distribution of reaction times is very skewed, and this motivated the author's development of a nonparametric test.

0	1	2	3
2.4	2.8	9.8	7.0
3.0	2.2	3.2	9.8
3.0	3.8	5.8	9.4
2.2	9.4	7.8	8.8
2.2	8.4	2.6	8.8
2.2	3.0	2.2	3.4
2.2	3.2	6.2	9.0
2.8	4.4	9.4	8.4
2.0	3.2	7.8	2.4
3.0	7.4	3.4	7.8

266. Testing homogeneity

Haldane, J.B.S. (1955) The rapid calculation of χ^2 as a test of homogeneity from a $2 \times n$ table. *Biometrika*, **42**, 519–520.

It is often required to test a $2 \times n$ contingency table for homogeneity. The author provides a rapid and accurate (though not exact) short method for calculating χ^2. As an example, the author cites an experiment in which the number of normal *Drosophila melanogaster* and *vestigial Drosophila* melanogaster were counted in 11 bottles. (On Mendelian grounds a ratio of 3 : 1 was expected; however, vestigial was known to have a high mortality. The question arose of whether the mortality had been significantly different in different bottles.)

	1	2	3	4	5	6	7	8	9	10	11	
Drosophila	25	80	38	52	9	21	33	24	30	51	56	419
v. *Drosophila*	1	15	12	8	0	7	6	2	7	7	3	68
	26	95	50	60	9	28	39	26	37	58	59	487

267. Random digits

Data provided by F. Daly, Open University.

In an experiment to compare six different computer programs, the following digit frequencies in sequences of 1000 random digits were observed.

	0	1	2	3	4	5	6	7	8	9
SC v.1.09	92	107	85	85	109	95	104	95	113	115
GW-Basic v.3.23	85	110	91	95	106	110	92	106	101	104
SPIDA v.5.50	110	94	86	97	101	94	113	133	84	88
Minitab v.7.20	112	93	96	87	108	84	103	120	111	86
Mathematica v.2	104	101	82	91	118	102	96	110	97	99
S-Plus v.2.3	101	86	97	110	100	88	87	105	100	126

268. Ice cream consumption

Koteswara Rao Kadiyala (1970) Testing for the independence of regression disturbances. *Econometrica*, **38**, 97–117.

Ice cream consumption (Y, pints per capita) was measured over 30 four-week periods from March 18th 1951 to July 11th 1953. It was thought that variables influencing consumption might include the price of ice cream per pint (X_1, \$), the weekly family income (X_2, \$) and the mean temperature (X_3, ° F). In fact it can be shown that of these three only X_3 has any real effect on consumption.

	Y	X_1	X_2	X_3
1	.386	.270	78	41
2	.374	.282	79	56
3	.393	.277	81	63
4	.425	.280	80	68
5	.406	.272	76	69
6	.344	.262	78	65
7	.327	.275	82	61
8	.288	.267	79	47
9	.269	.265	76	32
10	.256	.277	79	24
11	.286	.282	82	28
12	.298	.270	85	26
13	.329	.272	86	32
14	.318	.287	83	40
15	.381	.277	84	55
16	.381	.287	82	63
17	.470	.280	80	72
18	.443	.277	78	72
19	.386	.277	84	67
20	.342	.277	86	60
21	.319	.292	85	44
22	.307	.287	87	40
23	.284	.277	94	32
24	.326	.285	92	27
25	.309	.282	95	28
26	.359	.265	96	33
27	.376	.265	94	41
28	.416	.265	96	52
29	.437	.268	91	64
30	.548	.260	90	71

269. Hospital data

Brooks, D.G., Carroll, S.S. and Verdini, W.A. (1988) Characterizing the domain of a regression model. *American Statistician*, **42**, 187–190.

Data were collected on monthly man-hours (Y) associated with maintaining the anaesthesiology service for 12 naval hospitals in the United States; predictor variables were the number of surgical cases (X_1), the eligible population per thousand (X_2) and the number of operating rooms (X_3). In this paper a procedure is decribed for calculating the convex hull of the original twelve observations; then what to do if further observations are collected on predictor variables only, which may possibly lie outside the convex hull.

Y	X_1	X_2	X_3
304.37	89	25.5	4
2616.32	513	294.3	11
1139.12	231	83.7	4
285.43	68	30.7	2
1413.77	319	129.8	6
1555.68	276	180.8	6
383.78	82	43.4	4
2174.27	427	165.2	10
845.30	193	74.3	4
1125.28	224	60.8	5
3462.60	729	319.2	12
3682.33	951	376.2	12

270. Aquifer data

Cressie, N. (1989) Geostatistics. *American Statistician*, **43**, 197–202

Groundwater flow was measured in the regions surrounding a proposed site for high-level nuclear waste. The site will eventually contain several tens of thousands of waste canisters, buried deep underground about 30 feet apart in holes or trenches surrounded by salt, and covering an area of about two square miles. There were 85 data points, obtained (in principle — details are not provided) by drilling a narrow pipe into the aquifer and letting the water find its own level. The measurement units of water level, Z, are feet above sea level. The two location coordinates (x, y) are measured in miles from an arbitrary origin.

x	y	Z	x	y	Z
42.78275	127.62282	1464	103.26625	20.34239	1591
-27.39691	90.78732	2553	-14.31073	31.26545	2540
-1.16289	84.89600	2158	-18.13447	30.18118	2352
-18.61823	76.45199	2455	-18.12151	29.53241	2528
96.46549	64.58058	1756	-9.88796	38.14483	2575
108.56243	82.92325	1702	-12.16336	39.11081	2468
88.36356	56.45348	1805	11.65754	18.73347	2646
90.04213	39.25820	1797	61.69122	32.49406	1739
93.17269	33.05852	1714	69.57896	33.80841	1674
97.61099	56.27887	1466	66.72205	33.93264	1868
90.62946	35.08169	1729	-36.65446	150.91456	1865
92.55262	41.75238	1638	-19.55102	137.78404	1777
99.48996	59.15785	1736	-21.29791	131.82542	1579
-24.06744	184.76636	1476	-22.36166	137.13680	1771
-26.06285	114.07479	2200	21.14719	139.26199	1408
56.27842	26.84826	1999	7.68461	126.83751	1527
73.03881	18.88140	1680	-8.33227	107.77691	2003
80.26679	12.61593	1806	56.70724	171.26443	1386
80.23009	14.61795	1682	59.00052	164.54863	1089
68.83845	107.77423	1306	68.96893	177.24820	1384
76.39921	95.99380	1722	70.90225	161.38136	1030
64.46148	110.39641	1437	73.00243	162.98960	1092
43.39657	53.61499	1828	59.66237	170.10544	1161
39.07769	61.99805	2118	61.87429	174.30178	1415
112.80450	45.54766	1725	63.70810	173.91453	1231
54.25899	147.81987	1606	5.62706	79.08730	2300
6.13202	48.32772	2648	18.24739	77.39191	2238
-3.80469	40.40450	2560	85.68824	139.81701	1038
-2.23054	29.91113	2544	105.07646	132.03181	1332
-2.36177	33.82002	2386	-101.64278	10.65106	3510
-2.18890	33.68207	2400	-145.23654	28.02333	3490
63.22428	79.49924	1757	-73.99313	87.97270	2594
-10.77860	175.11346	1402	-94.48182	86.62606	2650
-18.98889	171.91694	1364	-88.84983	76.70991	2533
-38.57884	158.52742	1735	-120.25898	80.76485	3571
83.14496	159.11558	1376	-86.02454	54.36334	2811
-21.80248	15.02551	2729	-72.79097	43.09215	2728
-23.56457	9.41441	2766	-100.17372	42.89881	3136
-20.11299	22.09269	2736	-78.83539	40.82141	2553
-16.62654	17.25621	2432	-83.69063	46.50482	2798
29.90748	175.12875	1024	-95.61661	35.82183	2691
100.91568	22.97808	1611	-87.55480	29.39267	2946
101.29544	22.96385	1548			

271. Windmill data

Joglekar, G., Schuenemeyer, J.H. and LaRiccia, V. (1989) Lack-of-fit testing when replicates are not available. *American Statistician*, **43**, 135–143.

The authors consider two data sets for regression fits. In the first case (the windmill data) a reciprocal transformation is appropriate; in the second case (tensile strength) a quadratic in x is the appropriate model. For the windmill data direct current output (Y) was measured against wind velocity (x, miles per hour). There were 25 observations recorded. For the tensile-strength data, the tensile strength of Kraft paper (Y, psi) was measured against the percentage of hardwood in the batch of pulp from which the paper was produced (19 observations).

Wind velocity, Y	DC output, x
2.45	0.123
2.70	0.500
2.90	0.653
3.05	0.558
3.40	1.057
3.60	1.137
3.95	1.144
4.10	1.194
4.60	1.562
5.00	1.582
5.45	1.501
5.80	1.737
6.00	1.822
6.20	1.866
6.35	1.930
7.00	1.800
7.40	2.088
7.85	2.179
8.15	2.166
8.80	2.112
9.10	2.303
9.55	2.294
9.70	2.386
10.00	2.236
10.20	2.310

Tensile strength, Y	Hardwood concentration, x
6.3	1.0
11.1	1.5
20.0	2.0
24.0	3.0
26.1	4.0

30.0	4.5
33.8	5.0
34.0	5.5
38.1	6.0
39.9	6.5
42.0	7.0
46.1	8.0
53.1	9.0
52.0	10.0
52.5	11.0
48.0	12.0
42.8	13.0
27.8	14.0
21.9	15.0

272. Examination times and scores

Basak, I., Balch, W.R. and Basak, P. (1992) Skewness: asymptotic critical values for a test related to Pearson's measure. *Journal of Applied Statistics*, **19**, 479–487.

Final examination scores (out of 75) and corresponding exam completion times (seconds) were collected for 134 individuals.

49	49	70	55	52	55	61	65	57	71	49	48	49	69
2860	2063	2013	2000	1420	1934	1519	2735	2329	1590	1699	1816	1824	1899

44	53	49	52	53	36	61	68	67	53	33	64	57	56
1714	1741	1968	1721	2120	1435	1909	1707	1431	2024	1725	1634	1949	1278

41	40	42	40	51	53	62	61	49	54	57	71	45	70
1677	1945	1754	1200	1307	1895	1798	1375	2665	1743	1722	2562	2277	1579

58	62	28	72	37	67	51	55	68	58	61	43	60	53
1785	1068	1411	1162	1646	1489	1769	1550	1313	2472	2036	1914	1910	2730

51	51	60	64	66	52	45	48	51	73	63	32	59	68
2235	1993	1613	1532	2339	2109	1649	2238	1733	1981	1440	1482	1758	2540

35	64	62	51	52	44	64	65	56	52	59	66	42	67
1637	1779	1069	1929	2605	1491	1321	1326	1797	1158	1595	2105	1496	1301

48	56	47	68	58	59	45	31	47	56	38	47	65	61
2467	1265	3813	1216	1167	1767	1683	1648	1144	1162	1460	1726	1862	3284

45	63	66	44	57	56	56	54	61	58	46	62	68	58
1683	1654	2725	1992	1332	1840	1704	1510	3000	1758	1604	1475	1106	2040

47	66	61	58	45	55	54	54	54	41	65	66	38	51
1594	1215	1418	1828	2305	1902	2013	2026	1875	2227	2325	1674	2435	2715
49	49	51	42	61	69	42	53						
1773	1656	2320	1908	1853	1302	2161	1715						

273. Gamma irradiation

Singh, M., Kanji, G.K. and El-Bizri, K.S. (1992) A note on inverse estimation in non-linear models. *Journal of Applied Statistics*, 473–477.

The authors consider a regression model where the predictor variable is a dose of gamma irradiation (x, kiloRöntgen) and the response is the shoot length of plants emerging from seeds exposed to this dose (Y, cm, the mean shoot length of 5 plants of a chickpea cultivar ILC482). The fitted model in this case is

$$y = \frac{c}{1 + \exp[-b(x - m)]} \, .$$

x	0	10	20	30	40	50	60	70	80	90	100	110
y	8.85	9.40	9.18	8.70	7.53	6.43	5.85	4.73	3.98	3.50	3.10	2.80

274. Ramus heights

Elston, R.C. and Grizzle, J.E. (1962) Estimation of time-response curves and their confidence bands. *Biometrics*, **18**, 148–159.

Data were collected on the growth of the ramus bone (height in mm) for each of a cohort of 20 boys measured at ages 8, $8\frac{1}{2}$, 9, $9\frac{1}{2}$ years. The object of the study was to establish a normal growth curve for use by dentists. (The ramus is the ascending part of the jawbone.)

	8	$8\frac{1}{2}$	9	$9\frac{1}{2}$
1	47.8	48.8	49.0	49.7
2	46.4	47.3	47.7	48.4
3	46.3	46.8	47.8	48.5
4	45.1	45.3	46.1	47.2
5	47.6	48.5	48.9	49.3
6	52.5	53.2	53.3	53.7
7	51.2	53.0	54.3	54.5
8	49.8	50.0	50.3	52.7
9	48.1	50.8	52.3	54.4
10	45.0	47.0	47.3	48.3
11	51.2	51.4	51.6	51.9
12	48.5	49.2	53.0	55.5
13	52.1	52.8	53.7	55.0
14	48.2	48.9	49.3	49.8
15	49.6	50.4	51.2	51.8
16	50.7	51.7	52.7	53.3
17	47.2	47.7	48.4	49.5
18	53.3	54.6	55.1	55.3
19	46.2	47.5	48.1	48.4
20	46.3	47.6	51.3	51.8

275. A soil experiment

Johnson, D.E. and Graybill, F.A. (1972) An analysis of a two-way model with interaction and no replication. *Journal of the American Statistical Association*, **67**, 862–868.

These data are part of a larger experiment to determine the effectiveness of blast furnace slags as agricultural liming materials on three types of soil, sandy loam (I), sandy clay loam (II) and loamy sand (III). The treatments were all applied at 4000 lbs per acre, and what was measured was the corn yield in bushels per acre.

Treatment	I	II	III
None	11.1	32.6	63.3
Coarse slag	15.3	40.8	65.0
Medium slag	22.7	52.1	58.8
Agricultural slag	23.8	52.8	61.4
Agricultural limestone	25.6	63.1	41.1
Agricultural slag + minor elements	31.2	59.5	78.1
Agricultural limestone + minor elements	25.8	55.3	60.2

276. Rabbit foetuses

Jun Shao and Shein-Chung Chow (1990) Test for treatment effect based on binary data with random sample sizes. *Australian Journal of Statistics*, **32**, 53–70.

In this paper the following teratological experiment is described. A group of 12 female mated rabbits was subjected to a treatment (not described) from the 6th to the 18th days of gestation. On the 30th day the rabbits were sacrificed and the number of foetuses was counted for each rabbit. There was a control group (untreated) also of size 12. In the table below the variables are the number of live male foetuses found (x), the number of live foetuses (y) and the total number of foetuses (z) (so $x \leq y \leq z$). The purpose of the experiment was to explore the treatment effect on foetus survival.

Controls			Treated		
x	y	z	x	y	z
2	3	3	5	11	11
5	8	11	4	7	10
7	12	12	5	7	7
1	4	4	6	6	6
6	9	9	4	7	8
2	7	7	4	9	9
5	6	7	4	7	9
3	3	5	5	7	10
4	7	7	0	1	2
4	9	9	2	6	6
5	10	11	7	11	11
5	8	8	3	6	6

277. Blood fat concentration

Scott, D.W. (1992) *Multivariate density estimation*, Chichester: John Wiley & Sons, 275, Table B3.

Data were collected on the concentration of plasma cholesterol and plasma triglycerides (mg/dl) for 371 male patients evaluated for chest pain. For 51 patients there was no evidence of heart disease; for the remaining 320 there was evidence of narrowing of the arteries.

Original reference: Scott, D.W., Gotto, A.M., Cole, J.S. and Gorry, G.A. (1978) Plasma lipids as collateral risk factors in coronary artery disease: a study of 371 males with chest pain. *Journal of Chronic Diseases*, **31**, 337–345.

51 patients with no evidence of heart disease

195	348	237	174	205	158	201	171	190	85	180	82	193	210
170	90	150	167	200	154	228	119	169	86	178	166	251	211
234	143	222	284	116	87	157	134	194	121	130	64	206	99
158	87	167	177	217	114	234	116	190	132	178	157	265	73
219	98	266	486	190	108	156	126	187	109	149	146	147	95
155	48	207	195	238	172	168	71	210	91	208	139	160	116
243	101	209	97	221	156	178	116	289	120	201	72	168	100
168	227	207	160										

320 patients with narrowing of the arteries

184	145	263	142	185	115	271	128	173	56	230	304	222	151
215	168	233	340	212	171	221	140	239	97	168	131	231	145
221	432	131	137	211	124	232	258	313	256	240	221	176	166
210	92	251	189	175	148	185	256	184	222	198	149	198	333
208	112	284	245	231	181	171	165	258	210	164	76	230	492
197	87	216	112	230	90	265	156	197	158	230	146	233	142
250	118	243	50	175	489	200	68	240	196	185	116	213	130
180	80	208	220	386	162	236	152	230	162	188	220	200	101
212	130	193	188	230	158	169	112	181	104	189	84	180	202
297	232	232	328	150	426	239	154	178	100	242	144	323	196
168	208	197	291	417	198	172	140	240	441	191	115	217	327
208	262	220	75	191	115	119	84	171	170	179	126	208	149
180	102	254	153	191	136	176	217	283	424	253	222	220	172
268	154	248	312	245	120	171	108	239	92	196	141	247	137
219	454	159	125	200	152	233	127	232	131	189	135	237	400
319	418	171	78	194	183	244	108	236	148	260	144	254	170
250	161	196	130	298	143	306	408	175	153	251	117	256	271
285	930	184	255	228	142	171	120	229	242	195	137	214	223
221	268	204	150	276	199	165	121	211	91	264	259	245	446

227	146	197	265	196	103	193	170	211	122	185	120	157	59
224	124	209	82	223	80	278	152	251	152	140	164	197	101
172	106	174	117	192	101	221	179	283	199	178	109	185	168
181	119	191	233	185	130	206	133	210	217	226	72	219	267
215	325	228	130	245	257	186	273	242	85	201	297	239	137
179	126	218	123	279	317	234	135	264	269	237	88	162	91
245	166	191	90	207	316	248	142	139	173	246	87	247	91
193	290	332	250	194	116	195	363	243	112	271	89	197	347
242	179	175	246	138	91	244	177	206	201	191	149	223	154
172	207	190	120	144	125	194	125	105	36	201	92	193	259
262	88	211	304	178	84	331	134	235	144	267	199	227	202
243	126	261	174	185	100	171	90	222	229	231	161	258	328
211	306	249	256	209	89	177	133	165	151	299	93	274	323
219	163	233	101	220	153	348	154	194	400	230	137	250	160
173	300	260	127	258	151	131	61	168	91	208	77	287	209
308	260	227	172	168	126	178	101	164	80	151	73	165	155
249	146	258	145	194	196	140	99	187	390	171	135	221	156
294	135	167	80	208	201	208	148	185	231	159	82	222	108
266	164	217	227	249	200	218	207	245	322	242	180	262	169
169	158	204	84	184	182	206	148	198	124	242	248	189	176
260	98	199	153	207	150	206	107	210	95	229	296	232	583
267	192	228	149	187	115	304	149	140	102	209	376	198	105
270	110	188	148	160	125	218	96	257	402	259	240	139	54
213	261	178	125	172	146	198	103	222	348	238	156	273	146
131	96	233	141	269	84	170	284	149	237	194	272	142	111
218	567	194	278	252	233	184	184	203	170	239	38	232	161
225	240	280	218	185	110	163	156	216	101				

278. Snowfall data

Parzen, E. (1979) Nonparametric statistical data modelling. *Journal of the American Statistical Association*, **74**, 105–131.

The data give the annual snowfall in Buffalo, NY (inches) for the 63 years 1910 to 1972.

126.4	82.4	78.1	51.1	90.9	76.2	104.5	87.4	110.5	25.0	69.3	53.5
39.8	63.6	46.7	72.9	79.7	83.6	80.7	60.3	79.0	74.4	49.6	54.7
71.8	49.1	103.9	51.6	82.4	83.6	77.8	79.3	89.6	85.5	58.0	120.7
110.5	65.4	39.9	40.1	88.7	71.4	83.0	55.9	89.9	84.8	105.2	113.7
124.7	114.5	115.6	102.4	101.4	89.8	71.5	70.9	98.3	55.5	66.1	78.4
120.5	97.0	110.0									

279. Counting alpha-particles

Rutherford, E. and Geiger, M. (1910) The probability variations in the distribution of alpha-particles. *Philosophical Magazine, Series 6*, **20**, 698–704.

In this classic set of data Rutherford and Geiger counted the number of scintillations in 72 second intervals caused by the radioactive decay of a quantity of the element polonium. Altogether there were 10097 scintillations during 2608 intervals. Overall agreement with the Poisson distribution is excellent.

Count	0	1	2	3	4	5	6	7	8	9	10	11	12	13	14
Frequency	57	203	383	525	532	408	273	139	45	27	10	4	0	1	1

280. Eruptions of the Old Faithful geyser

Azzalini, A. and Bowman, A.W. (1990) A look at some data on the Old Faithful geyser. *Journal of the Royal Statistical Society, Series C*, **39**, 357–366.

The Old Faithful geyser at Yellowstone National Park, Wyoming, USA, was observed from August 1st to August 15th, 1985. During that time, data were collected on the duration of eruptions (not given here) and the waiting time between the starts of successive eruptions. There were 300 eruptions observed, so 299 waiting times (measured in minutes). There are two pronounced modes.

```
80 71 57 80 75 77  60 86 77 56 81 50 89 54 90 73 60 83
65 82 84 54 85 58  79 57 88 68 76 78 74 85 75 65 76 58
91 50 87 48 93 54  86 53 78 52 83 60 87 49 80 60 92 43
89 60 84 69 74 71 108 50 77 57 80 61 82 48 81 73 62 79
54 80 73 81 62 81  71 79 81 74 59 81 66 87 53 80 50 87
51 82 58 81 49 92  50 88 62 93 56 89 51 79 58 82 52 88
52 78 69 75 77 53  80 55 87 53 85 61 93 54 76 80 81 59
86 78 71 77 76 94  75 50 83 82 72 77 75 65 79 72 78 77
79 75 78 64 80 49  88 54 85 51 96 50 80 78 81 72 75 78
87 69 55 83 49 82  57 84 57 84 73 78 57 79 57 90 62 87
78 52 98 48 78 79  65 84 50 83 60 80 50 88 50 84 74 76
65 89 49 88 51 78  85 65 75 77 69 92 68 87 61 81 55 93
53 84 70 73 93 50  87 77 74 72 82 74 80 49 91 53 86 49
79 89 87 76 59 80  89 45 93 72 71 54 79 74 65 78 57 87
72 84 47 84 57 87  68 86 75 73 53 82 93 77 54 96 48 89
63 84 76 62 83 50  85 78 78 81 78 76 74 81 66 84 48 93
47 87 51 78 54 87  52 85 58 88 79
```

281. Starch films

Fong, D.K.H. (1992) Ranking and estimation of related means in the presence of a covariate — a Bayesian approach. *Journal of the American Statistical Association,* **87**, 1128–1136.

The author refers to an experiment first described in Freeman, H.A. (1942) *Industrial Statistics*, Wiley, New York. The breaking strength (y, in grams) was calculated against the thickness (x, inches/10000) of seven types of starch film.

281. Starch films

Wheat		Rice		Canna		Corn		Potato		Dasheen		S/Potato	
y	x	y	x	y	x	y	x	y	x	y	x	y	x
263.7	5.0	556.7	7.1	791.7	7.7	731.0	8.0	983.3	13.0	485.4	7.0	837.1	9.4
130.8	3.5	552.5	6.7	610.0	6.3	710.0	7.3	958.8	13.3	395.4	6.0	901.2	10.6
382.9	4.7	397.5	5.6	710.0	8.6	604.7	7.2	747.8	10.7	465.4	7.1	595.7	9.0
302.5	4.3	532.3	8.1	940.7	11.8	508.8	6.1	866.0	12.2	371.4	5.3	510.0	7.6
213.3	3.8	587.8	8.7	990.0	12.4	393.0	6.4	810.8	11.6	402.0	6.2		
132.1	3.0	520.9	8.3	916.2	12.0	416.0	6.4	950.0	9.7	371.9	5.8		
292.0	4.2	574.3	8.4	835.0	11.4	400.0	6.9	1282.0	10.8	430.0	6.6		
315.5	4.5	505.0	7.3	724.3	10.4	335.6	5.8	1233.8	10.1	380.0	6.6		
262.4	4.3	604.6	8.5	611.1	9.2	306.4	5.3	1660.0	12.7				
314.4	4.1	522.5	7.8	621.7	9.0	426.0	6.7	746.0	9.8				
310.8	5.5	555.0	8.0	735.4	9.5	382.5	5.8	650.0	10.0				
280.8	4.8	561.1	8.4	990.0	12.5	340.8	5.7	992.5	13.8				
331.7	4.8	862.7	11.7	436.7	6.1	896.7	13.3						
672.5	8.0	333.3	6.2	873.9	12.4								
496.0	7.4	382.3	6.3	924.4	12.2								
311.9	5.2	397.7	6.0	1050.0	14.1								
276.7	4.7	619.1	6.8	973.3	13.7								
325.7	5.4	857.3	7.9										
310.8	5.4	592.5	7.2										
288.0	5.4												
269.3	4.9												

282. Sheep diet

Street, A.P. and Street, D.J. (1988) Latin squares and agriculture: the other bicentennial. *Mathematical Scientist*, **13**, 48–55.

The authors report a paper of 1788 entitled "On the advantage and economy of feeding sheep in the house with roots", by M. Crette de Palluel, read before the Royal Agricultural Society of Paris in July of that year. The 1788 paper is a remarkably detailed account of an experiment which is, in fact, a 4 × 4 Latin square, assuming de Palluel randomized correctly. Sixteen sheep (four each of four different breeds) were fed on four different diets. They were weighed at the start of the experiment and then monthly thereafter. Each month, four sheep were killed immediately after their weights were recorded.

Food	p	potatoes
	t	turnips
	b	beets
	o	oats, barley and grey peas
Breed	i	Isle de France
	b	Beauce
	c	Champagne
	p	Picardy

Food	Breed	Start	Monthly weights (lb)			
p	i	69.75	79.75	-	-	-
	b	70.75	82.5	90.25	93	95
	c	69.25	83	82.5	84	-
	p	88	95	101	-	-
t	i	69	86	87	-	-
	b	71	86	-	-	-
	c	68.5	78.5	82.5	84	84.5
	p	79	95.5	97.5	97.5	-
b	i	72	83.25	90.5	94	-
	b	70.75	80.75	86	-	-
	c	77.25	90.5	-	-	-
	p	80	93.5	98.5	100.5	101
o	i	74	91	95.5	102	106
	b	73.5	84.25	91.5	96	-
	c	71	86.25	93	-	-
	p	71	87	-	-	-

283. Horse-kicks

Preece, D.A., Ross, G.J.S. and Kirby, S.P.J. (1988) Bortkewitsch's horse-kicks and the generalized linear model. *The Statistician*, **37**, 313–318.

The "horse-kicks" data are amongst the most well-known and least understood collections. They summarize the numbers of Prussian Militarpersonen killed by the kicks of a horse for each of 14 corps in each of 20 successive years 1875-1894. In this paper the data are revisited. Most often they appear as summary data (A: 196 deaths during 280 corps-years) and show moderately good agreement with the Poisson distribution. Bortkewitsch noted that four of the corps were less representative than the others. After removing these four (G, I, VI and XI) the summary data are as shown at Table B: 122 deaths during 200 corps-years. Here the Poisson agreement is very good indeed. Table C gives the full data broken down by corps and year, to which we may fit a generalized linear model with logarithmic link function and Poisson errors for the observations and with terms for corps and years. The authors fit a Poisson model compounded by a gamma distribution, resulting in a negative binomial distribution for the data.

(A) Summary data for 14 corps over 20 years:

Deaths	0	1	2	3	4	≥ 5
Frequency	144	91	32	11	2	0

(B) Summary data for 10 corps over 20 years:

Deaths	0	1	2	3	4	≥ 5
Frequency	109	65	22	3	1	0

(C) The full data set for 14 corps over 20 years:

	G	I	II	III	IV	V	VI	VII	VIII	IX	X	XI	XIV	XV	
1875	0	0	0	0	0	0	0	1	1	0	0	0	1	0	3
1876	2	0	0	0	1	0	0	0	0	0	0	0	1	1	5
1877	2	0	0	0	0	0	1	1	0	0	1	0	2	0	7
1878	1	2	2	1	1	0	0	0	0	0	1	0	1	0	9
1879	0	0	0	1	1	2	2	0	1	0	0	2	1	0	10
1880	0	3	2	1	1	1	0	0	0	2	1	4	3	0	18
1881	1	0	0	2	1	0	0	1	0	1	0	0	0	0	6
1882	1	2	0	0	0	0	1	0	1	1	2	1	4	1	14
1883	0	0	1	2	0	1	2	1	0	1	0	3	0	0	11
1884	3	0	1	0	0	0	0	1	0	0	2	0	1	1	9
1885	0	0	0	0	0	0	1	0	0	2	0	1	0	1	5
1886	2	1	0	0	1	1	1	0	0	1	0	1	3	0	11

1887	1	1	2	1	0	0	3	2	1	1	0	1	2	0	15
1888	0	1	1	0	0	1	1	0	0	0	0	1	1	0	6
1889	0	0	1	1	0	1	1	0	0	1	2	2	0	2	11
1890	1	2	0	2	0	1	1	2	0	2	1	1	2	2	17
1891	0	0	0	1	1	1	0	1	1	0	3	3	1	0	12
1892	1	3	2	0	1	1	3	0	1	1	0	1	1	0	15
1893	0	1	0	0	0	1	0	2	0	0	1	3	0	0	8
1894	1	0	0	0	0	0	0	0	1	0	1	1	0	0	4
	16	16	12	12	8	11	17	12	7	13	15	25	24	8	196

284. Pneumonia risk in smokers with chickenpox

Ellis, M.E., Neal, K.R. and Webb, A.K. (1987) Is smoking a risk factor for pneumonia in patients with chickenpox. *British Medical Journal*, **294**, 1002.

These data appeared in a paper which reported the carbon monoxide transfer factor levels in seven smokers with chickenpox with a view to determining their risk of contracting pneumonia. The measurements were taken on entry to hospital and one week after admission.

The main question is straightforward: has the carbon monoxide transfer factor changed significantly over one week?

The analysis is interesting. Box plots of the entry and one week transfer factors appear to show a difference and a two-sample t-test (which is, of course, incorrect for paired data such as these) seems to confirm this. However, a paired t-test produces a p-value above the conventional 5% level. Clearly the normality assumption required for the t-test is suspect and a Q-Q plot of sample quantiles of (entry – 1 week) against standard normal quantiles shows it to be untenable. A Wilcoxon signed rank test is appropriate.

| | CO transfer factor | |
Patient	Entry	One week
1	40	73
2	50	52
3	56	80
4	58	85
5	60	64
6	62	63
7	66	60

285. Anorexia data

Brian Everitt (private communication)

These are weights, in kg, of young girls receiving three different treatments for anorexia over a fixed period of time with the control group receiving the standard treatment.

There are three groups of paired data here with a wide variety of approaches. Clearly the main problem is to compare the methods of treatment. Whichever statistical technique is employed, it is instructive to look at the three scatterplots of after/before.

Cognitive behavioural treatment		Control		Family therapy	
Weight (kg)		Weight (kg)		Weight (kg)	
before	after	before	after	before	after
80.5	82.2	80.7	80.2	83.8	95.2
84.9	85.6	89.4	80.1	83.3	94.3
81.5	81.4	91.8	86.4	86.0	91.5
82.6	81.9	74.0	86.3	82.5	91.9
79.9	76.4	78.1	76.1	86.7	100.3
88.7	103.6	88.3	78.1	79.6	76.7
94.9	98.4	87.3	75.1	76.9	76.8
76.3	93.4	75.1	86.7	94.2	101.6
81.0	73.4	80.6	73.5	73.4	94.9
80.5	82.1	78.4	84.6	80.5	75.2
85.0	96.7	77.6	77.4	81.6	77.8
89.2	95.3	88.7	79.5	82.1	95.5
81.3	82.4	81.3	89.6	77.6	90.7
76.5	72.5	78.1	81.4	83.5	92.5
70.0	90.9	70.5	81.8	89.9	93.8
80.4	71.3	77.3	77.3	86.0	91.7
83.3	85.4	85.2	84.2	87.3	98.0
83.0	81.6	86.0	75.4		
87.7	89.1	84.1	79.5		
84.2	83.9	79.7	73.0		
86.4	82.7	85.5	88.3		
76.5	75.7	84.4	84.7		
80.2	82.6	79.6	81.4		
87.8	100.4	77.5	81.2		
83.3	85.2	72.3	88.2		
79.7	83.6	89.0	78.8		
84.5	84.6				
80.8	96.2				
87.4	86.7				

286. Breast development of Turkish girls

Neyzi, O., Alp, H. and Orhon, A. (1975) Breast development of 318 12-13 year old Turkish girls by socio-economic class of parents. *Annals of Human Biology*, 2, 1, 49–59.

The data are taken from an investigation into the effect of socio-economic class on physical development. Breast sizes were classified on a scale of 1 (no development) to 5 (fully developed) and the socio-economic class of their parents was assessed on a scale of 1 to 4.

The expected cell frequencies in the first column are too small and therefore the first two columns need to be pooled. A chi-squared test of independence does not yield a significant result but the story does not end there. Try pooling columns 1+2, 3 and 4 (after a suitable check that this is a reasonable thing to do, of course) and do another chi-squared test. What conclusions can you draw?

S-E class of parents	Breast development				
	1	2	3	4	5
1	2	14	28	40	18
2	1	21	25	25	9
3	1	12	12	12	2
4	6	17	34	33	6

287. Treatment of enuresis

Hills, M. and Armitage, P. (1979) The two-period cross-over clinical trial. *British Journal of Clinical Pharmacology*, 8, 7–20.

These data are from a clinical trial on 29 patients of a new drug for the treatment of enuresis. The patients were given the drug for 14 days and a placebo for a separate 14 days, the order of administration being chosen randomly for each patient. The data show the number of dry nights out of 14.

This is an interesting crossover trial which may be analysed via *t*-tests or, more powerfully, by regression. It turns out that the number of dry nights is significantly greater than zero for all patients. The order of administration is not important and the drug treatment proves ineffective.

Group A			Group B		
patient number (drug/placebo)	Period 1 (drug)	Period 2 (placebo)	patient number (placebo/drug)	Period 1 (placebo)	Period 2 (drug)
1	8	5	2	12	11
3	14	10	5	6	8
4	8	0	8	13	9
6	9	7	10	8	8
7	11	6	12	8	9

9	3	5
11	6	0
13	0	0
16	13	12
18	10	2
19	7	5
21	13	13
22	8	10
24	7	7
25	9	0
27	10	6
28	2	2

14	4	8
15	8	14
17	2	4
20	8	13
23	9	7
26	7	10
29	7	6

288. Births in the USA

Statistical Abstract of the U.S (1988), United States Department of Commerce, Bureau of the Census.

The data show the number of births per thousand population in different regions of the USA in 1985 and 1975.

There are two questions of interest, these being "Does the birth rate differ between regions?" and "Are the changes between 1975 and 1985 different for different regions?" The analysis needs care in checking the validity of statistical assumptions.

| | 1985 | | | | | 1975 | | | |
North East	Mid West	South Atlantic	South Central	West	North East	Mid West	South Atlantic	South Central	West
14.5	14.9	15.5	14.2	15.9	14.2	14.7	14.0	15.8	14.0
15.5	14.7	15.5	14.0	14.7	13.3	15.4	12.7	14.6	14.4
15.0	15.7	15.8	14.9	17.9	14.1	15.0	13.7	15.8	14.7
14.1	15.2	15.1	16.6	24.6	11.9	14.7	13.9	18.3	20.7
13.5	15.4	12.5	14.9	17.4	11.3	14.3	15.3	16.0	17.8
13.9	16.1	14.3	18.2	16.4	11.6	14.4	14.6	17.5	16.1
14.6	14.3	15.6	16.1	17.5	13.1	14.4	16.1	15.4	19.5
14.0	15.3	16.1	18.8	18.4	12.5	14.3	15.8	17.2	18.3
13.5	17.1	14.4	19.1	17.0	12.5	16.6	12.4	18.1	15.5
	17.1	22.8	18.6	25.7	16.5			17.3	
	15.9		14.9	16.4	15.4				
	16.2				14.6				

289. Flying bomb hits on London during World War II

Clarke, R.D. (1946) An application of the Poisson distribution. *Journal of the Institute of Actuaries*, **72**, 48.

The data are taken from a 36 sq km area of South London. The area was gridded into 1/4 km squares and the data comprise the numbers of squares which received no hits, 1 hit, 2 hits, etc. The question of concern to both military and civilian authorities was whether or not the weapon could be precisely aimed.

If the hits were random, a Poisson model should fit these data.

Number of hits	Number of 1/4 km squares
0	229
1	211
2	93
3	35
4	7
7	1

290. Brain and body weights of animals

Jerison, H.J. (1973) *Evolution of the Brain and Intelligence*, New York: Academic Press.

These data also appear in Weisberg, S. (1980) *Applied Linear Regression*, New York: John Wiley & Sons.

The data comprise brain and body weights of animals.

These data need transforming before they can even be plotted. Three obvious outliers are revealed. After fitting a linear model to the transformed data, a plot of residuals against fitted values reveals two further outliers.

Index	Species	Body weight (kg)	Brain weight (g)
1	Mountain beaver	1.350	8.100
2	Cow	465.000	423.000
3	Grey wolf	36.33	119.500
4	Goat	27.660	115.000
5	Guinea pig	1.040	5.500
6	Diplodocus	11700.000	50.000
7	Asian elephant	2547.000	4603.000

8	Donkey	187.100	419.000
9	Horse	521.000	655.000
10	Potar monkey	10.000	115.000
11	Cat	3.300	25.600
12	Giraffe	529.000	680.000
13	Gorilla	207.000	406.000
14	Human	62.000	1320.000
15	African elephant	6654.000	5712.000
16	Triceratops	9400.000	70.000
17	Rhesus monkey	6.800	179.000
18	Kangaroo	35.000	56.000
19	Hamster	0.120	1.000
20	Mouse	0.023	0.400
21	Rabbit	2.500	12.100
22	Sheep	55.500	175.000
23	Jaguar	100.000	157.000
24	Chimpanzee	52.160	440.000
25	Brachiosaurus	87000.000	154.500
26	Rat	0.280	1.900
27	Mole	0.122	3.000
28	Pig	192.000	180.000

291. Severe idiopathic respiratory distress syndrome

Van Vliet, P.K. and Gupta, J.M., (1973) Tham-v-sodium bicarbonate in idiopathic respiratory distress syndrome. *Archives of Disease in Childhood*, **48**, 249–255.

The data comprise birthweights (in kg) of infants with severe idiopathic respiratory distress syndrome. The children marked with asterisks died. Can children at risk be identified by their birthweights?

The data are skewed and need to be transformed before testing. It is interesting to compare with a nonparametric test.

1.050*	2.500*	1.890*	1.760	2.830
1.175*	1.030*	1.940*	1.930	1.410
1.230*	1.100*	2.200*	2.015	1.715
1.310*	1.185*	2.270*	2.090	1.720
1.500*	1.225*	2.440*	2.600	2.040
1.600*	1.262*	2.560*	2.700	2.200
1.720*	1.295*	2.730*	2.950	2.400
1.750*	1.300*	1.130	3.160	2.550
1.770*	1.550*	1.575	3.400	2.570
2.275*	1.820*	1.680	3.640	3.005

* = child died

292. Lowest temperatures for U.S. cities

Statistical Abstract of the U.S. (1988) United States Department of Commerce, Bureau of the Census.

The data are lowest temperatures (in °F) recorded in various months for cities in the U.S.A. Can the cities be characterized by these data? Can the number of variables be reduced so that the cities can be graphically represented?

It is interesting to fit a two component model to these data and plot them. This can be done by principal component analysis.

City	January	April	July	October
Atlanta	-8	26	53	28
Baltimore	-7	20	51	25
Bismarck	-44	-12	35	5
Boston	-12	16	54	28
Chicago	-27	7	40	17
Dallas	4	30	59	29
Denver	-25	-2	43	3
El Paso	-8	23	57	25
Honolulu	53	57	67	64
Houston	12	31	62	33
Juneau	-22	6	36	11
Los Angeles	23	39	49	41
Miami	30	46	69	51
Nashville	-17	23	51	26
New York	-6	12	52	28
Omaha	-23	5	44	13
Phoenix	17	32	61	34
Portland	-26	8	40	15
Reno	-16	13	33	8
San Francisco	24	31	43	34
Seattle	0	29	43	28
Washington	-5	24	55	29

293. Plasma citrate concentrations

Andersen, A.H., Jensen, E.B. and Schou, G. (1981) Two-way analysis of variance with correlated errors. *International Statistical Review*, **49**, 153–167.

The data are measurements on plasma citrate concentrations in micromols/litre obtained from 10 subjects at 8 am, 11 am, 2 pm, 5 pm and 8 pm. To what extent is there a normal profile for the level in the human body during the day?

Two-way analysis of variance may be used. The distribution of residuals should be looked at with a view to transforming the data.

8am	11am	2pm	5pm	8pm
93	121	112	117	121
116	135	114	98	135
125	137	119	105	102
144	173	148	124	122
105	119	125	91	133
109	83	109	80	104
89	95	88	91	116
116	128	122	107	119
151	149	141	126	138
137	139	125	109	107

294. Voting in congress

Romesburg, H.C. (1984) *Cluster Analysis for Researchers*, Belmont, CA: Lifetime Learning Publications.

The table shows the number of times 15 congressmen from New Jersey voted differently in the House of Representatives on 19 environmental bills. Can party affiliations be detected? Abstentions are not recorded, but two congressmen abstained more frequently than the others, these being Sandman (9 abstentions) and Thompson (6 abstentions).

This is an interesting data set because it can be used to demonstrate the powerful technique of multidimensional scaling. The two-dimensional plot obtained throws an interesting light upon one of the Republicans and the abstainers are seen to stand out from the rest.

Name (Party)	1	2	3	4	5	6	7	8	9	10	11	12	13	14	15
1 Hunt (R)	0	8	15	15	10	9	7	15	16	14	15	16	7	11	13
2 Sandman (R)	8	0	17	12	13	13	12	16	17	15	16	17	13	12	16
3 Howard (D)	15	17	0	9	16	12	15	5	5	6	5	4	11	10	7
4 Thompson (D)	15	12	9	0	14	12	13	10	8	8	8	6	15	10	7
5 Freylinghuysen (R)	10	13	16	14	0	8	9	13	14	12	12	12	10	11	11
6 Forsythe (R)	9	13	12	12	8	0	7	12	11	10	9	10	6	6	10
7 Widnall (R)	7	12	15	13	9	7	0	17	16	15	14	15	10	11	13
8 Roe (D)	15	16	5	10	13	12	17	0	4	5	5	3	12	7	6
9 Heltoski (D)	16	17	5	8	14	11	16	4	0	3	2	1	13	7	5
10 Rodino (D)	14	15	6	8	12	10	15	5	3	0	1	2	11	4	6
11 Minish (D)	15	16	5	8	12	9	14	5	2	1	0	1	12	5	5
12 Rinaldo (R)	16	17	4	6	12	10	15	3	1	2	1	0	12	6	4
13 Maraziti (R)	7	13	11	15	10	6	10	12	13	11	12	12	0	9	13
14 Daniels (D)	11	12	10	10	11	6	11	7	7	4	5	6	9	0	9
15 Patten (D)	13	16	7	7	11	10	13	6	5	6	5	4	13	9	0

295. Cortisol levels in psychotics

Rothschild, A.J., Schatzberg, A.F., Rosenbaum, A.H., Stahl, J.B. and Cole, J.O. (1982) The dexamethasone suppression test as a discriminator among subtypes of psychotic patients. *British Journal of Psychiatry*, **141**, 471–474.

Postdexamethasone 1600 hour cortisol levels (in micrograms/dl) in 31 control subjects and four groups of psychotics are to be compared.

Comparison of these samples by multiple boxplots shows that a transformation is necessary.

Control	Major Depression	Bipolar Depression	Schizophrenia	Atypical
1.0	1.0	1.0	0.5	0.9
1.0	3.0	1.0	0.5	1.3
1.0	3.5	1.0	1.0	1.5
1.5	4.0	1.5	1.0	4.8
1.5	10.0	2.0	1.0	
1.5	12.5	2.5	1.0	
1.5	14.0	2.5	1.5	
1.5	15.0	5.5	1.5	
1.5	17.5		1.5	
1.5	18.0		2.0	
1.5	20.0		2.5	
1.5	21.0		2.5	
1.5	24.5		5.5	
2.0	25.0		11.2	
2.0				
2.0				
2.0				
2.0				
2.0				
2.0				
2.0				
2.5				
2.5				
3.0				
3.0				
3.0				
3.5				
3.5				
4.0				
4.5				
10.0				

296. An historic data set: crime and drinking

These are Pearson's 1909 data on crime and drinking. The data are numbers of people convicted of the six crimes listed. Questions of interest are: are the types of crime drink-related and do the types of crime group together in any way?

Fraud is obviously different and a chi-squared test will confirm that crime is drink-related. It is interesting to test sub-groupings to show that drinking does not differentiate between certain types of crime.

Crime	Drinker	Abstainer
Arson	50	43
Rape	88	62
Violence	155	110
Stealing	379	300
Coining	18	14
Fraud	63	144

297. Depression in adolescents

Maag, J.W. and Behrens, J.T. (1989) Epidemiologic data on seriously emotionally disturbed and learning disabled adolescents: reporting extreme depressive symptomatology. *Behavioral Disorders*, **15**, No. 1.

Data from a study of seriously emotionally disturbed (SED) and learning disabled (LD) adolescents are shown in a four-way contingency table.

This could be an exercise in log-linear modelling. There is no possibility of fitting a main effects model and, in fact, three-way interactions have to be included. Logistic regression with low or high depression as the response seems a profitable line, but interaction terms will be needed.

Age	Group	Sex	Depression Low	High
	LD	Male	79	18
		Female	34	14
12-14		Male	14	5
	SED	Female	5	8

	LD	Male	63	10
		Female	26	11
15-16	SED	Male	32	3
		Female	15	7
	LD	Male	36	13
		Female	16	1
17-18	SED	Male	36	5
		Female	12	2

298. English and Greek teachers

Woods, N., Fletcher, P. and Hughes, A. (1986) *Statistics in Language Studies*, Cambridge: Cambridge University Press.

These are scores on each of 32 English sentences awarded by native English speaking teachers and by Greek teachers as a result of errors made by Greek-Cypriot learners of English. The scores are assessments made by the teachers of the quantity of errors made by the students.

The data are paired for each of 32 sentences so differences are to be tested for a mean of zero. A probability plot casts doubt upon the normality of the data and a power transformation is needed if a t-test is to be carried out.

Sentence	English	Greek	Sentence	English	Greek
1	22	36	17	23	39
2	16	9	18	18	19
3	42	29	19	30	28
4	25	35	20	31	41
5	31	34	21	20	25
6	36	23	22	21	17
7	29	25	23	29	26
8	24	31	24	22	37
9	29	35	25	26	34
10	18	21	26	20	28
11	23	33	27	29	33
12	22	13	28	18	24
13	31	22	29	23	37
14	21	29	30	25	33
15	27	25	31	27	39
16	32	25	32	11	20

299. Fun runners

Dale, G., Fleetwood, J.A., Weddell, A., Ellis, R.D. and Sainsbury, J.R.C. (1987) Beta endorphin: a factor in 'fun run' collapse? *British Medical Journal*, **294**, 1004.

Plasma beta concentrations (in pmol/l) of 11 normal fun runners before and after a Tyneside Great North Run, and another 11 runners after they had collapsed from exhaustion near the end.

The first two columns are paired data, the last one is from a different group of runners. Is there a relationship between plasma beta concentrations before and after the race? What can be said about the increase in concentration over the race? What can be deduced about collapsed runners? Beware of outliers.

Normal runner before race	Same runner after race	Collapsed runner after race
4.3	29.6	66
4.6	25.1	72
5.2	15.5	79
5.2	29.6	84
6.6	24.1	102
7.2	37.8	110
8.4	20.2	123
9.0	21.9	144
10.4	14.2	162
14.0	34.6	169
17.8	46.2	414

300. Olympic jumping events

World Almanac and Book of Facts (1988) New York: Pharos Books.

These data, supplied by the Amateur Athletics Association, are also quoted in Lunn, A.D. and McNeil, D.R. (1991) *Computer-Interactive Data Analysis,* Chichester: John Wiley & Sons, 86.

Distances (in metres) jumped by winners of Olympic jumping events from 1896 to 1988 are given in the table. Interest lies in whether a model can be built to predict future results accurately.

Year	High jump	Pole vault	Long jump	Triple jump
1896	1.81	3.30	6.34	13.72
1900	1.90	3.30	7.19	14.43
1904	1.80	3.51	7.34	14.33
1908	1.90	3.71	7.48	14.92
1912	1.93	3.95	7.60	14.76
1920	1.94	4.09	7.15	14.50
1924	1.98	3.95	7.45	15.53
1928	1.94	4.20	7.74	15.21
1932	1.97	4.31	7.64	15.72
1936	2.03	4.35	8.06	16.00
1948	1.98	4.30	7.82	15.40
1952	2.04	4.55	7.57	16.22
1956	2.11	4.56	7.83	16.34
1960	2.16	4.70	8.12	16.81
1964	2.18	5.10	8.07	16.85
1968	2.24	5.40	8.90	17.39
1972	2.23	5.50	8.24	17.35
1976	2.25	5.50	8.34	17.29
1980	2.36	5.78	8.54	17.35
1984	2.35	5.75	8.54	17.56
1988	2.38	5.90	8.72	17.61

301. Blood lactic acid

Afifi, A.A. and Azen, S.R. (1979) *Statistical Analysis, A Computer Oriented Approach*, New York: Academic Press, 125.

The data are to be used for calibration of an instrument to measure blood lactic acid concentration. The true concentration is compared with the concentration as measured by the instrument, with the true concentrations being designed (i.e. fixed in advance).

We expect a straight line through the origin, but there are two interesting features. Firstly, there are repeated measurements at each value of true concentration and, secondly, having calibrated the measuring instrument, we shall need the fit in order to deduce the correct x-values from measured y-values. Thus it is advisable to perform an inverse regression, i.e. use the true value as the response.

True concentration	Instrument reading	True concentration	Instrument reading
1	1.1	5	8.2
1	0.7	5	6.2
1	1.8	10	12.0
1	0.4	10	13.1
3	3.0	10	12.6
3	1.4	10	13.2
3	4.9	15	18.7
3	4.4	15	19.7
3	4.5	15	17.4
5	7.3	15	17.1

302. Skin cancers

Kleinbaum, D., Kupper, L. and Muller, K., (1989) *Applied Regression Analysis and other Multivariable Methods*, Boston, Massachusetts: PWS-Kent Publishing Company.

The data show the incidence of nonmelanoma skin cancer among women in Minneapolis (St. Paul) and Dallas (Fort Worth).

A regression model may be fitted by replacing the age groups with an indicator array. The response variable is a proportion and will need a folded transformation. Alternatively, a logistic regression may be fitted directly.

Number of cases	Town 0 = St. Paul 1 = Fort Worth	Age group	Population size
1	0	15-24	172675
16	0	25-34	123065
30	0	35-44	96216
71	0	45-54	92051
102	0	55-64	72159
130	0	65-74	54722
133	0	75-84	32185
40	0	85+	8328
4	1	15-24	181343
38	1	25-34	146207
119	1	35-44	121374
221	1	45-54	111353
259	1	55-64	83004
310	1	65-74	55932
65	1	85+	7583

303. Mental status of school-age children

Agresti, A. (1989) Tutorial on modelling ordered categorical response data. *Psychological Bulletin*, **105**, 290–301.

The table shows mental impairment in schoolchildren against socio-economic status of parents. Socio-economic status is represented by classifications 1 (lowest) to 6 (highest).

At first sight this appears to be a straightforward test of whether or not there is a relationship between mental status and parental socio-economic class. A fair amount of artistry is needed with a conventional chi-squared goodness-of-fit approach for the independence model if the data are to be exploited. Log-linear modelling will expose the relationship in detail.

		Socio-economic status of parents					
		1	2	3	4	5	6
Mental	Well	64	57	57	72	36	21
status	Mild symptoms	94	94	105	141	97	71
	Moderate symptoms	58	54	65	77	54	54
	Impaired funtioning	46	40	60	94	78	71

304. Diabetic mice

Dolkart, R.E., Halperin, B. and Perlman, J. (1971) Comparrison of antibody responses in normal and alloxan diabetic mice. *Diabetes*, **20**, 162–167.

The data shown are the amounts of nitrogen-bound bovine serum albumen produced by three groups of diabetic mice, these being normal, alloxan diabetic and alloxan diabetic treated with insulin.

Analysis of variance may be attempted, but closer investigation of the data shows that there is skewness and the variances are not plausibly equal. A simple power transformation solves these problems.

Normals	Alloxan	Alloxan+insulin
156	391	82
282	46	100
197	469	98
297	86	150
116	174	243
127	133	68
119	13	228
29	499	131
253	168	73
122	62	18
349	127	20
110	276	100
143	176	72
64	146	133
26	108	465
86	276	40
122	50	46
455	73	34
655		44
14		

305. Absorption of ions by potato

Steward, F.C. and Harrison, J.A. (1939) The absorption and accumulation of salts by living plants cells. *Annals of Botany*, New Series, **3**, 427–453.

This is a famous old data set which illustrates testing for parallel regression lines. Absorptions over time of rubidium and bromide ions by potato slices are given in mg per 1000 g of water in the tissue.

A scatterplot of rubidium and bromide ions absorbed against time shows two straight lines which are plausibly parallel. Regressions may be carried out and the slopes tested for equality.

Duration of immersion (h)	Rubidium mg/1000 g	Bromide mg/1000 g
21.7	7.2	0.7
46.0	11.4	6.4
67.0	14.2	9.9
90.2	19.1	12.8
95.5	20.0	15.8

306. Survival of cancer patients

Peto, R., Pike, M.C., Armitage, P., Breslow, N.E., Cox, D.R., Howard, S.V., Mantel, N., McPherson, K., Peto, J. and Smith, P.G. (1977) Design and analysis of randomised clinical trials requiring prolonged observation of each patient. *British Journal of Cancer*, **35**, 1–35.

This is a straightforward set of survival data on two groups of cancer patients, one of which received treatment, with 32% of the data being censored.

Treatment group	Survival time (days)	Failure status (1=Dead, 0=Alive)
1	8	1
2	180	1
2	632	1
1	852	0
1	52	1
2	2240	0
1	220	1
1	63	1
2	195	1

2	76	1
2	70	1
1	8	1
2	13	1
2	1990	0
1	1976	0
2	18	1
2	700	1
1	1296	0
1	1460	0
2	210	1
1	63	1
1	1328	0
2	1296	1
1	365	0
2	23	1

307. Major earthquakes since 1900: fatalities and intensities

World Almanac and Book of Facts (1977) New York: Pharos Books.

Numbers of fatalities and intensities of major earthquakes since 1900 are given below. The question of interest concerns possible relationships between number of deaths and intensity.

307. Major earthquakes since 1900: fatalities and intensities

Deaths	Intensity (Richter)	Year	Location	Deaths	Intensity (Richter)	Year	Location
452	8.3	1906	San Francisco	1250	6.8	1954	N Algeria
20000	8.6	1906	Valparaiso	2000	7.7	1956	N Afghanistan
83000	7.5	1908	Messina	2500	7.4	1957	N Iran
29980	7.5	1915	Avezzano	2000	7.1	1957	W Iran
100000	8.6	1920	Kansu	12000	5.8	1960	Agadir
99330	8.3	1923	Tokyo	5000	8.3	1960	S Chile
200000	8.3	1927	Nan-Shan	12230	7.1	1962	NW Iran
70000	7.6	1932	Kansu	1100	6.0	1963	Skopje
2990	8.9	1933	Japan	114	8.5	1964	Alaska
115	6.3	1933	Long Beach	2520	6.9	1966	E Turkey
10700	8.4	1934	Bihar-Nepal	12000	7.4	1968	NW Iran
30000	7.5	1935	Quetta	1086	7.4	1970	W Turkey
28000	8.3	1939	Chillan	66794	7.7	1970	N Peru
30000	7.9	1939	Erzincan	65	6.5	1971	S California
1300	6.0	1946	E Turkey	5057	6.9	1972	S Iran
2000	8.4	1946	Honshu	5000	6.2	1972	Nicaragua
5131	7.3	1948	Fukui	5200	6.3	1974	Pakistan
6000	6.8	1949	Pelileo	2312	6.8	1975	Turkey
1530	8.7	1950	Assam	22778	7.5	1976	Guatemala
1200	7.2	1953	NW Turkey	946	6.5	1976	NE Italy

308. Presidents, popes and monarchs

Lunn, A.D. and McNeil, D.R. (1991) *Computer-Interactive Data Analysis,* Chichester: John Wiley & Sons, 86.

Listed below are survival times in years from inauguration, election or coronation to death of US Presidents, Roman Catholic Popes and British Monarchs from 1690 to the present.

One question of interest is whether the survival times for the three groups differ in any marked way.

Presidents		Popes		Kings and Queens	
Washington	10	Alex VIII	2	James II	17
J. Adams	29	Innoc XII	9	Mary II	6
Jefferson	26	Clem XI	21	William III	13
Madison	28	Innoc XIII	3	Anne	12
Monroe	15	Ben XIII	6	George I	13
J.Q. Adams	23	Clem XII	10	George II	33
Jackson	17	Ben XIV	18	George III	59
Van Buren	25	Clem XIII	11	George IV	10
Harrison	0	Clem XIV	6	William IV	7
Tyler	20	Pius VI	25	Victoria	63
Polk	4	Pius VII	23	Edward VII	9
Taylor	1	Leo XII	6	George V	25
Filmore	24	Pius VIII	2	Edward VIII	36
Pierce	16	Greg XVI	15	George VI	15
Buchanan	12	Pius IX	32		
Lincoln	4	Leo XIII	25		
A. Johnson	10	Pius X	11		
Grant	17	Ben XV	8		
Hayes	16	Pius XI	17		
Garfield	0	Pius XII	19		
Arthur	7	John XXIII	5		
Cleveland	24	Paul VI	15		
Harrison	12	John Paul	0		
McKinley	4				
T. Roosevelt	18				
Taft	21				
Wilson	11				
Harding	2				
Coolidge	9				
Hoover	36				
F. Roosevelt	12				
Truman	28				
Kennedy	3				
Eisenhower	16				
L. Johnson	9				

309. Cervical cancer

Graham, S. and Shotz, W. (1979) Epidemiology of cancer of the cervix in Buffalo, New York. *Journal of the National Cancer Institute*, **63:1**, 23–27.

A case-control or retrospective study begins by selecting subjects with a disease and a disease-free comparison group. The null hypothesis is the so-called hypothesis of homogeneity. In the table, which shows age at first pregnancy by incidence of cervical cancer diagnosed in women aged 50-59, this corresponds to hypothesising that the probability of late first pregnancy is the same in both groups. Does age at first pregnancy affect the relative risk?

| | Disease status | |
Age at first pregnancy	Cervical cancer	Controls
≤ 25	42	203
> 25	7	114

310. Sickle cell disease

Anionwu, E., Watford, D., Brozovic, M. and Kirkwood, B. (1981) Sickle cell disease in a British urban community. *British Medical Journal*, **282**, 283–286.

The data are steady-state haemoglobin levels for patients with different types of sickle cell disease, these being HB SS, HB S/-thalassaemia and HB SC. One question of interest is whether the steady state haemoglobin levels differ significantly between patients with different types

HB SS	HB S/-thalassaemia	HB SC
7.2	8.1	10.7
7.7	9.2	11.3
8.0	10.0	11.5
8.1	10.4	11.6
8.3	10.6	11.7
8.4	10.9	11.8
8.4	11.1	12.0
8.5	11.9	12.1
8.6	12.0	12.3
8.7	12.1	12.6
9.1		12.6
9.1		13.3
9.1		13.3
9.8		13.8
10.1		13.9
10.3		

311. Men's olympic sprint times

World Almanac and Book of Facts (1988) New York: Pharos Books.

Quoted in Lunn, A.D. and McNeil, D.R. (1991) *Computer-Interactive Data Analysis*, Chichester: John Wiley & Sons, 127.

The data comprise times in seconds recorded by winners of men's Olympic sprint finals from 1900 to 1988 (there were no Olympic Games in 1916, 1940 and 1944). One question is whether a viable regression model can be fitted to these data and used to predict future winning times.

Year	100 m	200 m	400 m	800 m	1500 m
1900	10.80	22.20	49.40	121.40	246.00
1904	11.00	21.60	49.20	116.00	245.40
1908	10.80	22.40	50.00	112.80	243.40
1912	10.80	21.70	48.20	111.90	236.80
1920	10.80	22.00	49.60	113.40	241.80
1924	10.60	21.60	47.60	112.40	233.60
1928	10.80	21.80	47.80	111.80	233.20
1932	10.30	21.20	46.20	109.80	231.20
1936	10.30	20.70	46.50	112.90	227.80
1948	10.30	21.10	46.20	109.20	225.20
1952	10.40	20.70	45.90	109.20	225.20
1956	10.50	20.60	46.70	107.70	221.20
1960	10.20	20.50	44.90	106.30	215.60
1964	10.00	20.30	45.10	105.10	218.10
1968	9.95	19.83	43.80	104.30	214.90
1972	10.14	20.00	44.66	105.90	216.30
1976	10.06	20.23	44.26	103.50	219.20
1980	10.25	20.19	44.60	105.40	218.40
1984	9.99	19.80	44.27	103.00	212.50
1988	9.92	19.75	43.87	103.45	215.96

312. Women's olympic swimming

Lunn, A.D. and McNeil, D.R. (1991) *Computer-Interactive Data Analysis*, Chichester: John Wiley & Sons, 162.

Times in seconds recorded by winners of women's Olympic freestyle swimming finals from 1964 to 1988 are recorded below. One question is whether these data can be used to predict further values.

Year	100 m	200 m	400 m	800 m
1964	59.5		283.3	
1968	60	130.5	271.8	564
1972	58.6	123.6	259	533.7
1976	55.6	119.3	249.9	517.1
1980	54.8	118.3	248.8	508.9
1984	55.9	119.2	247.1	505
1988		117.6	243.8	500.2

313. A thrombosis study

van Oost, B.A., Veldhayzen, B., Timmermans, A.P.M. and Sixma, J.J. (1983) Increased urinary β-thromboglobulin excretion in diabetes assayed with a modified RIA kit-technique. *Thrombosis and Haemostasis*, **9**, 18–20.

The data are taken from a study in which urinary-thromboglobulin excretion in 12 normal and 12 diabetic patients was measured. In order to test for a difference between the groups, it is advisable first to look at boxplots, which expose the skewness. A suitable transformation is required, followed by a test for normality.

Normal	Diabetic	Normal	Diabetic
4.1	11.5	11.5	33.9
6.3	12.1	12.0	40.7
7.8	16.1	13.8	51.3
8.5	17.8	17.6	56.2
8.9	24.0	24.3	61.7
10.4	28.8	37.2	69.2

314. Deaths in USA in 1966

US Department of Health.

These data show the total number of deaths in the USA for each month in 1966. Interest lies in modelling the data to separate features of interest from chance fluctuations.

Month	Number	Month	Number
January	166761	July	159924
February	151296	August	145184
March	164804	September	141164
April	158973	October	154777
May	156455	November	150678
June	149251	December	163882

315. Comparison of experimental method

Sleigh, A., Hoff, R., Mott, K., Banneto, M., Maisk de Paiva, T., de Sousa Pedrosa, J. and Sherlock, I. (1982) Comparison of filtration staining (Bell) and thick smear (Kato) for the detection and quantitation of *Schistosoma mansoni* eggs in faeces. *Transactions of the Royal Society of Tropical Medicine and Hygiene*, **76**, 403–406.

The experiment was a comparison of Bell's method and the Kato-Katz method for detecting *Schistosoma mansoni* eggs in faeces; 315 samples were analysed in 2 sub-samples.

Bell (eggs detected)	Kato-Katz (eggs detected) +	−
+	184	54
−	14	63

316. Bat-to-prey detection distances

Griffin, D.R., Webster, F.A. and Michael, C.R. (1960) The echolocation of flying insects by bats. *Animal Behaviour*, **8**, 141–154.

Bats hunt insects by sending out high frequency sounds and listening for echoes. The data comprise detection distances, angle to time of flight and whether the prey was to the left or right for 11 catches.

Distance (cm)	62	52	68	23	34	45	27	42	83	56	40
Angle (degrees)	15	5	60	50	15	45	40	15	35	35	70
Left or right	L	R	R	R	L	R	R	L	R	R	R

317. Sexual performance of castrated mice

McGill, T.E. and Tucker, G.R. (1964) Genotype and sex drive in intact and castrated male mice. *Science*, **145**, 514–515.

The data shown are the result of a study of the sex drive of castrated male mice. 72 male mice from three different strains volunteered their services, along with 252 female mice. Strains 1 and 2 were pure-bred genetic stock, whilst strain 3 was hybrid. Each day, for 42 consecutive days, 36 females were brought to behavioural oestrus by hormone injections. Each male mouse was placed in a testing chamber with a female. He was allowed 10 minutes to begin mating and, if he showed no interest, the female was replaced. If three females were refused, he was scored negative for that day. Positive scores were only given to ejaculators and a mouse who ceased activity before ejaculation was given 40 minutes to resume. Each male had a maximum score of 42, one for each day of testing. At this point half of the males were castrated and the experiments were resumed, apart from two of the volunteers who did not survive the operation. Daily testing continued until the ejaculatory reflex was lost.

Does castration affect the mating ability of a mouse? Is there a genetic effect?

317. Sexual performance of castrated mice

Sexual performance of 24 intact males of each strain during 42 days:

Strain	Number of ejaculators	Number of ejaculations per ejaculator		Day of first ejaculation		Days between ejaculations	
		Median	Range	Median	Range	Median	Range
1	10	2	1–9	17	4–31	6	1–38
2	22	15	4–28	3	1–36	2	1–19
3	24	15	5–27	2	1–32	2	1–9

Sexual performance after castration:

Strain	Number of castrates	Number of pre-op ejaculators	Number of post-op ejaculators	Total ejaculations after castration	Day after castration on which last ejaculation occurred	
					Median	Range
1	11	4	0	0	0	0
2	11	10	3	3	3	3–8
3	12	12	9	42	28	3–60

318. The cloud point of a liquid

Draper, N.R. and Smith, H. (1966) *Applied Regression Analysis*, New York: John Wiley & Sons, 283.

The cloud point of a liquid is a measure of the degree of crystallization in a stock solution. It can be measured by the refractive index. Is it true that the percentage of I_8 in the basic stock is a good predictor of cloud point?

I_8 %	Cloud point	I_8 %	Cloud point
0	21.9	5	28.9
0	22.1	6	29.8
0	22.8	6	30.0
1	24.5	6	30.3
2	26.0	7	30.4
2	26.1	8	31.4
3	26.8	8	31.5
3	27.3	9	31.8
4	28.2	10	33.1
4	28.5		

319. Romano-British pottery

Tubb, A., Parker, A.J. and Nickless, G. (1980) The analysis of Romano-British pottery by atomic absorption spectrophotometry. *Archaeometry*, **22**, 153–171.

The data are the result of chemical analysis of 26 samples of pottery found at kiln sites in Wales, Gwent and the New Forest. The variables are the percentages of oxides of the various metals indicated, and the sites are L: Llanederyn, C: Caldicot, I: Island Thorns, A: Ashley Rails.

Data were gathered to see if a particular chemical composition could be linked to a particular site.

Site	Al	Fe	Mg	Ca	Na	Site	Al	Fe	Mg	Ca	Na
L	14.4	7.00	4.30	0.15	0.51	C	11.8	5.44	3.94	0.30	0.04
L	13.8	7.08	3.43	0.12	0.17	C	11.6	5.39	3.77	0.29	0.06
L	14.6	7.09	3.88	0.13	0.20	I	18.3	1.28	0.67	0.03	0.03
L	11.5	6.37	5.64	0.16	0.14	I	15.8	2.39	0.63	0.01	0.04
L	13.8	7.06	5.34	0.20	0.20	I	18.0	1.50	0.67	0.01	0.06
L	10.9	6.26	3.47	0.17	0.22	I	18.0	1.88	0.68	0.01	0.04
L	10.1	4.26	4.26	0.20	0.18	I	20.8	1.51	0.72	0.07	0.10
L	11.6	5.78	5.91	0.18	0.16	A	17.7	1.12	0.56	0.06	0.06
L	11.1	5.49	4.52	0.29	0.30	A	18.3	1.14	0.67	0.06	0.05
L	13.4	6.92	7.23	0.28	0.20	A	16.7	0.92	0.53	0.01	0.05
L	12.4	6.13	5.69	0.22	0.54	A	14.8	2.74	0.67	0.03	0.05
L	13.1	6.64	5.51	0.31	0.24	A	19.1	1.64	0.60	0.10	0.03
L	12.7	6.69	4.45	0.20	0.22						
L	12.5	6.44	3.94	0.22	0.23						

320. A split plot field trial

This is a classic set involving an experiment to investigate the effect of manure (nitrogen) on the yield of barley. Six blocks of three whole plots were used along with three varieties of barley, each whole plot being devoted to one variety only. The whole plots were each divided into 4 subplots to cater for four levels of manure (0, 0.01, 0.02, and 0.04 tons per acre).

Block	Variety	Manure	Yield	Block	Variety	Manure	Yield
1	1	0	111	4	1	0	74
1	1	1	130	4	1	1	89
1	1	2	157	4	1	2	81
1	1	4	174	4	1	4	122
1	2	0	117	4	2	0	64
1	2	1	114	4	2	1	103
1	2	2	161	4	2	2	132
1	2	4	141	4	2	4	133
1	3	0	105	4	3	0	70
1	3	1	140	4	3	1	89
1	3	2	118	4	3	2	104
1	3	4	156	4	3	4	117
2	1	0	61	5	1	0	62
2	1	1	91	5	1	1	90
2	1	2	97	5	1	2	100
2	1	4	100	5	1	4	116
2	2	0	70	5	2	0	80
2	2	1	108	5	2	1	82
2	2	2	126	5	2	2	94
2	2	4	149	5	2	4	126
2	3	0	96	5	3	0	63
2	3	1	124	5	3	1	70
2	3	2	121	5	3	2	109
2	3	4	144	5	3	4	99
3	1	0	68	6	1	0	53
3	1	1	64	6	1	1	74
3	1	2	112	6	1	2	118
3	1	4	86	6	1	4	113
3	2	0	60	6	2	0	89
3	2	1	102	6	2	1	82
3	2	2	89	6	2	2	86
3	2	4	96	6	2	4	104
3	3	0	89	6	3	0	97
3	3	1	129	6	3	1	99
3	3	2	132	6	3	2	119
3	3	4	124	6	3	4	121

321. Olympic triple jump distances

World Almanac and Book of Facts (1988), New York: Pharos Books.

The data show Olympic triple jump winning distances, in metres, for the years 1908 to 1988.

Year	Jump	Year	Jump	Year	Jump
1908	13.72	1936	15.21	1964	16.85
1912	14.43	1940	15.72	1968	17.39
1916	14.33	1944	16.00	1972	17.35
1920	14.92	1948	15.40	1976	17.29
1924	14.76	1952	16.22	1980	17.35
1928	14.50	1956	16.34	1984	17.56
1932	15.53	1960	16.81	1988	17.61

322. Species of flea beetle

Lubischew, A.A. (1962) On the use of discriminant functions in taxonomy. *Biometrics*, **18**, 455–477.

The genus of flea beetle *Chaetocnema* contains three species that are difficult to distinguish from one another and, indeed, were confused for a long time. The data comprise the maximal width of aedeagus in the forepart (in microns) and the front angle of the aedeagus (1 unit = 7.5 degrees). The objective is to use these data to form a classification rule.

Chaetocnema concinna		*Chaetocnema heikertingeri*		*Chaetocnema heptapotamica*	
Width	Angle	Width	Angle	Width	Angle
150	15	120	14	145	8
147	13	123	16	140	11
144	14	130	14	140	11
144	16	131	16	131	10
153	13	116	16	139	11
140	15	122	15	139	10
151	14	127	15	136	12
143	14	132	16	129	11
144	14	125	14	140	10
142	15	119	13	137	9
141	13	122	13	141	11
150	15	120	15	138	9
148	13	119	14	143	9
154	15	123	15	142	11

147	14	125	15	144	10
137	14	125	14	138	10
134	15	129	14	140	10
157	14	130	13	130	9
149	13	129	13	137	11
147	13	122	12	137	10
148	14	129	15	136	9
		124	15	140	10
		120	13		
		119	16		
		119	14		
		133	13		
		121	15		
		128	14		
		129	14		
		124	13		
		129	14		

323. Survival times of cancer patients

Cameron, E. and Pauling, L. (1978) Supplemental ascorbate in the supportive treatment of cancer: re-evaluation of prolongation of survival times in terminal human cancer. *Proceedings of the National Academy of Science USA*, **75**, 4538–4542.

Patients with advanced cancer of the stomach, bronchus, colon, ovary or breast were treated with ascorbate. Do the survival times differ with the organ affected? The possibility of a transformation of the data before analysis should be considered.

Stomach	Bronchus	Colon	Ovary	Breast
124	81	248	1234	1235
42	461	377	89	24
25	20	189	201	1581
45	450	1843	356	1166
412	246	180	2970	40
51	166	537	456	727
1112	63	519		3808
46	64	455		791
103	155	406		1804
876	859	365		3460
146	151	942		719
340	166	776		
396	37	372		
	223	163		
	138	101		
	72	20		
	245	283		

324. Birthweights of Poland China pigs

Snedecor, G.W. (1956) *Statistical Methods*, 5th edition, Ames: Iowa State College Press.

Birthweights of Poland China pigs (lb) are given for eight litters. Are the birthweights different for different litters? Is there a difference between large and small litters?

				Litter			
1	**2**	**3**	**4**	**5**	**6**	**7**	**8**
2.0	3.5	3.3	3.2	2.6	3.1	2.6	2.5
2.8	2.8	3.6	3.3	2.6	2.9	2.2	2.4
3.3	3.2	2.6	3.2	2.9	3.1	2.2	3.0
3.2	3.5	3.1	2.9	2.0	2.5	2.5	1.5
4.4	2.3	3.2	3.3	2.0		1.2	
3.6	2.4	3.3	2.5	2.1		1.2	
1.9	2.0	2.9	2.6				
3.3	1.6	3.4	2.8				
2.8		3.2					
1.1		3.2					

325. Performance of various computer CPUs

Ein-Dor, P. and Feldmesser, J. (1987): Attributes of the performance of central processing units: a relative performance prediction model. *Communications of the ACM*, **30**, 308–317.

The data are characteristics, performance measures and relative performance measures of 209 CPUs. The relative performance measures are relative to an IBM 370/158-3. What factors affect performance and relative performance?

325. Performance of various computer CPUs

Make and Model	Cycle time(ns)	Minimum memory(kb)	Maximum memory(kb)	Cache size(kb)	Minimum channels	Maximum channels	Relative performance	Estimated relative performance
ADVISOR 32/60	125	256	6000	256	16	128	198	199
AMDAHL 470V/7	29	8000	32000	32	8	32	269	253
AMDAHL 470/7A	29	8000	32000	32	8	32	220	253
AMDAHL 470V/7B	29	8000	32000	32	8	32	172	253
AMDAHL 470V/7C	29	8000	16000	32	8	16	132	132
AMDAHL 470V/8	26	8000	32000	64	8	32	318	290
AMDAHL 580-5840	23	16000	32000	64	16	32	367	381
AMDAHL 580-5850	23	16000	32000	64	16	32	489	381
AMDAHL 580-5860	23	16000	64000	64	16	32	636	749
AMDAHL 580-5880	23	32000	64000	128	32	64	1144	1238
APOLLO DN320	400	1000	3000	0	1	2	38	23
APOLLO DN420	400	512	3500	4	1	6	40	24
BASF 7/65	60	2000	8000	65	1	8	92	70
BASF 7/68	50	4000	16000	65	1	8	138	117
BTI 5000	350	64	64	0	1	4	10	15
BTI 8000	200	512	16000	0	4	32	35	64
BURROUGHS B1955	167	524	2000	8	4	15	19	23
BURROUGHS B2900	143	512	5000	0	7	32	28	29
BURROUGHS B2925	143	1000	2000	0	5	16	31	22
BURROUGHS B4955	110	5000	5000	142	8	64	120	124
BURROUGHS B5900	143	1500	6300	0	5	32	30	35
BURROUGHS B5920	143	3100	6200	0	5	20	33	39
BURROUGHS B6900	143	2300	6200	0	6	64	61	40

BURROUGHS B6925	110	3100	6200	0	6	64	76	45
C.R.D. 68/10-80	320	128	6000	0	1	12	23	28
C.R.D. UNIVERSE 2203T	320	512	2000	4	1	3	69	21
C.R.D. UNIVERSE 68	320	256	6000	0	1	6	33	28
C.R.D. UNIVERSE 68/05	320	256	3000	4	1	3	27	22
C.R.D. UNIVERSE 68/137	320	512	5000	4	1	5	77	28
C.R.D. UNIVERSE 68/37	320	256	5000	4	1	6	27	27
CDC CYBER 170/750	25	1310	2620	131	12	24	274	102
CDC CYBER 170/760	25	1310	2620	131	12	24	368	102
CDC CYBER 170/815	50	2620	10480	30	12	24	32	74
CDC CYBER 170/825	50	2620	10480	30	12	24	63	74
CDC CYBER 170/835	56	5240	20970	30	12	24	106	138
CDC CYBER 170/845	64	5240	20970	30	12	24	208	136
CDC OMEGA 480-I	50	500	2000	8	1	4	20	23
CDC OMEGA 480-II	50	1000	4000	8	1	5	29	29
CDC OMEGA 480-III	50	2000	8000	8	1	5	71	44
CAMBEX 1636-1	50	1000	4000	8	3	5	26	30
CAMBEX 1636-10	50	1000	8000	8	3	5	36	41
CAMBEX 1641-1	50	2000	16000	8	3	5	40	74
CAMBEX 1641-11	50	2000	16000	8	3	6	52	74
CAMBEX 1651-1	50	2000	16000	8	3	6	60	74
DEC DECSYS 10 1091	133	1000	12000	9	3	12	72	54
DEC DECSYS 20 2060	133	1000	8000	9	3	12	72	41
DEC MICROVAX-1	810	512	512	8	1	1	18	18
DEC VAX 11/730	810	1000	5000	0	1	1	20	28
DEC VAX 11/750	320	512	8000	4	1	5	40	36
DEC VAX 11/780	200	512	8000	8	1	8	62	38
DG ECLIPSE C/350	700	384	8000	0	1	1	24	34
DG ECLIPSE M/600	700	256	2000	0	1	1	24	19
DG ECLIPSE MV/1000	140	1000	16000	16	1	3	138	72

DG ECLIPSE MV/4000	200	1000	8000	0	1	2	36	36
DG ECLIPSE MV/6000	110	1000	4000	16	1	2	26	30
DG ECLIPSE MV/8000	110	1000	12000	16	1	2	60	56
DG ECLIPSE MV/8000 II	220	1000	8000	16	1	2	71	42
FORMATION F4000/100	800	256	8000	0	1	4	12	34
FORMATION F4000/200	800	256	8000	0	1	4	14	34
FORMATION F4000/200AP	800	256	8000	0	1	4	20	34
FORMATION F4000/300	800	256	8000	0	1	4	16	34
FORMATION F4000/300AP	800	256	8000	0	1	4	22	34
FOUR PHASE 2000/260	125	512	1000	0	8	20	36	19
GOULD CONCEPT 32/8705	75	2000	8000	64	1	38	144	75
GOULD CONCEPT 32/8750	75	2000	16000	64	1	38	144	113
GOULD CONCEPT 32/8780	75	2000	16000	128	1	38	259	157
HP 3000/30	90	256	1000	0	3	10	17	18
HP 3000/40	105	256	2000	0	3	10	26	20
HP 3000/44	105	1000	4000	0	3	24	32	28
HP 3000/48	105	2000	4000	8	3	19	32	33
HP 3000/64	75	2000	8000	8	3	24	62	47
HP 3000/88	75	3000	8000	8	3	48	64	54
HP 3000/III	175	256	2000	0	3	24	22	20
HARRIS 100	300	768	3000	0	6	24	36	23
HARRIS 300	300	768	3000	6	6	24	44	25
HARRIS 500	300	768	12000	6	6	24	50	52
HARRIS 600	300	768	4500	0	1	24	45	27
HARRIS 700	300	384	12000	6	1	24	53	50
HARRIS 80	300	192	768	6	6	24	36	18
HARRIS 800	180	768	12000	6	1	31	84	53
HONEYWELL DPS 6/35	330	1000	3000	0	2	4	16	23
HONEYWELL DPS 6/92	300	1000	4000	8	3	64	38	30
HONEYWELL DPS 6/96	300	1000	16000	8	2	112	38	73

HONEYWELL DPS 7/35	330	1000	2000	0	1	2	16	20
HONEYWELL DPS 7/45	330	1000	4000	0	3	6	22	25
HONEYWELL DPS 7/55	140	2000	4000	0	3	6	29	28
HONEYWELL DPS 7/65	140	2000	4000	0	4	8	40	29
HONEYWELL DPS 8/44	140	2000	4000	8	1	20	35	32
HONEYWELL DPS 8/49	140	2000	32000	32	1	20	134	175
HONEYWELL DPS 8/50	140	2000	8000	32	1	54	66	57
HONEYWELL DPS 8/52	140	2000	32000	32	1	54	141	181
HONEYWELL DPS 8/62	140	2000	32000	32	1	54	189	181
HONEYWELL DPS 8/20	140	2000	4000	8	1	20	22	32
IBM 3033 S	57	4000	16000	1	6	12	132	82
IBM 3033 U	57	4000	24000	64	12	16	237	171
IBM 3081	26	16000	32000	64	16	24	465	361
IBM 3081 D	26	16000	32000	64	8	24	465	350
IBM 3083 B	26	8000	32000	0	8	24	277	220
IBM 3083 E	26	8000	16000	0	8	16	185	113
IBM 370/125-2	480	96	512	0	1	1	6	15
IBM 370/148	203	1000	2000	0	1	5	24	21
IBM 370/158-3	115	512	6000	16	1	6	45	35
IBM 38/3	1100	512	1500	0	1	1	7	18
IBM 38/4	1100	768	2000	0	1	1	13	20
IBM 38/5	600	768	2000	0	1	1	16	20
IBM 38/7	400	2000	4000	0	1	1	32	28
IBM 38/8	400	4000	8000	0	1	1	32	45
IBM 4321	900	1000	1000	0	1	2	11	18
IBM 4331-1	900	512	1000	0	1	2	11	17
IBM 4331-11	900	1000	4000	4	1	2	18	26
IBM 4331-2	900	1000	4000	8	1	2	22	28
IBM 4341	900	2000	4000	0	3	6	37	28
IBM 4341-1	225	2000	4000	8	3	6	40	31

IBM 4341-10	225	2000	4000	8	3	6	34	31
IBM 4341-11	180	2000	8000	8	1	6	50	42
IBM 4341-12	185	2000	16000	16	1	6	76	76
IBM 4341-2	180	2000	16000	16	1	6	66	76
IBM 4341-9	225	1000	4000	2	3	6	24	26
IBM 4361-4	25	2000	12000	8	1	4	49	59
IBM 4361-5	25	2000	12000	16	3	5	66	65
IBM 4381-1	17	4000	16000	8	6	12	100	101
IBM 4381-2	17	4000	16000	32	6	12	133	116
IBM 8130 A	1500	768	1000	0	0	0	12	18
IBM 8130 B	1500	768	2000	0	0	0	18	20
IBM 8140	800	768	2000	0	0	0	20	20
IPL 4436	50	2000	4000	0	3	6	27	30
IPL 4443	50	2000	8000	8	3	6	45	44
IPL 4445	50	2000	8000	8	1	6	56	44
IPL 4446	50	2000	16000	24	1	6	70	82
IPL 4460	50	2000	16000	24	1	6	80	82
IPL 4480	50	8000	16000	48	1	10	136	128
MAGNUSON M80/30	100	1000	8000	0	2	6	16	37
MAGNUSON M80/31	100	1000	8000	24	2	6	26	46
MAGNUSON M80/32	100	1000	8000	24	3	6	32	46
MAGNUSON M80/42	50	2000	16000	12	3	16	45	80
MAGNUSON M80/43	50	2000	16000	24	6	16	54	88
MAGNUSON M80/44	50	2000	16000	24	6	16	65	88
MICRODATA SEQ.MS/3200	150	512	4000	0	8	128	30	33
NAS AS/3000	115	2000	8000	16	1	3	50	46
NAS AS/3000 N	115	2000	4000	2	1	5	40	29
NAS AS/5000	92	2000	8000	32	1	6	62	53
NAS AS/5000 E	92	2000	8000	32	1	6	60	53
NAS AS/5000 N	92	2000	8000	4	1	6	50	41

NAS AS/6130	75	4000	16000	16	1	6	66	86
NAS AS/6150	60	4000	16000	32	1	6	86	95
NAS AS/6620	60	2000	16000	64	5	8	74	107
NAS AS/6630	60	4000	16000	64	5	8	93	117
NAS AS/6650	50	4000	16000	64	5	10	110	119
NAS AS/7000	72	4000	16000	64	8	16	143	120
NAS AS/7000 N	72	2000	8000	16	6	8	105	48
NAS AS/8040	40	8000	16000	32	8	16	214	126
NAS AS/8050	40	8000	32000	64	8	24	277	266
NAS AS/8060	35	8000	32000	64	8	24	370	270
NAS AS/9000 DPC	38	16000	32000	128	16	32	510	426
NAS AS/9000 N	48	4000	24000	32	8	24	214	151
NAS AS/9040	38	8000	32000	64	8	24	326	267
NAS AS/9060	30	16000	32000	256	16	24	510	603
NCR V8535 II	112	1000	1000	0	1	4	8	19
NCR V8545 II	84	1000	2000	0	1	6	12	21
NCR V8555 II	56	1000	4000	0	1	6	17	26
NCR V8565 II	56	2000	6000	0	1	8	21	35
NCR V8565 II E	56	2000	8000	0	1	8	24	41
NCR V8575 II	56	4000	8000	0	1	8	34	47
NCR V8585 II	56	4000	12000	0	1	8	42	62
NCR V8595 II	56	4000	16000	0	1	8	46	78
NCR V8635	38	4000	8000	32	16	32	51	80
NCR V8650	38	4000	8000	32	16	32	116	80
NCR V8635	38	8000	16000	64	4	8	100	142
NCR V8665	38	8000	24000	160	4	8	140	281
NCR V8670	38	4000	16000	128	16	32	212	190
NIXDORF 8890/30	200	1000	2000	0	1	2	25	21
NIXDORF 8890/50	200	1000	4000	0	1	4	30	25
NIXDORF 8890/70	200	2000	8000	64	1	5	41	67

PERKIN-ELMER 3205	250	512	4000	0	1	7	25	24
PERKIN-ELMER 3210	250	512	4000	0	4	7	50	24
PERKIN-ELMER 3230	250	1000	16000	1	1	8	50	64
PRIME 50-2250	160	512	4000	2	1	5	30	25
PRIME 50-250 II	160	512	2000	2	3	8	32	20
PRIME 50-550 II	160	1000	4000	8	1	14	38	29
PRIME 50-750 II	160	1000	8000	16	1	14	60	43
PRIME 50-850 II	160	2000	8000	32	1	13	109	53
SIEMENS 7.521	240	512	1000	8	1	3	6	19
SIEMENS 7.531	240	512	2000	8	3	5	11	22
SIEMENS 7.536	105	2000	4000	8	6	8	22	31
SIEMENS 7.541	105	2000	6000	16	4	16	33	41
SIEMENS 7.551	105	2000	8000	16	4	14	58	47
SIEMENS 7.561	52	4000	16000	32	6	12	130	99
SIEMENS 7.865-2	70	4000	12000	8	6	8	75	67
SIEMENS 7.870-2	59	4000	12000	32	6	12	113	81
SIEMENS 7.872-2	59	8000	16000	64	12	24	188	149
SIEMENS 7.875-2	26	8000	24000	32	8	16	173	183
SIEMENS 7.880-2	26	8000	32000	64	12	16	248	275
SIEMENS 7.881-2	26	8000	32000	128	24	32	405	382
SPERRY 1100/61 H1	116	2000	8000	32	5	28	70	56
SPERRY 1100/81	50	2000	32000	24	6	26	114	182
SPERRY 1100/82	50	2000	32000	48	26	52	208	227
SPERRY 1100/83	50	2000	32000	112	52	104	307	341
SPERRY 1100/84	50	4000	32000	112	52	104	397	360
SPERRY 1100/93	30	8000	64000	96	12	176	915	919
SPERRY 1100/94	30	8000	64000	128	12	176	1150	978
SPERRY 80/3	180	262	4000	0	1	3	12	24
SPERRY 80/4	180	512	4000	0	1	3	14	24
SPERRY 80/5	180	262	4000	0	1	3	18	24

SPERRY 80/6	180	512	4000	0	1	3	21	24
SPERRY 80/8	124	1000	8000	0	1	8	42	37
SPERRY 90/80 MODEL 3	98	1000	8000	32	2	8	46	50
STRATUS 32	125	2000	8000	0	2	14	52	41
WANG VS10	480	512	8000	32	0	0	67	47
WANG VS 90	480	1000	4000	0	0	0	45	25

326. Monthly deaths from lung diseases in the UK

Diggle, P.J. (1990) *Time Series: A Biostatistical Introduction*, Oxford: Oxford University Press, Table A.3 combining both sexes.

The table gives monthly deaths from bronchitis, emphysema and asthma in the UK from 1974 to 1979 for both sexes. What can be said about the way the numbers of deaths depend upon the time of year? Has anything changed between 1974 and 1979?

Year						Month						
	1	**2**	**3**	**4**	**5**	**6**	**7**	**8**	**9**	**10**	**11**	**12**
1974	3035	2552	2704	2554	2014	1655	1721	1524	1596	2074	2199	2512
1975	2933	2889	2938	2497	1870	1726	1607	1545	1396	1787	2076	2837
1976	2787	3891	3179	2011	1636	1580	1489	1300	1356	1653	2013	2823
1977	2996	2523	2540	2520	1994	1641	1691	1479	1596	1877	2032	2484
1978	2899	2990	2890	2379	1933	1734	1617	1495	1440	1777	1970	2745
1979	2841	3535	3010	2091	1667	1589	1518	1349	1392	1619	1954	2633

327. Convictions for drunkenness

Cook, T. (1971) *New Society*, 20 May, 1971.

At the time of publication of these data, T. Cook was the Director of the Alcoholics Recovery Unit at Camberwell. The table shows the number of people convicted for drunkenness at Tower Bridge and Lambeth Magistrates' Courts during the six month period from 1 January to 27 June, 1970, classified by age and sex of the offender. Can the age distribution be considered the same for both sexes? If not, what are the possible reasons? Is there anything which needs to be taken into account over and above the information given in the table?

	Age group				
	0-29	**30-39**	**40-49**	**50-59**	**≥ 60**
Number of males	185	207	260	180	71
Number of females	4	13	10	7	10

328. Ear infections in swimmers

Val Gebski, from a private communication from Cameron Kirton of the New South Wales Water Board, Sydney, Australia.

The data come from the 1990 Pilot Surf/Health Study of NSW Water Board. The first column takes values 1 or 2 according to the recruit's perception of whether (s)he is a Frequent Ocean Swimmer, the second column has values 1 or 4 according to recruit's usually chosen swimming location (1 for non-beach, 4 for beach), the third column has values 2 (aged 15–19), 3 (aged 20–25) or 4 (aged 25–29), the fourth column has values 1 (male) or 2 (female) and, finally, the fifth column has the number of self-diagnozed ear infections that were reported by the recruit.

The objective of the study was to determine, in particular, whether beach swimmers run a greater risk of contracting ear infections than non-beach swimmers.

328. Ear infections in swimmers

Frq	Loc	Age grp	Sex	Ill		Frq	Loc	Age grp	Sex	Ill		Frq	Loc	Age grp	Sex	Ill		Frq	Loc	Age grp	Sex	Ill
1	1	2	1	0		2	1	2	1	0		1	4	2	1	0		2	4	2	1	0
1	1	2	1	0		2	1	2	1	0		1	4	2	1	0		2	4	2	1	0
1	1	2	1	0		2	1	2	1	0		1	4	2	1	0		2	4	2	1	0
1	1	2	1	0		2	1	2	1	0		1	4	2	1	0		2	4	2	1	0
1	1	2	1	0		2	1	2	1	0		1	4	2	1	0		2	4	2	1	0
1	1	2	1	0		2	1	2	1	0		1	4	2	1	0		2	4	2	1	0
1	1	2	1	0		2	1	2	1	0		1	4	2	1	0		2	4	2	1	0
1	1	2	1	0		2	1	2	1	0		1	4	2	1	0		2	4	2	1	0
1	1	2	1	0		2	1	2	1	0		1	4	2	1	0		2	4	2	1	0
1	1	2	1	0		2	1	2	1	0		1	4	2	1	0		2	4	2	1	0
1	1	2	1	1		2	1	2	1	0		1	4	2	1	1		2	4	2	1	1
1	1	2	1	1		2	1	2	1	1		1	4	2	1	1		2	4	2	1	1
1	1	2	1	1		2	1	2	1	1		1	4	2	1	2		2	4	2	1	2
1	1	2	1	1		2	1	2	1	2		1	4	2	1	2		2	4	2	1	2
1	1	2	1	2		2	1	2	1	2		1	4	2	1	3		2	4	2	1	2
1	1	2	1	2		2	1	2	1	2		1	4	2	2	4		2	4	2	2	2
1	1	2	1	2		2	1	2	1	2		1	4	2	2	4		2	4	2	2	5
1	1	2	1	2		2	1	2	1	3		1	4	2	2	9		2	4	2	2	0
1	1	2	1	2		2	1	2	1	3		1	4	2	2	0		2	4	2	2	0

0 0 0 0 0 1 2 2 2 3 10 0 0 0 0 1 1 5 0 0 0 0 0 0 1 1 1 2
2 2 2 2 2 2 2 2 2 2 2 1 1 1 1 1 1 1 2 2 2 2 2 2 2 2 2 2
2 2 2 2 2 2 2 2 2 2 3 3 3 3 3 3 3 3 3 3 3 3 3 3 3 3 3 3
4 4 4 4 4 4 4 4 4 4 4 4 4 4 4 4 4 4 4 4 4 4 4 4 4 4 4 4
2 2 2 2 2 2 2 2 2 2 2 2 2 2 2 2 2 2 2 2 2 2 2 2 2 2 2 2

0 0 0 1 2 3 3 6 9 0 0 0 0 0 0 1 0 0 0 0 0 1 1 2 3 0 0 0
2 2 2 2 2 2 2 2 2 1 1 1 1 1 1 1 1 2 2 2 2 2 2 2 2 1 1 1
2 2 2 2 2 2 2 2 3 3 3 3 3 3 3 3 3 3 3 3 3 3 3 3 4 4 4
4 4 4 4 4 4 4 4 4 4 4 4 4 4 4 4 4 4 4 4 4 4 4 4 4 4 4 4
1 1 1 1 1 1 1 1 1 1 1 1 1 1 1 1 1 1 1 1 1 1 1 1 1 1 1 1

3 3 3 4 4 4 5 0 0 0 6 0 0 0 0 0 0 0 1 1 2 2 3 3 0 0 0 0
1 1 1 1 1 1 1 2 2 2 2 1 1 1 1 1 1 1 1 1 1 1 1 1 2 2 2 2
2 2 2 2 2 2 2 2 2 2 3 3 3 3 3 3 3 3 3 3 3 3 3 3 3 3 3 3
1 1 1 1 1 1 1 1 1 1 1 1 1 1 1 1 1 1 1 1 1 1 1 1 1 1 1 1
2 2 2 2 2 2 2 2 2 2 2 2 2 2 2 2 2 2 2 2 2 2 2 2 2 2 2 2

3 3 3 3 4 4 6 11 16 0 0 4 10 0 0 0 0 1 2 2 2 3 3 5 17 0 0 0
1 1 1 1 1 1 1 1 1 2 2 2 2 1 1 1 1 1 1 1 1 1 1 1 1 2 2 2
2 2 2 2 2 2 2 2 2 2 2 2 3 3 3 3 3 3 3 3 3 3 3 3 3 3 3 3
1 1 1 1 1 1 1 1 1 1 1 1 1 1 1 1 1 1 1 1 1 1 1 1 1 1 1 1
1 1 1 1 1 1 1 1 1 1 1 1 1 1 1 1 1 1 1 1 1 1 1 1 1 1 1 1

0 0 0 0 0 0 0 0 0 0 0 1 2 2 0 0 0 2 2 2

1 1 1 1 1 1 1 1 1 1 1 1 1 1 2 2 2 2 2 2

4 4 4 4 4 4 4 4 4 4 4 4 4 4 4 4 4 4 4 4

4 4 4 4 4 4 4 4 4 4 4 4 4 4 4 4 4 4 4 4

2 2 2 2 2 2 2 2 2 2 2 2 2 2 2 2 2 2 2 2

0 0 0 0 0 0 0 1 1 2 2 9 0 0 0 1 3 4 5 6 0

1 1 1 1 1 1 1 1 1 1 1 1 2 2 2 2 2 2 2 2 1

4 4 4 4 4 4 4 4 4 4 4 4 4 4 4 4 4 4 4 2

4 4 4 4 4 4 4 4 4 4 4 4 4 4 4 4 4 4 4 4

1 1 1 1 1 1 1 1 1 1 1 1 1 1 1 1 1 1 1 2

0 1 1 2 2 3 0 0 0 0 1 1 3 3 0 0 0 1 0 0 0 0

2 2 2 2 2 2 1 1 1 1 1 1 1 2 2 2 2 1 1 1 1

3 3 3 3 3 3 4 4 4 4 4 4 4 4 4 2 2 2 2

1 1 1 1 1 1 1 1 1 1 1 1 1 1 1 1 4 4 4 4

2 2 2 2 2 2 2 2 2 2 2 2 2 2 2 2 1 1 1 1

0 0 1 2 3 3 4 1 1 2 3 4 4 10 0 0 2 2 0 0 0

2 2 2 2 2 2 2 1 1 1 1 1 1 2 2 2 2 1 1 1

3 3 3 3 3 3 3 4 4 4 4 4 4 4 4 4 4 2 2 2

1 1 1 1 1 1 1 1 1 1 1 1 1 1 1 1 1 1 1 1

1 1 1 1 1 1 1 1 1 1 1 1 1 1 1 1 1 1 2 2 2

329. A perception experiment

Lederman, S.J., Klatzky, R.L. and Barber, B.P.O. (1985) Spatial and movement-based heuristics for encoding pattern information through touch. *Journal of Experimental Psychology: General*, **114**, 33–49.

The data comprise mean signed errors in judgements of angular position in a psychological perception experiment in which 24 normal blindfolded subjects were asked to walk around an irregularly curved track. At various positions they were asked to estimate the angular position of the shortest line to their starting point. Does the error depend upon the true angle? Does the error depend upon whether the subject has moved clockwise or anti-clockwise?

True angle in degrees	Error in degrees	Path (1=clockwise 2=anti)
–30	50	1
–15	22	1
–8	16	1
–7	10	1
14	8	1
24	14	1
28	–18	1
39	–2	1
50	–3	1
52	–11	1
60	–11	1
97	–31	1
82	22	2
120	0	2
129	8	2
132	7	2
142	–4	2
153	22	2
156	–17	2
165	–13	2
188	–32	2
190	–23	2
199	–47	2
212	–53	2

330. Boiling points in the Alps

Atkinson, A.C. (1985) *Plots, Transformations and Regression*, Oxford: Clarendon Press, 4.

These data also appear in Weisberg, S. (1980) *Applied Linear Regression*, New York: John Wiley & Sons.

These are the well known Forbes' data on the boiling point of water in the Alps. There are 17 observations on boiling point (°F) and barometric pressure (in inches of mercury).

Boiling point	Pressure	Boiling point	Pressure
194.5	20.79	201.3	24.01
194.3	20.79	203.6	25.14
197.9	22.40	204.6	26.57
198.4	22.67	209.5	28.49
199.4	23.15	208.6	27.76
199.9	23.35	210.7	29.04
200.9	23.89	211.9	29.88
201.1	23.99	212.2	30.06
201.4	24.02		

331. Remission times of leukaemia patients

Gehan, E.A. (1965) A generalized Wilcoxon test for comparing arbitrarily single-censored samples. *Biometrika*, **52**, 203–233.

These data are originally from a study described by Freireich, E.J. *et al* (1963) The effect of 6-mercaptopurine on the duration of steroid-induced remissions in acute leukemia. *Blood*, **21**, 699–716. They also appear in Cox, D.R. and Oakes, D. (1984) *Analysis of Survival Data*, London: Chapman and Hall.

In a clinical trial of 42 leukaemia patients, some were treated with the drug 6-mercaptopurine and the rest were controls. Censoring is represented by 0. The trial was designed with matched pairs, although Gehan, and Cox and Oakes and several other authors who have re-analysed the data chose to ignore the pairing. Are the remission times best fitted by an exponential, Weibull, log-normal or log-logistic model? Are the remission times for the two groups equal?

Pair	Time	Censoring	Treatment	Pair	Time	Censoring	Treatment
1	1	1	control	11	11	0	6-MP
1	10	1	6-MP	12	5	1	control
2	22	1	control	12	20	0	6-MP
2	7	1	6-MP	13	4	1	control
3	3	1	control	13	19	0	6-MP
3	32	0	6-MP	14	15	1	control
4	12	1	control	14	6	1	6-MP
4	23	1	6-MP	15	8	1	control
5	8	1	control	15	17	0	6-MP
5	22	1	6-MP	16	23	1	control
6	17	1	control	16	35	0	6-MP
6	6	1	6-MP	17	5	1	control
7	2	1	control	17	6	1	6-MP
7	16	1	6-MP	18	11	1	control
8	11	1	control	18	13	1	6-MP
8	34	0	6-MP	19	4	1	control
9	8	1	control	19	9	0	6-MP
9	32	0	6-MP	20	1	1	control
10	12	1	control	20	6	0	6-MP
10	25	0	6-MP	21	8	1	control
11	2	1	control	21	10	0	6-MP

332. Foster feeding rats of different genotype

Scheffe, H. (1959) *The Analysis of Variance*, New York: John Wiley & Sons, 140.

The data are from a foster feeding experiment with rat mothers and litters of four different genotypes: A, B, I and J. The measurement is the litter weight (in g) after a trial feeding period. This is an example of unbalanced double classification.

Genotype of litter	Genotype of mother	Weight (g) of litter	Genotype of litter	Genotype of mother	Weight (g) of litter
A	A	61.5	B	J	40.5
A	A	68.2	I	A	37.0
A	A	64.0	I	A	36.3
A	A	65.0	I	A	68.0
A	A	59.7	I	B	56.3
A	B	55.0	I	B	69.8
A	B	42.0	I	B	67.0
A	B	60.2	I	I	39.7
A	I	52.5	I	I	46.0
A	I	61.8	I	I	61.3
A	I	49.5	I	I	55.3
A	I	52.7	I	I	55.7
A	J	42.0	I	J	50.0
A	J	54.0	I	J	43.8
A	J	61.0	I	J	54.5
A	J	48.2	J	A	59.0
A	J	39.6	J	A	57.4
B	A	60.3	J	A	54.0
B	A	51.7	J	A	47.0
B	A	49.3	J	B	59.5
B	A	48.0	J	B	52.8
B	B	50.8	J	B	56.0
B	B	64.7	J	I	45.2
B	B	61.7	J	I	57.0
B	B	64.0	J	I	61.4
B	B	62.0	J	J	44.8
B	I	56.5	J	J	51.5
B	I	59.0	J	J	53.0
B	I	47.2	J	J	42.0
B	I	53.0	J	J	54.0
B	J	51.3			

333. Record times for Scottish hill races

Atkinson, A.C. (1986) Comment: Aspects of diagnostic regression analysis. *Statistical Science*, **1**, 397–402.

These are the record times in 1984 for 35 Scottish hill races. It seems that there is some evidence in favour of transforming the response variable (time), but care should be taken to inspect these data for influential observations.

Location	Distance (miles)	Total height gained (feet)	Time (minutes)
Greenmantle New Year Dash	2.5	650	16.083
Carnethy 5 Hill Race	6.0	2500	48.350
Craig Dunain	6.0	900	33.650
Ben Rha	7.5	800	45.600
Ben Lomond	8.0	3070	62.267
Goatfell	8.0	2866	73.217
Bens of Jura	16.0	7500	204.617
Cairnpapple	6.0	800	36.367
Scolty	5.0	800	29.750
Traprain Law	6.0	650	39.750
Lairig Ghru	28.0	2100	192.667
Dollar	5.0	2000	43.050
Lomonds of Fife	9.5	2200	65.000
Cairn Table	6.0	500	44.133
Eildon Two	4.5	1500	26.933
Cairngorm	10.0	3000	72.250
Seven Hills of Edinburgh	14.0	2200	98.417
Knock Hill	3.0	350	78.650
Black Hill	4.5	1000	17.417
Creag Beag	5.5	600	32.567
Kildcon Hill	3.0	300	15.950
Meall Ant-Suidhe	3.5	1500	27.900
Half Ben Nevis	6.0	2200	47.633
Cow Hill	2.0	900	17.933
N Berwick Law	3.0	600	18.683
Creag Dubh	4.0	2000	26.217
Burnswark	6.0	800	34.433
Largo Law	5.0	950	28.567
Criffel	6.5	1750	50.500
Achmony	5.0	500	20.950
Ben Nevis	10.0	4400	85.550
Knockfarrel	6.0	600	32.550
Two Breweries	18.0	5200	170.250
Cockleroi	4.5	850	28.100
Moffat Chase	20.0	5000	159.833

334. Hardness of timber

Williams, E.J. (1959) *Regression Analysis*, New York: John Wiley & Sons, 43, Table 3.7.

Janka hardness is an important structural property of Australian timbers which is difficult to measure directly. However, it is related to the density of the timber which is comparatively easy to measure. Therefore it is desirable to fit a model enabling the Janka hardness to be predicted from the density. The Janka hardness and density of 36 Australian eucalypt hardwoods are given in the table.

A polynomial regression may be a reasonable model. It may be appropriate to consider transforming the hardness.

Density	Hardness	Density	Hardness	Density	Hardness
24.7	484	39.4	1210	53.4	1880
24.8	427	39.9	989	56.0	1980
27.3	413	40.3	1160	56.5	1820
28.4	517	40.6	1010	57.3	2020
28.4	549	40.7	1100	57.6	1980
29.0	648	40.7	1130	59.2	2310
30.3	587	42.9	1270	59.8	1940
32.7	704	45.8	1180	66.0	3260
35.6	979	46.9	1400	67.4	2700
38.5	914	48.2	1760	68.8	2890
38.8	1070	51.5	1710	69.1	2740
39.3	1020	51.5	2010	69.1	3140

335. Dimensions of jellyfish

Lunn, A.D. and McNeil, D.R. (1991) *Computer-Interactive Data Analysis*, Chichester: John Wiley & Sons, 308.

Dimensions in millimetres are given of two samples of jellyfish from Hawkesbury River in New South Wales, Australia. One of the samples is from Dangar Island and the other from Salamander Bay. The dimensions measured were length and width. What can one learn from graphing the two principal components? Try graphing principal components of the logarithms of the measurements. Can the dimensions determine the location?

Dangar Island		Salamander Bay	
Breadth	**Length**	**Breadth**	**Length**
6.5	8.0	12.0	14.0
6.0	9.0	15.0	16.0
6.5	9.0	14.0	16.5
7.0	9.0	13.0	17.0
8.0	9.5	15.0	17.0
7.0	10.0	15.0	18.0
8.0	10.0	15.0	18.0
8.0	10.0	16.0	18.0
7.0	11.0	14.0	19.0
8.0	11.0	15.0	19.0
9.0	11.0	16.0	19.0
10.0	13.0	16.5	19.0
11.0	13.0	18.0	19.0
12.0	13.0	18.0	19.0
11.0	14.0	16.0	20.0
11.0	14.0	16.0	20.0
13.0	14.0	17.0	20.0
14.0	16.0	18.0	20.0
15.0	16.0	19.0	20.0
15.0	16.0	15.0	21.0
15.0	19.0	16.0	21.0
16.0	16.0	21.0	21.0
		19.0	22.0
		20.0	22.0

336. Malaria parasites

Wang C.C. (1970) Multiple invasion of erythrocyte by malaria parasites, *Transactions of the Royal Society of Tropical Medicine and Hygiene*, **64**, 268–270.

Multiple invasion of erythrocytes is thought to be more likely to occur with *Plasmodium falciparum* than with other malaria parasites. Do these parasites attack certain red blood cells preferentially, or are multiple infections random? The data are observed frequencies of the number of parasites per erythrocyte in the blood film of a patient with *Plasmodium falciparum* malaria. If multiple infections are due to chance, the distribution of the number of parasites per cell will be Poisson.

Number of parasites per erythrocyte	Observed frequency
0	40000
1	8621
2	1259
3	99
4	21
5+	0

337. The New York Choral Society

Chambers, J.M., Cleveland, W.S., Kleiner, B. and Tukey, P.A. (1983) *Graphical Methods for Data Analysis*, Boston: Duxbury Press, 350.

The data give heights in inches of male singers in the New York Choral Society. The singers are grouped according to voice parts with the vocal range decreasing in pitch going from Tenor 1 to Bass 2. Is there a difference in height for singers of different pitch? Is there a trend according to pitch? How may the heights of the groups best be compared?

Tenor 1	Tenor 2	Bass 1	Bass 2	Tenor 1	Tenor 2	Bass 1	Bass 2
69	68	72	72	64	69	71	70
72	73	70	75			70	69
71	69	72	67			74	72
66	71	69	75			70	71
76	69	73	74			75	74
74	76	71	72			75	75
71	71	72	72			69	
66	69	68	74			72	
68	71	68	72			71	
67	66	71	72			70	
70	69	66	74			71	
65	71	68	70			68	
72	71	71	66			70	
70	71	73	68			75	
68	69	73	75			72	
64	70	70	68			66	
73	69	68	70			72	
66	68	70	72			70	
68	70	75	67			69	
67	68	68	70				

338. Testing crash helmets

Silverman, B.W. (1985) Some aspects of the spline smoothing approach to non-parametric curve fitting. *Journal of the Royal Statistical Society, Series B*, **47**, 1–52.

The data are taken from a simulated motorcycle accident used to test crash helmets and comprise a series of measurements of head acceleration (in units of g) and times after impact (in milliseconds). The objective is to fit a smooth curve through the data points.

Time after impact	Head acceleration	Time after impact	Head acceleration	Time after impact	Head acceleration
2.4	0.0	16.8	−91.1	30.2	36.2
2.6	−1.3	16.8	−77.7	31.0	75.0
3.2	−2.7	17.6	−37.5	31.2	8.1
3.6	0.0	17.6	−85.6	32.0	54.9
4.0	−2.7	17.6	−123.1	32.0	48.2
6.2	−2.7	17.6	−101.9	32.8	46.9
6.6	−2.7	17.8	−99.1	33.4	16.0
6.8	−1.3	17.8	−104.4	33.8	45.6
7.8	−2.7	18.6	−112.5	34.4	1.3
8.2	−2.7	18.6	−50.8	34.8	75.0
8.8	−1.3	19.2	−123.1	35.2	−16.0
8.8	−2.7	19.4	−85.6	35.2	−54.9
9.6	−2.7	19.4	−72.3	35.4	69.6
10.0	−2.7	19.6	−127.2	35.6	34.8
10.2	−5.4	20.2	−123.1	35.6	32.1
10.6	−2.7	20.4	−117.9	36.2	−37.5
11.0	−5.4	21.2	−134.0	36.2	22.8
11.4	0.0	21.4	−101.9	38.0	46.9
13.2	−2.7	21.8	−108.4	38.0	10.7
13.6	−2.7	22.0	−123.1	39.2	5.4
13.8	0.0	23.2	−123.1	39.4	−1.3
14.6	−13.3	23.4	−128.5	40.0	−21.5
14.6	−5.4	24.0	−112.5	40.4	−13.3
14.6	−5.4	24.2	−95.1	41.6	30.8
14.6	−9.3	24.2	−81.8	41.6	−10.7
14.6	−16.0	24.6	−53.5	42.4	29.4
14.6	−22.8	25.0	−64.4	42.8	0.0
14.8	−2.7	25.0	−57.6	42.8	−10.7
15.4	−22.8	25.4	−72.3	43.0	14.7
15.4	−32.1	25.4	−44.3	44.0	−1.3
15.4	−53.5	25.6	−26.8	44.4	0.0
15.4	−54.9	26.0	−5.4	45.0	10.7
15.6	−40.2	26.2	−107.1	46.6	10.7
15.6	−21.5	26.2	−21.5	47.8	−26.8
15.8	−21.5	26.4	−65.6	47.8	−14.7
15.8	−50.8	27.0	−16.0	48.8	−13.3
16.0	−42.9	27.2	−45.6	50.6	0.0
16.0	−26.8	27.2	−24.2	52.0	10.7
16.2	−21.5	27.2	9.5	53.2	−14.7
16.2	−50.8	27.6	4.0	55.0	−2.7
16.2	−61.7	28.2	12.0	55.0	10.7
16.4	−5.4	28.4	−21.5	55.4	−2.7
16.4	−80.4	28.4	37.5	57.6	10.7
16.6	−59.0	28.6	46.9		
16.8	−71.0	29.4	−17.4		

339. The speed of light

Weekes, A.J. *A Genstat Primer*, London: Edward Arnold, Table 1.1.

These are the classical data of Michelson and Morley measurements of the speed of light. Five experiments were performed, each consisting of 20 consecutive runs. The response is the speed of light measurement in suitable units. These data may be thought of as a randomized block with experiment and run as factors. Run could also be viewed as a possible covariate to account for changes in the measurement over the course of a single experiment, in which case one might look at a polynomial regression.

| | Experiment | | | | |
Run	1	2	3	4	5
1	850	960	880	890	890
2	740	940	880	810	840
3	900	960	880	810	780
4	1070	940	860	820	810
5	930	880	720	800	760
6	850	800	720	770	810
7	950	850	620	760	790
8	980	880	860	740	810
9	980	900	970	750	820
10	880	840	950	760	850
11	1000	830	880	910	870
12	980	790	910	920	870
13	930	810	850	890	810
14	650	880	870	860	740
15	760	880	840	880	810
16	810	830	840	720	940
17	1000	800	850	840	950
18	1000	790	840	850	800
19	960	760	840	850	810
20	960	800	840	780	870

340. Murder-suicides through aircraft crashes

Phillips, D.P. (1978) Airplane accident fatalities increase just after newspaper stories about murder and suicide. *Science*, **201**, 748–750.

These data were used to investigate the hypothesis that newspaper and TV publicity of murder-suicides through deliberate crashing of private aircraft triggers further similar murder-suicides. The amount of publicity following 17 crashes known to be murder-suicides is represented by an index and the number of fatal crashes during the week immediately following each is recorded in the second column.

Index of newspaper coverage	number of multi-fatality crashes	Index of newspaper coverage	number of multi-fatality crashes
376	8	63	2
347	5	44	7
322	8	40	4
104	4	5	3
103	6	5	2
98	4	0	4
96	8	0	3
85	6	0	2
82	4		

341. Average monthly temperatures for Nottingham, 1920-1939

Anderson, O.D. (1976) *Time Series Analysis and Forecasting: The Box-Jenkins approach*. London and Boston: Butterworths, 166.

The data comprise average air temperatures (°F)at Nottingham Castle for 20 years. The objective is to perform a time-series analysis.

						Month						
Year	1	2	3	4	5	6	7	8	9	10	11	12
1920	40.6	40.8	44.4	46.7	54.1	58.5	57.7	56.4	54.3	50.5	42.9	39.8
1921	44.2	39.8	45.1	47.0	54.1	58.7	66.3	59.9	57.0	54.2	39.7	42.8
1922	37.5	38.7	39.5	42.1	55.7	57.8	56.8	54.3	54.3	47.1	41.8	41.7
1923	41.8	40.1	42.9	45.8	49.2	52.7	64.2	59.6	54.4	49.2	36.3	37.6
1924	39.3	37.5	38.3	45.5	53.2	57.7	60.8	58.2	56.4	49.8	44.4	43.6
1925	40.0	40.5	40.8	45.1	53.8	59.4	63.5	61.0	53.0	50.0	38.1	36.3
1926	39.2	43.4	43.4	48.9	50.6	56.8	62.5	62.0	57.5	46.7	41.6	39.8
1927	39.4	38.5	45.3	47.1	51.7	55.0	60.4	60.5	54.7	50.3	42.3	35.2
1928	40.8	41.1	42.8	47.3	50.9	56.4	62.2	60.5	55.4	50.2	43.0	37.3
1929	34.8	31.3	41.0	43.9	53.1	56.9	62.5	60.3	59.8	49.2	42.9	41.9
1930	41.6	37.1	41.2	46.9	51.2	60.4	60.1	61.6	57.0	50.9	43.0	38.8
1931	37.1	38.4	38.4	46.5	53.5	58.4	60.6	58.2	53.8	46.6	45.5	40.6
1932	42.4	38.4	40.3	44.6	50.9	57.0	62.1	63.5	56.3	47.3	43.6	41.8
1933	36.2	39.3	44.5	48.7	54.2	60.8	65.5	64.9	60.1	50.2	42.1	35.8
1934	39.4	38.2	40.4	46.9	53.4	59.6	66.5	60.4	59.2	51.2	42.8	45.8
1935	40.0	42.6	43.5	47.1	50.0	60.5	64.6	64.0	56.8	48.6	44.2	36.4
1936	37.3	35.0	44.0	43.9	52.7	58.6	60.0	61.1	58.1	49.6	41.6	41.3
1937	40.8	41.0	38.4	47.4	54.1	58.6	61.4	61.8	56.3	50.9	41.4	37.1
1938	42.1	41.2	47.3	46.6	52.4	59.0	59.6	60.4	57.0	50.7	47.8	39.2
1939	39.4	40.9	42.4	47.8	52.4	58.0	60.7	61.8	58.2	46.7	46.6	37.8

342. Insurance premiums

National Roads and Motorists Association magazine (Australia), The Open Road, June 1985, 14.

Age specific term life premium rates for a sum insured of $50,000 are given in the table. The four separate sets of points may be plotted and cubic spline regression used to fit them.

| | Male | | Female | |
Age	Smoker	Non-smoker	Smoker	Non-smoker
33	130	100	110	95
34	135	105	110	95
35	140	105	115	100
36	145	110	120	100
37	155	110	125	105
38	160	115	130	105
39	170	120	140	110
40	180	125	145	115
41	195	130	155	120
42	210	140	165	130
43	230	145	175	135
44	250	155	190	145
45	270	170	205	155
46	295	180	225	165
47	325	200	245	180
48	360	215	265	195
49	395	235	290	210
50	435	260	320	230
51	485	285	350	250
52	535	315	380	275
53	590	350	420	305
54	650	390	460	335
55	715	435	505	370

343. Testing authorship

Brinegar, C.S. (1963) Mark Twain and the Q.C.S. letters — a statistical test of authorship, *Journal of the Royal Statistical Association*, **58**, 85–96.

In 1861 the New Orleans Crescent published a set of ten letters signed Quintus Curtius Snodgrass. Although the events discussed in the letters seem to have occurred, there is no record of anyone of that name. It has been claimed that the author was, in fact, none other than Mark Twain. One way of settling authorship disputes is to compare distributions of word lengths.

Number of occurrences

Word length	Mark Twain	Q.C.S.
1	312	424
2	1146	2685
3	1394	2752
4	1177	2302
5	661	1431
6	442	992
7	367	896
8	231	638
9	181	465
10	109	276
11	50	152
12	24	101
13+	12	61

344. The rise of the planet Venus

World Almanac and Book of Facts (1986). New York: Pharos Books.

The data are the times, in minutes after midnight, that Venus rose at various latitudes in the Northern hemisphere at ten-day intervals during November and December, 1986. The difficulty here lies in deciding upon the right model.

Number of minutes past midnight (GMT)	Latitude in degrees north	Days after October 31, 1986
300	20	16
251	20	26
263	30	26
277	40	26
297	50	26
231	30	36
244	40	36
262	50	36
290	60	36
227	40	46
246	50	46
275	60	46
243	50	56
276	60	56

345. Pulse rates of Peruvian Indians

Ryan, T.A. Jr., Joiner, B.L. and Ryan, B.F. (1985) *The Minitab Student Handbook*, Boston: Duxbury Press 317–318.

The data show the pulse rates of Peruvian Indians. What distribution provides a plausible probability model? What is a typical pulse rate and how should a confidence interval for it be constructed?

88	76	84	64	60
64	60	64	68	74
68	68	72	76	72
52	72	64	60	56
72	88	80	76	64
72	60	76	88	72
64	60	60	72	92
80	72	64	68	

346. Vocabulary of children

Weiner, B. (1977) *Discovering Psychology*, Chicago: Science Research Association, 97.

The average oral vocabulary size of children at various ages is given in the table. A scatterplot shows a clear relationship, but is it linear? Is there an outlier?

Age in years	Number of words
1.0	3
1.5	22
2.0	272
2.5	446
3.0	896
3.5	1222
4.0	1540
4.5	1870
5.0	2072
6.0	2562

347. Educating cats

Weiner, B. (1977) *Discovering Psychology*, Chicago: Science Research Association, 27.

In an experiment to determine the learning curve of a cat, cats were required to master the task of escaping from a maze. The response is the rate at which cats escape from the maze (in escapes per minute), whilst the stimulus is the number of trials taken to learn the task.

Number of trials	Rate of escape	Number of trials	Rate of escape
1	0.38	12	3.00
2	1.50	14	6.00
3	0.67	16	6.00
4	1.09	18	7.50
6	2.00	20	6.00
8	1.77	22	10.00
10	6.00		

348. Shocking rats

Bond, N.W. (1979) Impairment of shuttlebox avoidance-learning following repeated alcohol withdrawal episodes in rats. *Pharmacology, Biochemistry and Behavior*, **11**, 589–591.

In an experiment to investigate the speed of learning of rats, times were recorded for a rat to get through a shuttlebox in successive attempts. If the time taken exceeded 5 seconds, the rat received an electric shock for the duration of the next attempt. The data are the number of shocks received and the average time for all attempts between shocks.

Number of shocks	Average time	Number of shocks	Average time
0	11.4	8	5.7
1	11.9	9	4.4
2	7.1	10	4.0
3	14.2	11	2.8
4	5.9	12	2.6
5	6.1	13	2.4
6	5.4	14	5.2
7	3.1	15	2.0

349. Smoking habits

Sterling, T.D. and Weinkam, J.J. (1976) Smoking characteristics by type of employment. *Journal of Occupational Medicine*, **18**, 743–753.

Proportions of blue collar, professional and other white males are tabulated by smoking habits. There are many ways of displaying and reporting these data, but it is necessary to be careful in deciding upon the most informative and valid visual display.

Daily average	Occupation			
	Blue collar	Professional	Other	Total
20+	44.1	6.2	49.7	8951
10-19	42.7	5.8	51.5	2589
1-9	38.6	7.6	53.8	1572
Former smoker	30.9	11.0	58.1	8509
Never smoked	30.5	11.8	58.7	9694

350. Mothers of schizophrenics

Werner, M., Stabenau, J.B. and Pollin, W. (1970) TAT method for the differentiation of families of schizophrenics, delinquents and normals. *Journal of Abnormal Psychology*, **75**, 139–145.

The TAT (Thematic Apperception Test) is a personality test in which subjects are shown ten pictures and asked to invent a story about each picture. An experiment was performed to determine whether mothers of schizophrenics respond differently from the mothers of normal children. The data give the number of stories out of ten exhibiting a positive parent-child relationship. Clearly the distributions of scores in the two groups need to be compared.

Normal	Schizophrenic	Normal	Schizophrenic
8	2	2	0
4	1	1	2
6	1	1	4
3	3	4	2
1	2	3	3
4	7	3	3
4	2	2	0
6	1	6	1
4	3	3	2
2	1	4	2

351. Unemployment and suicide rates

Smith, D. (1977) *Patterns in Human Geography*, Canada: Douglas David and Charles Ltd., 158.

The data are the percentage of unemployed and suicide rate per million for 11 cities in the USA. Is there a significant relationship and, if so, what is it?

City	Percentage unemployed	Suicide rate
New York	3.0	72
Los Angeles	4.7	224
Chicago	3.0	82
Philadelphia	3.2	92
Detroit	3.8	104
Boston	2.5	71
San Francisco	4.8	235
Washington	2.7	81
Pittsburgh	4.4	86
St. Louis	3.1	102
Cleveland	3.5	104

352. Main US budget items for 1978 and 1979

Newsweek Magazine, 30 January, 1978.

The figures for the eight main items in the US budget for 1978 and 1979 are in billions of dollars.

	1978	1979
Military spending	107.6	117.8
Social security	103.9	115.1
Health care	44.3	49.7
Debt service	43.8	49.0
Welfare	43.7	44.9
Education	27.5	30.4
Energy	19.9	21.8
Veteran's benefits	18.9	19.3

353. Damage to ships

McCullagh, P. and Nelder, J.A. (1983) *Generalized Linear Models*, London: Chapman and Hall, Section 6.3.2, 137.

The table gives the number of damage incidents and aggregate months of service by ship type, year of construction, and period of operation. The ship types are coded A to E, the years of construction are 1960-64, 1965-69, 1970-74 and 1975-79, coded as 60, 65, 70, 75 respectively, the period of operation is either 1960-74 or 1975-79, coded 60 or 75 respectively and the service is recorded in aggregate months of service. A Poisson regression model may be fitted to these data.

Type	Year	Period	Service	Incidents
A	60	60	127	0
A	60	75	63	0
A	65	60	1095	3
A	65	75	1095	4
A	70	60	1512	6
A	70	75	3353	18
A	75	60	0	0
A	75	75	2244	11
B	60	60	44882	39
B	60	75	17176	29
B	65	60	28609	58
B	65	75	20370	53
B	70	60	7064	12
B	70	75	13099	44
B	75	60	0	0
B	75	75	7117	18
C	60	60	1179	1
C	60	75	552	1
C	65	60	781	0
C	65	75	676	1
C	70	60	783	6
C	70	75	1948	2
C	75	60	0	0
C	75	75	274	1
D	60	60	251	0
D	60	75	105	0
D	65	60	288	0
D	65	75	192	0
D	70	60	349	2
D	70	75	1208	11

D	75	60	0	0
D	75	75	2051	4
E	60	60	45	0
E	60	75	0	0
E	65	60	789	7
E	65	75	437	7
E	70	60	1157	5
E	70	75	2161	12
E	75	60	0	0
E	75	75	542	1

354. Methadone treatment of heroin addicts

Caplehorn, J. (1991) Methadone dosage and retention of patients in maintenance treatment. *Medical Journal of Australia*, **154**, 195–199.

The data are the times, in days, that heroin addicts spend in a clinic. There are two clinics and the covariates are believed to affect the times spent in the clinic by addicts. A 1 in the column Prison? indicates a prison record. This can be used as an interesting example of stratified Cox regression where stratification is done by clinic.

Clinic is coded 1 and 2, Status is 0 = censored, 1 = departed from clinic, Time is in days and Dose is in mg/day:

ID	Clinic	Status	Time	Prison?	Dose
1	1	1	428	0	50
2	1	1	275	1	55
3	1	1	262	0	55
4	1	1	183	0	30
5	1	1	259	1	65
6	1	1	714	0	55
7	1	1	438	1	65
8	1	0	796	1	60
9	1	1	892	0	50
10	1	1	393	1	65
11	1	0	161	1	80
12	1	1	836	1	60
13	1	1	523	0	55
14	1	1	612	0	70
15	1	1	212	1	60
16	1	1	399	1	60
17	1	1	771	1	75
18	1	1	514	1	80
19	1	1	512	0	80
21	1	1	624	1	80
22	1	1	209	1	60

23	1	1	341	1	60
24	1	1	299	0	55
25	1	0	826	0	80
26	1	1	262	1	65
27	1	0	566	1	45
28	1	1	368	1	55
30	1	1	302	1	50
31	1	0	602	0	60
32	1	1	652	0	80
33	1	1	293	0	65
34	1	0	564	0	60
36	1	1	394	1	55
37	1	1	755	1	65
38	1	1	591	0	55
39	1	0	787	0	80
40	1	1	739	0	60
41	1	1	550	1	60
42	1	1	837	0	60
43	1	1	612	0	65
44	1	0	581	0	70
45	1	1	523	0	60
46	1	1	504	1	60
48	1	1	785	1	80
49	1	1	774	1	65
50	1	1	560	0	65
51	1	1	160	0	35
52	1	1	482	0	30
53	1	1	518	0	65
54	1	1	683	0	50
55	1	1	147	0	65
57	1	1	563	1	70
58	1	1	646	1	60
59	1	1	899	0	60
60	1	1	857	0	60
61	1	1	180	1	70
62	1	1	452	0	60
63	1	1	760	0	60
64	1	1	496	0	65
65	1	1	258	1	40
66	1	1	181	1	60
67	1	1	386	0	60
68	1	0	439	0	80
69	1	0	563	0	75
70	1	1	337	0	65
71	1	0	613	1	60
72	1	1	192	1	80
73	1	0	405	0	80
74	1	1	667	0	50
75	1	0	905	0	80
76	1	1	247	0	70
77	1	1	821	0	80
78	1	1	821	1	75
79	1	0	517	0	45
80	1	0	346	1	60
81	1	1	294	0	65

82	1	1	244	1	60
83	1	1	95	1	60
84	1	1	376	1	55
85	1	1	212	0	40
86	1	1	96	0	70
87	1	1	532	0	80
88	1	1	522	1	70
89	1	1	679	0	35
90	1	0	408	0	50
91	1	0	840	0	80
92	1	0	148	1	65
93	1	1	168	0	65
94	1	1	489	0	80
95	1	0	541	0	80
96	1	1	205	0	50
97	1	0	475	1	75
98	1	1	237	0	45
99	1	1	517	0	70
100	1	1	749	0	70
101	1	1	150	1	80
102	1	1	465	0	65
103	2	1	708	1	60
104	2	0	713	0	50
105	2	0	146	0	50
106	2	1	450	0	55
109	2	0	555	0	80
110	2	1	460	0	50
111	2	0	53	1	60
113	2	1	122	1	60
114	2	1	35	1	40
118	2	0	532	0	70
119	2	0	684	0	65
120	2	0	769	1	70
121	2	0	591	0	70
122	2	0	769	1	40
123	2	0	609	1	100
124	2	0	932	1	80
125	2	0	932	1	80
126	2	0	587	0	110
127	2	1	26	0	40
128	2	0	72	1	40
129	2	0	641	0	70
131	2	0	367	0	70
132	2	0	633	0	70
133	2	1	661	0	40
134	2	1	232	1	70
135	2	1	13	1	60
137	2	0	563	0	70
138	2	0	969	0	80
143	2	0	1052	0	80
144	2	0	944	1	80
145	2	0	881	0	80
146	2	1	190	1	50
148	2	1	79	0	40
149	2	0	884	1	50

150	2	1	170	0	40
153	2	1	286	0	45
156	2	0	358	0	60
158	2	0	326	1	60
159	2	0	769	1	40
160	2	1	161	0	40
161	2	0	564	1	80
162	2	1	268	1	70
163	2	0	611	1	40
164	2	1	322	0	55
165	2	0	1076	1	80
166	2	0	2	1	40
168	2	0	788	0	70
169	2	0	575	0	80
170	2	1	109	1	70
171	2	0	730	1	80
172	2	0	790	0	90
173	2	0	456	1	70
175	2	1	231	1	60
176	2	1	143	1	70
177	2	0	86	1	40
178	2	0	1021	0	80
179	2	0	684	1	80
180	2	1	878	1	60
181	2	1	216	0	100
182	2	0	808	0	60
183	2	1	268	1	40
184	2	0	222	0	40
186	2	0	683	0	100
187	2	0	496	0	40
188	2	1	389	0	55
189	1	1	126	1	75
190	1	1	17	1	40
192	1	1	350	0	60
193	2	0	531	1	65
194	1	0	317	1	50
195	1	0	461	1	75
196	1	1	37	0	60
197	1	1	167	1	55
198	1	1	358	0	45
199	1	1	49	0	60
200	1	1	457	1	40
201	1	1	127	0	20
202	1	1	7	1	40
203	1	1	29	1	60
204	1	1	62	0	40
205	1	0	150	1	60
206	1	1	223	1	40
207	1	0	129	1	40
208	1	0	204	1	65
209	1	1	129	1	50
210	1	1	581	0	65
211	1	1	176	0	55
212	1	1	30	0	60
213	1	1	41	0	60

214	1	0	543	0	40
215	1	0	210	1	50
216	1	1	193	1	70
217	1	1	434	0	55
218	1	1	367	0	45
219	1	1	348	1	60
220	1	0	28	0	50
221	1	0	337	0	40
222	1	0	175	1	60
223	2	1	149	1	80
224	1	1	546	1	50
225	1	1	84	0	45
226	1	0	283	1	80
227	1	1	533	0	55
228	1	1	207	1	50
229	1	1	216	0	50
230	1	0	28	0	50
231	1	1	67	1	50
232	1	0	62	1	60
233	1	0	111	0	55
234	1	1	257	1	60
235	1	1	136	1	55
236	1	0	342	0	60
237	2	1	41	0	40
238	2	0	531	1	45
239	1	0	98	0	40
240	1	1	145	1	55
241	1	1	50	0	50
242	1	0	53	0	50
243	1	0	103	1	50
244	1	0	2	1	60
245	1	1	157	1	60
246	1	1	75	1	55
247	1	1	19	1	40
248	1	1	35	0	60
249	2	0	394	1	80
250	1	1	117	0	40
251	1	1	175	1	60
252	1	1	180	1	60
253	1	1	314	0	70
254	1	0	480	0	50
255	1	0	325	1	60
256	2	1	280	0	90
257	1	1	204	0	50
258	2	1	366	0	55
259	2	0	531	1	50
260	1	1	59	1	45
261	1	1	33	1	60
262	2	1	540	0	80
263	2	0	551	0	65
264	1	1	90	0	40
266	1	1	47	0	45

355. Smoking and motherhood

Freeman, D.H. (1987) *Applied Categorical Data Analysis*, New York: Marcel Dekker, 211.

The obvious question is whether or not a mother's smoking puts a new-born baby at risk. However, several more questions are of interest. Are smokers more likely to have premature babies? Are older mothers more likely to have premature babies? Obviously premature babies are less likely to survive than those who have gone a full term, but how does smoking affect premature babies?

		Premature died in 1st year	Premature alive at year 1	Full term died in 1st year	Full term alive at year 1
Young mothers	Non-smokers	50	315	24	4012
	Smokers	9	40	6	459
Older mothers	Non-smokers	41	147	14	1594
	Smokers	4	11	1	124

356. Crime rates in the USA

Statistical Abstract of the U.S. (1988) Department of Commerce, USA, 159, Table 265.

Numbers are given of offences known to police per 100,000 residents of fifty states of the United States plus the District of Columbia for the year 1986.

Crime rates are given in offences per 100,000:

State	Murder	Rape	Robbery	Aggravated assault	Burglary	Larceny/ theft	Motor vehicle theft
ME	2	14.8	28	102	803	2347	164
NH	2.2	21.5	24	92	755	2208	228
VT	2	21.8	22	103	949	2697	181
MA	3.6	29.7	193	331	1071	2189	906
RI	3.5	21.4	119	192	1294	2568	705
CT	4.6	23.8	192	205	1198	2758	447
NY	10.7	30.5	514	431	1221	2924	637
NJ	5.2	33.2	269	265	1071	2822	776
PA	5.5	25.1	152	176	735	1654	354
OH	5.5	38.6	142	235	988	2574	376
IN	6	25.9	90	186	887	2333	328
IL	8.9	32.4	325	434	1180	2938	628
MI	11.3	67.4	301	424	1509	3378	800
WI	3.1	20.1	73	162	783	2802	254
MN	2.5	31.8	102	148	1004	2785	288
IA	1.8	12.5	42	179	956	2801	158

MO	9.2	29.2	170	370	1136	2500	439
ND	1	11.6	7	32	385	2049	120
SD	4	17.7	16	87	554	1939	99
NE	3.1	24.6	51	184	748	2677	168
KS	4.4	32.9	80	252	1188	3008	258
DE	4.9	56.9	124	241	1042	3090	272
MD	9	43.6	304	476	1296	2978	545
DC	31	52.4	754	668	1728	4131	975
VA	7.1	26.5	106	167	813	2522	219
WV	5.9	18.9	41	99	625	1358	169
NC	8.1	26.4	88	354	1225	2423	208
SC	8.6	41.3	99	525	1340	2846	277
GA	11.2	43.9	214	319	1453	2984	430
FL	11.7	52.7	367	605	2221	4373	598
KY	6.7	23.1	83	222	824	1740	193
TN	10.4	47	208	274	1325	2126	544
AL	10.1	28.4	112	408	1159	2304	267
MS	11.2	25.8	65	172	1076	1845	150
AR	8.1	28.9	80	278	1030	2305	195
LA	12.8	40.1	224	482	1461	3417	442
OK	8.1	36.4	107	285	1787	3142	649
TX	13.5	51.6	240	354	2049	3987	714
MT	2.9	17.3	20	118	783	3314	215
ID	3.2	20	21	178	1003	2800	181
WY	5.3	21.9	22	243	817	3078	169
CO	7	42.3	145	329	1792	4231	486
NM	11.5	46.9	130	538	1845	3712	343
AZ	9.3	43	169	437	1908	4337	419
UT	3.2	25.3	59	180	915	4074	223
NV	12.6	64.9	287	354	1604	3489	478
WA	5	53.4	135	244	1861	4267	315
OR	6.6	51.1	206	286	1967	4163	402
CA	11.3	44.9	343	521	1696	3384	762
AK	8.6	72.7	88	401	1162	3910	604
HI	4.8	31	106	103	1339	3759	328

357. Olympic decathlon, 1988

International Athletics Federation, London. Quoted in Lunn, A.D. and McNeil, D.R. (1991) *Computer-Interactive Data Analysis*, Chichester: John Wiley & Sons, 276.

Results for the men's decathlon in the 1988 Olympics may be analysed with the usual multivariate techniques, although there is an obvious outlier to watch out for. Principal components analysis and/or factor analysis will show how the events group together. It is interesting to look at the different ways of expressing the multivariate distances apart of individuals and comparing these with the final points scores.

357. Olympic decathlon, 1988

Timed events have times in seconds, distance measurements are in metres:

Athlete	100 m	Long jump	Shot	High jump	400 m	110m hurd.	disc.	Pole vlt.	Jav.	1500 m	Score
Schenk-GDR	11.25	7.43	15.48	2.27	48.90	15.13	49.28	4.7	61.32	268.95	8488
Voss-GDR	10.87	7.45	14.97	1.97	47.71	14.46	44.36	5.1	61.76	273.02	8399
Steen-CAN	11.18	7.44	14.20	1.97	48.29	14.81	43.66	5.2	64.16	263.20	8328
Thompson-GB	10.62	7.38	15.02	2.03	49.06	14.72	44.80	4.9	64.04	285.11	8306
Blondel-FRA	11.02	7.43	12.92	1.97	47.44	14.40	41.20	5.2	57.46	256.64	8286
Plaziat-FRA	10.83	7.72	13.58	2.12	48.34	14.18	43.06	4.9	52.18	274.07	8272
Bright-USA	11.18	7.05	14.12	2.06	49.34	14.39	41.68	5.7	61.60	291.20	8216
De Wit-HOL	11.05	6.95	15.34	2.00	48.21	14.36	41.32	4.8	63.00	265.86	8189
Johnson-USA	11.15	7.12	14.52	2.03	49.15	14.66	42.36	4.9	66.46	269.62	8180
Tarnovetsky-URS	11.23	7.28	15.25	1.97	48.60	14.76	48.02	5.2	59.48	292.24	8167
Keskitalo-FIN	10.94	7.45	15.34	1.97	49.94	14.25	41.86	4.8	66.64	295.89	8143
Gaehwiler-SWI	11.18	7.34	14.48	1.94	49.02	15.11	42.76	4.7	65.84	256.74	8114
Szabo-HUN	11.02	7.29	12.92	2.06	48.23	14.94	39.54	5.0	56.80	257.85	8093
Smith-CAN	10.99	7.37	13.61	1.97	47.83	14.70	43.88	4.3	66.54	268.97	8083
Shirley-AUS	11.03	7.45	14.20	1.97	48.94	15.44	41.66	4.7	64.00	267.48	8036
Poelman-NZ	11.09	7.08	14.51	2.03	49.89	14.78	43.20	4.9	57.18	268.54	8021
Olander-SWE	11.46	6.75	16.07	2.00	51.28	16.06	50.66	4.8	72.60	302.42	7869
Freimuth-GDR	11.57	7.00	16.60	1.94	49.84	15.00	46.66	4.9	60.20	286.04	7860
Warming-DEN	11.07	7.04	13.41	1.94	47.97	14.96	40.38	4.5	51.50	262.41	7859
Hraban-CZE	10.89	7.07	15.84	1.79	49.68	15.38	45.32	4.9	60.48	277.84	7781
Werthner-AUT	11.52	7.36	13.93	1.94	49.99	15.64	38.82	4.6	67.04	266.42	7753
Gugler-SWI	11.49	7.02	13.80	2.03	50.60	15.22	39.08	4.7	60.92	262.93	7745

Penalver-ESP	11.38	7.08	14.31	2.00	50.24	14.97	46.34	4.4	55.68	272.68	7743
Kruger-GB	11.30	6.97	13.23	2.15	49.98	15.38	38.72	4.6	54.34	277.84	7623
Lee Fu-An-TPE	11.00	7.23	13.15	2.03	49.73	14.96	38.06	4.5	52.82	285.57	7579
Mellado-ESA	11.33	6.83	11.63	2.06	48.37	15.39	37.52	4.6	55.42	270.07	7517
Moser-SWI	11.10	6.98	12.69	1.82	48.63	15.13	38.04	4.7	49.52	261.90	7505
Valenta-CZE	11.51	7.01	14.17	1.94	51.16	15.18	45.84	4.6	56.28	303.17	7422
O'Connell-IRL	11.26	6.90	12.41	1.88	48.24	15.61	38.02	4.4	52.68	272.06	7310
Richards-GB	11.50	7.09	12.94	1.82	49.27	15.56	42.32	4.5	53.50	293.85	7237
Gong-CHN	11.43	6.22	13.98	1.91	51.25	15.88	46.18	4.6	57.84	294.99	7231
Miller-FIJ	11.47	6.43	12.33	1.94	50.30	15.00	38.72	4.0	57.26	293.72	7016
Kwang-lk-KOR	11.57	7.19	10.27	1.91	50.71	16.20	34.36	4.1	54.94	269.98	6907
Kunwar-NEP	12.12	5.83	9.71	1.70	52.32	17.05	27.10	2.6	39.10	281.24	5339

358. Measurements of dog mandibles

Higham, C.F.W., Kijngam, A. and Manly, B.F.J. (1980) An analysis of prehistoric canid remains from Thailand. *Journal of Archaeological Science*, **7**, 148–165.

These data also appear in Manly, B.F.J. (1986) *Multivariate Statistical Methods: A Primer*. London: Chapman and Hall.

The ancestry of the prehistoric dog is not clear and it is thought that it could descend from the golden jackal or the wolf. Prehistoric sites in Thailand have revealed dog bones from around 3500BC to the present and, in order to clarify the situation, measurements were made on these and common wild dogs. Since the wolf is not native to Thailand, the nearest indigenous wolves were used along with the dingo, which may have originated in India, the cuon, which is indigenous to South East Asia, and modern village dogs from Thailand.

It is interesting to examine the various useful multivariate distances between the varieties of canine. The results are, perhaps, surprising.

Measurements in the table are in millimetres and are as follows:

1 = breadth of mandible,
2 = height of mandible below first molar,
3 = length of first molar,
4 = breadth of first molar,
5 = length from first to third molars inclusive,
6 = length from first to fourth premolars inclusive.

	Mean measurements					
	1	**2**	**3**	**4**	**5**	**6**
Modern dog	9.7	21.0	19.4	7.7	32.0	36.5
Golden jackal	8.1	16.7	18.3	7.0	30.3	32.9
Chinese wolf	13.5	27.3	26.8	10.6	41.9	48.1
Indian wolf	11.5	24.3	24.5	9.3	40.0	44.6
Cuon	10.7	23.5	21.4	8.5	28.8	37.6
Dingo	9.6	22.6	21.1	8.3	34.4	43.1
Prehistoric dog	10.3	22.1	19.1	8.1	32.3	35.0

359. Children with Down's syndrome

Moran, P.A.P. (1974) Are there two maternal age groups in Down's syndrome? *British Journal of Psychiatry*, **124**, 453–455.

These data also appear in Andrews, D.F. and Herzberg, A.M. (1985) *Data: A Collection of Problems from Many Fields for the Student and Research Worker*, New York: Springer-Verlag, 221.

The number of mothers of children with Down's syndrome for births in Australia from 1942-1952 are given in the first column out of the total number of births in the second column. The third column gives the mothers' age group. The objective is to model the risk of a child having the syndrome given the age of the mother at the birth.

The data are interesting because different models apply for younger and older mothers. This may be handled by an extra covariate.

Number of mothers	Total number of births	Age group (years)
15	35555	< 20
128	207931	20-24
208	253450	25-29
194	170970	30-34
297	86046	35-39
240	24498	40-44
37	1707	> 44

360. Protein consumption in European countries

Weber A. (1973) Agrarpolitik im Spannungsfeld der internationalen Ernaehrungspolitik, Institut fuer Agrarpolitik und marktlehre, Kiel.

These data also appear in Gabriel, K.R. (1981) Biplot display of multivariate matrices for inspection of data and diagnosis. In *Interpreting Multivariate Data* (Ed. V. Barnett), New York: John Wiley & Sons, 147–173.

Protein consumption in twenty-five European countries for nine food groups is given in the table. It is possible to use multivariate methods to determine whether there are groupings of countries and whether meat consumption is related to that of other foods.

360. Protein consumption in European countries

Country	Red meat	White meat	Eggs	Milk	Fish	Cereals	Starchy foods	Pulses, nuts, oil-seeds	Fruits, vegetables
Albania	10.1	1.4	0.5	8.9	0.2	42.3	0.6	5.5	1.7
Austria	8.9	14.0	4.3	19.9	2.1	28.0	3.6	1.3	4.3
Belgium	13.5	9.3	4.1	17.5	4.5	26.6	5.7	2.1	4.0
Bulgaria	7.8	6.0	1.6	8.3	1.2	56.7	1.1	3.7	4.2
Czechoslovakia	9.7	11.4	2.8	12.5	2.0	34.3	5.0	1.1	4.0
Denmark	10.6	10.8	3.7	25.0	9.9	21.9	4.8	0.7	2.4
E Germany	8.4	11.6	3.7	11.1	5.4	24.6	6.5	0.8	3.6
Finland	9.5	4.9	2.7	33.7	5.8	26.3	5.1	1.0	1.4
France	18.0	9.9	3.3	19.5	5.7	28.1	4.8	2.4	6.5
Greece	10.2	3.0	2.8	17.6	5.9	41.7	2.2	7.8	6.5
Hungary	5.3	12.4	2.9	9.7	0.3	40.1	4.0	5.4	4.2
Ireland	13.9	10.0	4.7	25.8	2.2	24.0	6.2	1.6	2.9
Italy	9.0	5.1	2.9	13.7	3.4	36.8	2.1	4.3	6.7
Netherlands	9.5	13.6	3.6	23.4	2.5	22.4	4.2	1.8	3.7
Norway	9.4	4.7	2.7	23.3	9.7	23.0	4.6	1.6	2.7
Poland	6.9	10.2	2.7	19.3	3.0	36.1	5.9	2.0	6.6
Portugal	6.2	3.7	1.1	4.9	14.2	27.0	5.9	4.7	7.9
Romania	6.2	6.3	1.5	11.1	1.0	49.6	3.1	5.3	2.8
Spain	7.1	3.4	3.1	8.6	7.0	29.2	5.7	5.9	7.2
Sweden	9.9	7.8	3.5	24.7	7.5	19.5	3.7	1.4	2.0
Switzerland	13.1	10.1	3.1	23.8	2.3	25.6	2.8	2.4	4.9
UK	17.4	5.7	4.7	20.6	4.3	24.3	4.7	3.4	3.3
USSR	9.3	4.6	2.1	16.6	3.0	43.6	6.4	3.4	2.9
W Germany	11.4	12.5	4.1	18.8	3.4	18.6	5.2	1.5	3.8
Yugoslavia	4.4	5.0	1.2	9.5	0.6	55.9	3.0	5.7	3.2

361. Male Egyptian skulls

Thomson, A. and Randall-Maciver, R. (1905) *Ancient Races of the Thebaid*, Oxford: Oxford University Press.

Measurements on male Egyptian skulls from 5 epochs are to be analysed with a view to deciding whether there are any differences between the epochs and if they show any changes with time. A steady change of head shape with time would indicate interbreeding with immigrant populations. These data are analysed in detail by Manly, B.F.J. (1986) *Multivariate Statistical Methods*, New York: Chapman and Hall.

Measurements are Maximum Breadth, Basibregmatic Height, Basialveolar Length and Nasal Height.

361. Male Egyptian skulls

| c 4000 BC | | | | c 3300 BC | | | | c 1850 BC | | | | c 200 BC | | | | c AD 150 | | | |
MB	BH	BL	NH	MB	BH	BL	NH	MB	BH	BL	NH	MB	BH	BL	NH	MB	BH	BL	NH
131	138	89	49	124	138	101	48	137	141	96	52	137	134	107	54	137	123	91	50
125	131	92	48	133	134	97	48	129	133	93	47	141	128	95	53	136	131	95	49
131	132	99	50	138	134	98	45	132	138	87	48	141	130	87	49	128	126	91	57
119	132	96	44	148	129	104	51	130	134	106	50	135	131	99	51	130	134	92	52
136	143	100	54	126	124	95	45	134	134	96	45	133	120	91	46	138	127	86	47
138	137	89	56	135	136	98	52	140	133	98	50	131	135	90	50	126	138	101	52
139	130	108	48	132	145	100	54	138	138	95	47	140	137	94	60	136	138	97	58
125	136	93	48	133	130	102	48	136	145	99	55	139	130	90	48	126	126	92	45
131	134	102	51	131	134	96	50	136	131	92	46	140	134	90	51	132	132	99	55
134	134	99	51	133	125	94	46	126	136	95	56	138	140	100	52	139	135	92	54
129	138	95	50	133	136	103	53	137	129	100	53	132	133	90	53	143	120	95	51
134	121	95	53	131	139	98	51	137	139	97	50	134	134	97	54	141	136	101	54
126	129	109	51	131	136	99	56	136	126	101	50	135	135	99	50	135	135	95	56
132	136	100	50	138	134	98	49	137	133	90	49	133	136	95	52	137	134	93	53
141	140	100	51	130	136	104	53	129	142	104	47	136	130	99	55	142	135	96	52
131	134	97	54	131	128	98	45	135	138	102	55	134	137	93	52	139	134	95	47
135	137	103	50	138	129	107	53	129	135	92	50	131	141	99	55	138	125	99	51
132	133	93	53	123	131	101	51	134	125	90	60	129	135	95	47	137	135	96	54
139	136	96	50	130	129	105	47	138	134	96	51	136	128	93	54	133	125	92	50
132	131	101	49	134	130	93	54	136	135	94	53	131	125	88	48	145	129	89	47
126	133	102	51	137	136	106	49	132	130	91	52	139	130	94	53	138	136	92	46
135	135	103	47	126	131	100	48	133	131	100	50	144	124	86	50	131	129	97	44
134	124	93	53	135	136	97	52	138	137	94	51	141	131	97	53	143	126	88	54

134	124	91	55
132	127	97	52
137	125	85	57
129	128	81	52
140	135	103	48
147	129	87	48
136	133	97	51
130	131	98	53
133	128	92	51
138	126	97	54
131	142	95	53
136	138	94	55
132	136	92	52
135	130	100	51
130	127	99	45
136	133	91	49
134	123	95	52
136	137	101	54
133	131	96	49
138	133	100	55
138	133	91	46
129	126	91	50
134	139	101	49
131	134	90	53
132	130	104	50
130	132	93	52
135	132	98	54
130	128	101	51
128	134	103	50
130	130	104	49
138	135	100	55
128	132	93	53
127	129	106	48
131	136	114	54
124	138	101	46

362. Olympic women's heptathlon 1988

International Athletics Federation, London. Quoted in Lunn, A.D. and McNeil, D.R. (1981) *Computer-Interative Data Analysis*, Chichester: John Wiley & Sons Ltd, 308.

Results in the women's heptathlon in the 1988 Olympics may be analysed with the usual multivariate techniques, after removal of obvious outliers.

Timed events have times in seconds, distances are measured in metres.

Athlete	100 m	High jump	Shot	200 m	Long jump	Javelin	800 m	Score
Joyner-Kersee-USA	12.69	1.86	15.80	22.56	7.27	45.66	128.51	7291
John-GDR	12.85	1.80	16.23	23.65	6.71	42.56	126.12	6897
Behmer-GDR	13.20	1.83	14.20	23.10	6.68	44.54	124.20	6858
Sablovskaite-URS	13.61	1.80	15.23	23.92	6.25	42.78	132.24	6540
Choubenkova-URS	13.51	1.74	14.76	23.93	6.32	47.46	127.90	6540
Schulz-GDR	13.75	1.83	13.50	24.65	6.33	42.82	125.79	6411
Fleming-AUS	13.38	1.80	12.88	23.59	6.37	40.28	132.54	6351
Greiner-USA	13.55	1.80	14.13	24.48	6.47	38.00	133.65	6297
Lajbnerova-CZE	13.63	1.83	14.28	24.86	6.11	42.20	136.05	6252
Bouraga-URS	13.25	1.77	12.62	23.59	6.28	39.06	134.74	6252
Wijnsma-HOL	13.75	1.86	13.01	25.03	6.34	37.86	131.49	6205
Dimitrova-BUL	13.24	1.80	12.88	23.59	6.37	40.28	132.54	6171
Schneider-SWI	13.85	1.86	11.58	24.87	6.05	47.50	134.93	6137
Braun-FRG	13.71	1.83	13.16	24.78	6.12	44.58	142.82	6109
Ruotsalainen-FIN	13.79	1.80	12.32	24.61	6.08	45.44	137.06	6101
Yuping-CHN	13.93	1.86	14.21	25.00	6.40	38.60	146.67	6087
Hagger-GB	13.47	1.80	12.75	25.47	6.34	35.76	138.48	5975
Brown-USA	14.07	1.83	12.69	24.83	6.13	44.34	146.43	5972
Mulliner-GB	14.39	1.71	12.68	24.92	6.10	37.76	138.02	5746
Hautenauve-BEL	14.04	1.77	11.81	25.61	5.99	35.68	133.90	5734
Kytola-FIN	14.31	1.77	11.66	25.69	5.75	39.48	133.35	5686
Geremias-BRA	14.23	1.71	12.95	25.50	5.50	39.64	144.02	5508
Hui-Ing-TAI	14.85	1.68	10.00	25.23	5.47	39.14	137.30	5290
Jeong-Mi-KOR	14.53	1.71	10.83	26.61	5.50	39.26	139.17	5289
Launa-PNG	16.42	1.50	11.78	26.16	4.88	46.38	163.43	4566

363. Percentages employed in different industries in Europe

Euromonitor (1979), *European Marketing Data and Statistics*, Euromonitor Publications, London, 76–77.

These data also appear in Manly, B.F.J. (1986) *Multivariate Statistical Methods: A Primer*, 11. New York: Chapman and Hall.

Multivariate techniques may be used to examine groupings of countries with a similar employment structure and to investigate relationships between the countries.

Industry groups are: Agriculture, Mining, Manufacturing, Power Supplies, Construction, Service Industries, Finance, Social and Personal Services, Transport and Communications. Figures are percentages employed.

Country	Agr	Min	Man	PS	Con	SI	Fin	SPS	TC
Belgium	3.3	0.9	27.6	0.9	8.2	19.1	6.2	26.6	7.2
Denmark	9.2	0.1	21.8	0.6	8.3	14.6	6.5	32.2	7.1
France	10.8	0.8	27.5	0.9	8.9	16.8	6.0	22.6	5.7
W Germany	6.7	1.3	35.8	0.9	7.3	14.4	5.0	22.3	6.1
Ireland	23.2	1.0	20.7	1.3	7.5	16.8	2.8	20.8	6.1
Italy	15.9	0.6	27.6	0.5	10.0	18.1	1.6	20.1	5.7
Luxembourg	7.7	3.1	30.8	0.8	9.2	18.5	4.6	19.2	6.2
Netherlands	6.3	0.1	22.5	1.0	9.9	18.0	6.8	28.5	6.8
UK	2.7	1.4	30.2	1.4	6.9	16.9	5.7	28.3	6.4
Austria	12.7	1.1	30.2	1.4	9.0	16.8	4.9	16.8	7.0
Finland	13.0	0.4	25.9	1.3	7.4	14.7	5.5	24.3	7.6
Greece	41.4	0.6	17.6	0.6	8.1	11.5	2.4	11.0	6.7
Norway	9.0	0.5	22.4	0.8	8.6	16.9	4.7	27.6	9.4
Portugal	27.8	0.3	24.5	0.6	8.4	13.3	2.7	16.7	5.7
Spain	22.9	0.8	28.5	0.7	11.5	9.7	8.5	11.8	5.5
Sweden	6.1	0.4	25.9	0.8	7.2	14.4	6.0	32.4	6.8
Switzerland	7.7	0.2	37.8	0.8	9.5	17.5	5.3	15.4	5.7
Turkey	66.8	0.7	7.9	0.1	2.8	5.2	1.1	11.9	3.2
Bulgaria	23.6	1.9	32.3	0.6	7.9	8.0	0.7	18.2	6.7
Czechoslovakia	16.5	2.9	35.5	1.2	8.7	9.2	0.9	17.9	7.0
E Germany	4.2	2.9	41.2	1.3	7.6	11.2	1.2	22.1	8.4
Hungary	21.7	3.1	29.6	1.9	8.2	9.4	0.9	17.2	8.0
Poland	31.1	2.5	25.7	0.9	8.4	7.5	0.9	16.1	6.9
Rumania	34.7	2.1	30.1	0.6	8.7	5.9	1.3	11.7	5.0
USSR	23.7	1.4	25.8	0.6	9.2	6.1	0.5	23.6	9.3
Yugoslavia	48.7	1.5	16.8	1.1	4.9	6.4	11.3	5.3	4.0

364. Ages at death of English rulers

Gebski, V., Leung, O., McNeil, D.R. and Lunn, A.D. (1992) *The SPIDA User's Manual*. Sydney: Statistical Computing Laboratory.

The table gives the ages at which English rulers died.

William I	60	Henry VI	49	James II	68
William II	43	Edward IV	41	William III	51
Henry I	67	Edward V	13	Mary II	33
Stephen	50	Richard III	35	Anne	49
Henry II	56	Henry VII	53	George I	67
Richard I	42	Henry VIII	56	George II	77
John	50	Edward VI	16	George III	81
Henry III	65	Mary I	43	George IV	67
Edward I	68	Elizabeth I	69	William IV	71
Edward II	43	James I	59	Victoria	81
Edward III	65	Charles I	48	Edward VII	68
Richard II	34	Cromwell 'I'	59	George V	70
Henry IV	47	Cromwell 'II'	86	Edward VIII	77
Henry V	34	Charles II	55	George VI	56

365. Deaths in coal mining disasters

Jarrett, R.G. (1979) A note on the intervals between coal-mining disasters. *Biometrika*, **66**, 191–193.

These data also appear in Andrews, D.F. and Herzberg, A.M. (1985) *Data: A Collection of Problems from Many Fields for the Student and Research Worker*, New York: Springer-Verlag, 51–56.

The table gives the number of fatalities in coal mining disasters in Britain which claimed ten or more lives. A good graphical display is needed to report these data and they could be tested for differences in the average numbers of the fatalities during the different periods.

Years						
1860-69	**1870-79**	**1880-89**	**1890-99**	**1900-19**	**1920-39**	**1940-59**
13	30	62	176	81	12	11
76	19	120	87	16	39	10
13	19	164	10	11	27	12
12	20	101	10	33	52	16
142	26	25	112	119	13	22
22	38	48	139	25	13	57
13	19	74	290	14	14	12

10	70	37	13	10	27	13
13	26	13	57	75	10	15
47	11	32	20	168	45	15
16	27	43	63	27	11	104
59	34	20	19	136	27	21
26	18	68		344	11	12
39	54	14		88	14	83
13	15	14		439	265	17
15	23	42		12	10	45
26	17	178		39	19	19
34	43	81		27	58	
30	16	143		52	30	
12	23	10		13	79	
38	18	22		35		
24	10	28				
361	18	39				
91	36	73				
14	207	30				
178	17	23				
12	43	20				
10	23	64				
62	189					
26	268					
37	63					
53	21					
59	28					
11						
27						

366. A study on obesity

Beckles, G.L.A., Miller, G.J., Alexis, S.D., Price, S.G.L., Kirkwood, B.R., Carson, D.C. and Byam, M.T.A. (1985) Obesity in women in an urban Trinidadian community. Prevalence and associated characteristics. *International Journal of Obesity*, **9**, 127–135.

The data are from a study on obesity in women, giving the relationship between triceps skinfold and early menarche. The proportion of women who had early menarche seems to have increased with triceps skinfold size. A chi-squared test for trend or an ordinal regression may be used.

	Triceps skinfold group		
Age at menarche	Small	Intermediate	Large
< 12 years	15	29	36
12+ years	156	197	150

367. The star cluster CYG OB1

Humphreys, R.M. (1978) Studies of luminous stars in nearby galaxies. I. supergiants and O stars in the milky way, *Astrophysics Journal, Supplementary Series*, **38**, 309–350.

These data also appear in Rousseeuw, P.J. and Leroy, A.M. (1987) *Robust Regression and Outlier Detection*, New York: John Wiley & Sons.

The data are calibrated according to Vanisma, F. and De Greve, J.P. (1972): Close binary systems before and after mass transfer. *Astrophysics and Space Science*, **87**, 377–401.

Data from the Hertzprung-Russel Diagram of Star Cluster CYG OB1 were taken from Humphreys (1978) by C. Doom, who calibrated them according to Vanisma and De Greve (1982). The objective is to find a relationship between light intensity and surface temperature.

Index	Log surface temperature	Log light intensity	Index	Log surface temperature	Log light intensity
1	4.37	5.23	25	4.38	5.02
2	4.56	5.74	26	4.42	4.66
3	4.26	4.93	27	4.29	4.66
4	4.56	5.74	28	4.38	4.90
5	4.30	5.19	29	4.22	4.39
6	4.46	5.46	30	3.48	6.05
7	3.84	4.65	31	4.38	4.42
8	4.57	5.27	32	4.56	5.10
9	4.26	5.57	33	4.45	5.22
10	4.37	5.12	34	3.49	6.29
11	3.49	5.73	35	4.23	4.34
12	4.43	5.45	36	4.62	5.62
13	4.48	5.42	37	4.53	5.10
14	4.01	4.05	38	4.45	5.22
15	4.29	4.26	39	4.53	5.18
16	4.42	4.58	40	4.43	5.57
17	4.23	3.94	41	4.38	4.62
18	4.42	4.18	42	4.45	5.06
19	4.23	4.18	43	4.50	5.34
20	3.49	5.89	44	4.45	5.34
21	4.29	4.38	45	4.55	5.54
22	4.29	4.22	46	4.45	4.98
23	4.42	4.42	47	4.42	4.50
24	4.49	4.85			

368. Frequency of breast self-examination

Senie, R.T., Rosen, P.P., Lesser, M.L. and Kinne, D.W. (1981) Breast self-examinations and medical examination relating to breast cancer stage. *American Journal of Public Health*, **71**, 583–590.

The cross-sectional study is an important type of design which relies upon sample surveys of large populations. The investigator is interested in associations among the responses and this is illustrated by the data on breast self-examination by age of diagnosis of breast cancer. Are age and breast self-examination frequency related?

Frequency of breast self-examination

Age	Monthly	Occasionally	Never
< 45	91	90	51
45-59	150	200	155
60+	109	198	172

369. Annual wages of production line workers in the USA

Dyer, D. (1981) Structural probability bounds for the strong Pareto law, *Canadian Journal of Statistics*, **9**, 71.

Annual wages (multiples of US$100) of 30 production line workers in a large American firm are given.

112	125	108
154	119	105
119	128	158
108	132	104
112	107	119
156	151	111
123	103	101
103	104	157
115	116	112
107	140	115

370. One way to kill a cat

Snedecor, G. and Cochran, W. (1967) *Statistical Methods*, 6th edition, Ames: Iowa State University Press, 457–459, Table 15.4.2, contributed by Vos, B.J. and Dawson, W.T.

The table gives lethal doses of ouabain needed to kill a cat, injected at four different rates of injection.   The rates are in (mg/kg/min)/1045.75.   A scatterplot shows plausible linearity with, possibly, some falling off at a rate of 8. Regression of the mean dosage on the rate is a nice example of weighted least squares.

Rates (mg/kg/min)/1045.75

1	2	4	8
5	3	34	51
9	6	34	56
11	22	38	62
13	27	40	63
14	27	46	70
16	28	58	73
17	28	60	76
20	37	60	89
22	40	65	92
28	42		
31	50		
31			

371. Viral lesions on tobacco leaves

Youden, W.J. and Beale, H.P. (1934) *Contrib. Boyce Thompson Inst.*, **6**, 437.

These data also appear in Snedecor, G. and Cochran, W. (1967) *Statistical Methods*, 6th edition, Ames: Iowa State University Press, 95, Table 4.3.1.

Two virus preparations were soaked into cheesecloth and each was rubbed onto different halves of a tobacco leaf. Numbers of local lesions appearing (lesions appear as small, dark rings) on each half were counted for eight leaves. Do the two extracts produce different effects?

ID	Prep 1	Prep 2
1	31	18
2	20	17
3	18	14
4	17	11
5	9	10
6	8	7
7	10	5
8	7	6

372. Finger ridges of identical twins

Newman, H.H., Freeman, F.N. and Holzinger, K.J. (1937) *Twins*, Chicago: University of Chicago Press.

These data also appear in Snedecor, G. and Cochran, W. (1967) *Statistical Methods*, 6th edition, Ames: Iowa State University Press, 295, Table 10.20.1.

The data are counts of the numbers of finger ridges of individuals for 12 pairs of identical twins. Numbers of finger ridges are obviously similar within the pairs, but vastly different between different sets of twins. The interest lies in estimating the correlation between identical twins and obtaining a confidence interval for the correlation.

ID	Twin 1	Twin 2	ID	Twin 1	Twin 2
1	71	71	7	114	113
2	79	82	8	57	44
3	105	99	9	114	113
4	115	114	10	94	91
5	76	70	11	75	83
6	83	82	12	76	72

373. Statures of brother and sister

Pearson, K. and Lee, A. (1902-3) On the laws of inheritance in man. *Biometrika*, **2**, 357.

An historic data set on the heights (in inches) of brothers and sisters, published in the second volume of *Biometrika*. Pearson was interested in the correlation between statures of siblings.

Family	Height of brother	Height of sister
1	71	69
2	68	64
3	66	65
4	67	63
5	70	65
6	71	62
7	70	65
8	73	64
9	72	66
10	65	59
11	66	62

374. Candidates in the 1992 British Election

The Independent, Friday 27th March, 1992.

The table below is a cross classification of geographical region by political party by sex of the numbers of candidates who contested the British General Election which took place on 9th April 1992. The format is males/females.

The article in *The Independent* based on these data discussed the difficulties women found in breaking through the prejudices of constituency parties. Thus questions of interest for this data set are whether the proportion of women candidates differs between parties and geographical regions.

| | | Political party | | |
Region	Cons	Labour	Lib-Dem	Green	Other
South East	101/8	84/25	81/28	42/15	86/27
South West	45/3	36/12	35/13	21/6	61/11
Greater London	76/8	57/27	63/19	37/13	93/21
East Anglia	19/1	16/4	16/4	6/4	23/8
East Midlands	39/3	35/7	36/6	8/3	19/7
Wales	36/2	34/4	30/8	7/0	44/10
Scotland	63/9	67/5	51/21	14/6	87/17
West Midlands	50/8	43/15	49/9	11/4	30/5
Yorks and Humbers	51/3	45/9	42/12	22/3	22/6
North West	65/8	57/16	61/12	17/5	75/20
North	32/4	34/2	32/4	7/1	6/3

375. Papers presented at the first ten ICPRs

Newsletter of the International Association for Pattern Recognition, **13**, No. 3, October 1990.

The data give the numbers of papers presented at the International Conference on Pattern Recognition, for the years 1973 to 1990 for the ten countries most heavily represented.

Papers have usually been attributed to the country of the first author. After 1984 numbers decreased because the refereeing process was made stricter. Questions of interest relate to the relative performance of different countries. Can sudden changes be detected and do any countries behave anomalously?

Country	1973	1974	1976	1978	1980	1982	1984	1986	1988	1990
USA	52	52	81	55	134	63	154	65	62	150
Japan	13	14	17	89	40	52	52	55	60	40
France	2	15	14	22	28	25	34	58	37	23

West Germany	4	13	10	12	23	77	23	22	8	11
China	0	0	0	1	7	6	14	40	86	10
Canada	0	6	4	7	8	12	37	16	19	26
UK	0	12	7	4	9	8	11	6	8	8
Italy	4	7	5	8	4	8	10	10	10	6
Sweden	0	4	4	5	9	4	3	8	7	5
Netherlands	1	4	4	3	6	8	6	6	3	3

376. Body mass index data

These data were provided by Dr S.L. Channon, Middlesex Hospital, with permission.

Body mass index (bmi) is defined as weight (in kilograms) divided by the square of height (in metres). The figures below show body mass indices for twenty patients (1) on admission to hospital for treatment for anorexia, (2) calculated in terms of the patient's stated preferred weight at the time of admission, and (3) at follow up after discharge.

Questions of interest here relate to the change of bmi between admission (1) and discharge (3) and to the relationship between this change and the bmi based on the preferred weight (2).

Patient	(1)	(2)	(3)
1	16.84	18.01	24.03
2	16.26	19.22	18.50
3	14.33	18.65	16.61
4	14.30	17.54	16.57
5	11.98	12.05	18.99
6	13.59	17.38	15.62
7	15.03	19.16	25.96
8	13.95	12.31	13.82
9	14.02	17.96	21.09
10	12.22	15.63	21.38
11	12.0	16.84	12.34
12	16.05	18.08	13.17
13	15.86	19.22	17.32
14	11.55	15.06	16.07
15	15.76	21.10	17.22
16	14.12	17.46	13.06
17	14.44	18.15	14.72
18	18.12	18.54	21.16
19	12.60	18.01	18.92
20	15.52	15.89	18.01

377. Citation rates

Members of British University Psychology Departments with 60 or more citations in the 1985 Social Science Citation Index.

What function might we use to model the distribution of numbers of papers produced by the authors?

Psychologist	Number of citations	Psychologist	Number of citations
H.J. Eysenck	813	O.J. Braddick	83
J.A. Gray	251	S.J. Cooper	82
E.K. Warrington	180	D.N. Lee	84
N.J. Mackintosh	176	G.D. Wilson	84
J.M. Argyle	170	C.B. Trevarthan	73
M. Coltheart	164	H. Giles	70
P.B. Warr	120	H.R. Schaffer	67
D.A. Booth	101	P.H. Venables	68
M.R. Trimble	97	T.W. Robbins	69
J. Sandler	97	L. Weiskranz	64
S.B.G. Eysenck	91	A.F. Furnham	62
M.W. Eysenck	90	E.T. Rolls	61
T.G.R. Bower	85	W. Yule	60

378. A World Heavyweight Boxing Championship match

The Independent, 27th February, 1989.

These data compare the performances of Mike Tyson and Frank Bruno in their 1989 World Heavyweight Championship match. Tyson won.

Interest focusses on a comparison between the two boxers. Who is more accurate (thrown punches connect), who is more powerful, etc.

	Tyson Round					Bruno Round				
	1	2	3	4	5	1	2	3	4	5
Total punches	43	39	28	37	55	55	42	35	18	20
Punches connected	13	14	13	15	34	14	8	5	3	7
Jabs thrown	6	9	5	13	8	14	24	16	8	6
Jabs connected	0	1	1	3	3	2	1	0	0	1
Power punches	37	30	23	24	47	41	18	19	10	14
Power connected	13	13	12	12	31	12	7	5	3	6
Knockdowns	1	0	0	0	0	0	0	0	0	0

379. Assessment of conference papers

Papers submitted to COMPSTAT-88, the 1988 conference on computational statistics, were sent to a number of judges for assessment. The ratings were combined and discussed to decide which papers should be accepted for presentation at the meeting. Here are the ratings for a subsample of 30 papers from eight judges. A is the highest rating, E the lowest.

The main interest here is the inter-rater reliability of the judges, and how one might produce a ranking of the papers.

	Judge							
	1	2	3	4	5	6	7	8
1	A	A	E	A	A	A	A	B
2	A	E	A	A	A	D	B	B
3	B	B	E	A	A	C	A	A
4	A	E	B	A	A	C	B	A
5	B	C	C	A	A	B	A	A
6	B	A	C	A	A	C	A	C
7	B	A	B	B	B	B	A	A
8	A	B	A	E	B	B	B	A
9	B	A	C	A	B	B	A	B
10	B	A	B	B	B	C	A	A
11	B	B	A	A	B	C	A	B
12	C	B	B	A	A	B	A	B
13	B	A	E	C	B	B	A	A
14	B	E	A	B	C	A	A	B
15	B	A	B	A	B	A	C	C
16	A	B	B	B	B	A	B	B
17	B	B	B	A	C	B	A	B
18	B	A	B	B	B	C	A	B
19	B	E	B	A	B	B	C	A
20	B	C	B	E	A	B	A	B
21	B	E	B	B	B	A	A	C
22	B	E	B	B	B	A	A	C
23	B	B	C	C	B	A	A	B
24	A	C	B	A	B	C	B	B
25	B	B	C	A	B	B	A	C
26	B	A	B	C	B	A	C	B
27	C	A	B	B	C	B	A	B
28	B	A	C	A	C	E	B	B
29	B	C	B	C	B	D	A	A
30	C	A	B	C	D	B	A	B

380. Clinical trial for sprains

David Hand: private communication.

These data arose during a clinical trial of a treatment for sprains. The first value for each patient is the patient number, and the next four are tenderness scores on a scale 0=none, 1=slight, 2=moderate, and 3=severe on consecutive weeks. A * represents a missing value.

Primary interest centres on whether the treatment produces better results than the control. The analysis is complicated by missing values and truncated series of observations. Note that the condition is one that would be expected to disappear spontaneously, so it is rate of recovery which is the key concern.

Treatment group Patient	Scores	Control group Patient	Scores
1	211	17	3000
2	21*0	18	11*0
3	1100	19	2220
4	110	20	1100
5	2100	21	2210
6	21	22	110
7	210	23	2222
8	2100	24	210
9	200	25	2100
10	2110	26	3221
11	3221	27	22*0
12	3221	28	222
13	3210	29	2211
14	3233	30	2211
15	332		
16	3211		

381. Jeans sales in the UK

Conrad, S. (1989) *Assignments in Applied Statistics*, Chichester: John Wiley & Sons, 78.

The data show the estimated monthly sales of pairs of jeans, in 1000s, in the UK over six years. Interest here is in forecasting future sales. Techniques of time series analysis are appropriate.

Month	1980	1981	1982	1983	1984	1985
January	1998	1924	1969	2149	2319	2137
February	1968	1959	2044	2200	2352	2130
March	1937	1889	2100	2294	2476	2154
April	1827	1819	2103	2146	2296	1831
May	2027	1824	2110	2241	2400	1899
June	2286	1979	2375	2369	3126	2117
July	2484	1919	2030	2251	2304	2266
August	2266	1845	1744	2126	2190	2176
September	2107	1801	1699	2000	2121	2089
October	1690	1799	1591	1759	2032	1817
November	1808	1952	1770	1947	2161	2162
December	1927	1956	1950	2135	2289	2267

382. The 13 most common psychiatric disorders in the UK

Dunn, G. (1986) Patterns of psychiatric diagnosis in general practice: the Second National Morbidity Survey. *Psychological Medicine*, **16**, 573–581, Table 1.

For each of 22 medical practices, each patient was asked if they had had one or more episodes of each psychiatric disorder. The number of patients who answered 'yes' is recorded in the appropriate cell of the table. The first column gives the number of patients registered with the practice over the six years of the survey.

Key to disorders: 130, anxiety neurosis; 134, depressive neurosis; 150, unclassified symptoms; 135, physical disorders of presumably psychogenic origin; 146, insomnia; 147, tension headache; 126, affective psychosis; 136, neurasthenia; 148, enuresis; 132, phobic neurosis; 131, hysterical neurosis; 125, schizophrenia; 139, alcoholism and drug dependence.

This data matrix is quite complex and one might wish to reduce it to a manageable form. Various multivariate techniques might be considered, including principal components analysis and correspondence analysis. One might also be interested in exploring if there are different types of practice, in terms of their profile of psychiatric illnesses. Note that the definition of the data means that there could be a lack of independence between observations. One could discuss how this might influence the analysis.

Disorder type

Practice Size	130	134	150	135	146	147	126	136	148	132	131	125	139
2519	221	68	83	108	70	36	58	1	17	20	7	7	6
1504	281	184	11	3	69	17	0	2	9	1	1	7	0
2161	234	206	112	55	108	63	3	28	18	2	5	3	3
4187	478	608	40	402	156	108	1	1	64	25	12	11	9
1480	76	251	173	159	23	37	4	30	14	1	9	6	3
2125	386	142	9	16	113	90	0	0	10	7	15	11	4
6514	1122	742	197	435	81	172	330	193	66	33	22	16	21
1820	208	398	45	14	13	35	5	1	3	4	10	12	9
2671	409	282	512	168	119	45	252	225	41	27	39	5	17
4220	314	429	105	53	65	37	20	1	29	28	23	14	11
2377	72	148	425	120	56	116	3	0	17	6	6	5	2
5009	566	305	123	174	71	99	5	16	28	15	13	9	7
2037	241	207	4	4	37	29	0	1	11	0	3	0	0
1759	390	277	56	68	160	45	17	15	26	32	23	3	2
1767	248	178	61	89	122	70	13	11	16	19	5	6	3
3443	218	185	317	198	164	75	45	89	26	21	13	22	27
2200	212	210	115	104	64	70	85	7	11	12	15	5	1
2639	280	155	286	180	53	74	36	4	27	9	16	4	12
1897	288	224	132	153	79	51	14	15	7	9	9	3	5
2278	331	251	170	439	76	84	9	24	20	8	5	5	7
2242	254	303	177	152	148	138	21	53	16	15	16	9	5
2497	330	290	186	215	119	121	21	7	16	26	13	13	9

383. Randomized experiment on tomato yield

Box, G.E.P., Hunter, W.G. and Hunter, J.S. (1978) *Statistics for experimenters*, New York: John Wiley & Sons, Table 4.1.

To compare the effects of two fertilizers on the yield of tomatoes a gardener selected cards from a shuffled pack to determine in which order in the row to use the two fertilizers. The results and the associated tomato yield (in pounds) are shown below. The question is whether fertilizer B (a supposedly improved fertilizer) is better than fertilizer A (the standard one).

Position in row	1	2	3	4	5	6	7	8	9	10	11
Fert'r	A	A	B	B	A	B	B	B	A	A	B
Yield	29.9	11.4	26.6	23.7	25.3	28.5	14.2	17.9	16.5	21.1	24.3

384. Screening for psychiatric disorder

Below are given the scores of two groups of subjects on 25 questions. The first group have been classified by a psychiatrist as being healthy and the second group as suffering from mild psychiatric illness. The responses to the questions are coded from 1 to 4, in order of increasing feeling of severity of tension, inadequacy, etc; 0 signifies a missing value.

The key question of interest is whether one can use these data to build a classification rule to identify people who are likely to be suffering from mild psychiatric illness on the basis of their responses to the questions. If so this will be much easier, cheaper, and quicker than interview by a psychiatrist and can be used as a first stage of psychiatric health screening. A subsequent question is to estimate likely future performance.

	Group 1		Group 2
Subject	Scores on 25 items	Subject	Scores on 25 items
1	2222232122211212222222222	1	2112221122211211222222222
2	2221222212111211222221222	2	3121212212121211212121241
3	1121233322211211333331233	3	3223222222223211332333300
4	2221222111111211122121221	4	2122222123121213222211221
5	1121222111111211112120000	5	2321232111211333232232333
6	1121122112111211111221222	6	1121222312311211223331212
7	2222222222221212232322232	7	2121123112222221222212212
8	1121222111111211232211232	8	1121322121111211121112211
9	1121222111111211112111221	9	2222222211111221122111211
10	2122222111111211222121232	10	3134444444441400000000000
11	2221232212121211222232222	11	2221212112121121111121121
12	2121222112111211222121221	12	3222232232213343322333223
13	1122232113211210112121232	13	3223322212221211122232213
14	1121222111111211112121221	14	1222111213221212211111131
15	3323341222231212232232223	15	2243342223421212323343343

16	2222242211121211112211222	16	32223302333323240333333323
17	11111111111111111111111121	17	4433343334334413333344444
18	312123212322121123322233	18	33322221234322123232333322
19	1221222121221211112122222	19	424444433444231434343434444
20	112123211321121133122112	20	212222222221121133232232

385. Salaries in different occupations

The Guardian, April 6, 1992.

The data below show the salaries offered in 72 randomly selected advertisements in *The Guardian* on days specializing in different occupations. The units are pounds sterling. When ranges are given in the advertisements, the mid-point of the range is reproduced below.

Interest here concerns whether there is a difference between advertised starting salaries in the two occupations. Thus various two group tests might be considered. Questions such as asymmetry of distributions could be looked at and one could transform or perform nonparametric tests.

Key
(1) Creative, media, and marketing
(2) Education

(1)			(2)		
17703	13796	12000	25899	17378	19236
42000	22958	22900	21676	15594	18780
18780	10750	13440	15053	17375	12459
15723	13552	17574	19461	20111	22700
13179	21000	22149	22485	16799	35750
37500	18245	17547	17378	12587	20539
22955	19358	9500	15053	24102	13115
13000	22000	25000	10998	12755	13605
13500	12000	15723	18360	35000	20539
13000	16820	12300	22533	20500	16629
11000	17709	10750	23008	13000	27500
12500	23065	11000	24260	18066	17378
13000	18693	19000	25899	35403	15053
10500	14472	13500	18021	17378	20594
12285	12000	32000	17970	14855	9866
13000	20000	17783	21074	21074	21074
16000	18900	16600	15053	19401	25598
15000	14481	18000	20739	15053	15053
13944	35000	11406	15053	15083	31530
23960	18000	23000	30800	10294	16799
11389	30000	15379	37000	11389	15053
12587	12548	21458	48000	11389	14359
17000	17048	21262	16000	26544	15344
9000	13349	20000	20147	14274	31000

386. Whisky prices

Chance (1991), **4**, No. 1, 29.

In the 1960s in the United States of America 16 states owned the retail liquor stores while in 26 the stores were privately owned. (Some states were omitted for technical reasons.) The table shows the prices in dollars of a fifth of Seagram 7 Crown Whisky in the two sets of states in 1961.

Interest here centres on a comparison between the two groups. This may include comparisons of means and medians, but also of dispersions. The two groups are the entire populations of scores, and not a sample from them.

16 monopoly states:

4.65,	4.55,	4.11,	4.15,	4.20,	4.55,	3.80,	4.00,	4.19,	4.75,
4.74,	4.50,	4.10,	4.00,	5.05,	4.20				

26 private-ownership states:

4.82,	5.29,	4.89,	4.95,	4.55,	4.90,	5.25,	5.30,	4.29,	4.85,
4.54,	4.75,	4.85,	4.85,	4.50,	4.75,	4.79,	4.85,	4.79,	4.95,
4.95,	4.75,	5.20,	5.10,	4.80,	4.29				

387. 1992 London Marathon

The Independent, Monday 13th April, 1992.

The data below give the finishing times in hours, minutes, and seconds of the first 50 men and women to finish in the 1992 London marathon.

Men		Women	
2.10.02	2.15.34	2.29.39	2.43.43
2.10.07	2.15.46	2.29.59	2.44.33
2.10.08	2.15.48	2.31.33	2.44.39
2.10.10	2.16.26	3.34.02	2.44 47
2.10.36	2.16.28	2.34.29	2.47.12
2.10.49	2.16.31	2.34.38	2.47.30
2.10.55	2.16.36	2.34.39	2.47.36
2.11.25	2.16.42	2.35.21	2.47.41

2.11.28	2.16.48	2.37.21	2.47.48
2.12.02	2.17.07	2.37.40	2.48.17
2.12.09	2.17.10	2.37.52	2.48.32
2.12.29	2.13.22	2.39.06	2.49.46
2.12.45	2.17.34	2.39.22	2.50.10
2.13.06	2.18.21	2.40.00	2.50.52
2.13.14	2.18.25	2.40.09	2.51.29
2.13.22	2.18.39	2.40.26	2.51.44
2.13.33	2.18.47	2.40.39	2.52.45
2.14.11	2.18.53	2.40.40	2.54.03
2.14.22	2.19.19	2.40.47	2.54.29
2.14.23	2.19.22	2.41.15	2.54.33
2.14.25	2.19.25	2.41.18	2.54.38
2.14.46	2.19.48	2.41.35	2.55.00
2.14.49	2.19.49	2.42.10	2.55.56
2.14.56	2.20.05	2.42.25	2.56.46
2.15.31	2.20.08	2.43.14	2.56.46

388. Predicting cervical cancer

Cuzick, J., Terry, G., Ho, L., Hollingworth, T. and Anderson, M. (1992) Human papillomavirus type 16 DNA in cervical smears as predictor of high-grade cervical cancer. *The Lancet*, **339**, 959–960.

About 5% of the cervical smears conducted each year in the UK show evidence of abnormality. Most of these show mild or moderate dyskaryosis, where the underlying pathology is highly variable: biopsy shows that about a third of these have high-grade disease (cervical intraepithelial neoplasia 3, CIN3), a third have CIN1/2, and the others have either normal cervices or less severe indications. A non-invasive test which could accurately predict which are the ones with CIN3 would be very valuable. A new test using the polymerase chain reaction for detecting human papillomavirus type 16 DNA in cervical smears has been proposed. The table shows a cross-classification of the results of this test by CIN grade for 30 women referred because of abnormal smears.

	Histological diagnosis			
New test result	Normal	CIN1	CIN2	CIN3
Negative or low	6	1	2	10
Intermediate	0	0	0	0
High	1	0	0	10

389. Treatments for head and neck cancer

Efron, B. (1988) Three examples of computer intensive statistical inference. *Sankhya, Series A*, **50**, 338–362.

These data arose in a clinical trial of cancer of the head and neck, comparing radiation only with radiation plus chemotherapy. Subjects were randomly allocated to the two treatment groups. *y* is the observed number of days following treatment before relapse. *d*=1 if the relapse was observed and *d*=0 if the experiment was terminated before the relapse was observed.

Radiation alone					Radiation plus chemotherapy			
y	*d*	*y*	*d*		*y*	*d*	*y*	*d*
7	1	185	0		37	1	519	1
34	1	218	1		84	1	528	0
42	1	225	1		92	1	547	0
63	1	241	1		94	1	613	0
64	1	248	1		110	1	633	1
74	0	273	1		112	1	725	1
83	1	277	1		119	1	759	0
84	1	279	0		127	1	817	1
91	1	297	1		130	1	1092	0
108	1	319	0		133	1	1245	0
112	1	405	1		140	1	1331	0
129	1	417	1		146	1	1557	1
133	1	420	1		155	1	1642	0
133	1	440	1		159	1	1771	0
139	1	523	0		169	0	1776	1
140	1	523	1		173	1	1897	0
140	1	583	1		179	1	2023	0
146	1	594	1		194	1	2146	0
149	1	1101	1		195	1	2297	0
154	1	1116	0		209	1		
157	1	1146	1		249	1		
160	1	1226	0		281	1		
160	1	1349	0		319	1		
165	1	1412	0		339	1		
173	1	1417	1		432	1		
176	1				469	1		

390. Tau particle decay

Efron, B. (1988) Three examples of computer intensive statistical inference. *Sankhya, Series A*, **50**, 338–362.

Soon after a tau particle is produced it decays into collections of other particles. Some times this decay produces just a single charged particle. The data below show 13 independent estimates of the proportion of times that a decay of this nature occurs. Also given are standard deviations for each estimate, as calculated by the laboratory making that measurement.

Issues of interest in the original papers describing these data were to find confidence intervals for the proportion of times that decay of this type occurs.

Proportion	84.0	86.0	85.2	85.2	85.1	87.8	84.7	86.7
S.D.	2.0	2.2	1.7	2.9	3.1	4.1	1.9	0.7

Proportion	86.9	86.1	87.9	87.2	84.7
S.D.	0.4	1.0	1.3	0.9	1.0

391. Factors influencing depression

Everitt, B.S. and Smith, A.M.R. (1979) Interactions in contingency tables: a brief discussion of alternative definitions. *Psychological Medicine*, **9**, 581–583.

Two tables are given showing the number of women from a sample in Camberwell, South London, who developed depression in a one year period cross-classified by two factors.

The primary research questions to be addressed here were whether or not there was an interactive effect between the two column variables in each table in the proportions suffering from depression.

(a)

	Lack of intimacy with husband/boyfriend?			
	Yes		No	
	Severe life event?		Severe life event?	
	Yes	No	Yes	No
Depression	24	2	9	2
No depression	52	60	79	191

(b)

	Three children less than 14?			
	Yes		No	
	Severe life event?		Severe life event?	
	Yes	No	Yes	No
Depression	9	0	24	4
No depression	12	20	119	231

392. Vocabulary scores of university students

These data were kindly provided by Dr S.L. Channon of the Middlesex Hospital, London.

Fifty-four University students (male and female) agreed to participate in an experiment on problem solving. They were tested on the vocabulary subtest of the WAIS-R.

The overall population produces a distribution of scores which has mean 10 and sd 3. A glance at the table shows that the mean of this sample is above 10. Is the standard deviation of the sample consistent with a population value of 3? What about normality and symmetry?

14	11	13	13	13	15	11	16	10
13	14	11	13	12	10	14	10	14
16	14	14	11	11	11	13	12	13
11	11	15	14	16	12	17	9	16
11	19	14	12	12	10	11	12	13
13	14	11	11	15	12	16	15	11

393. Two groups drug comparison

Crowder, M.J. and Hand, D.J. (1990) *Analysis of repeated measures*. London: Chapman and Hall, 9.

Two drugs, both in tablet form, were each given to the same five volunteer subjects in a pilot trial. Drug A was given first and drug B second, with a washout period in between. In each phase antibiotic blood serum levels were measured 1, 2, 3, and 6 hours after medication, as shown in the table.

These data are suitable for repeated measures analysis.

	Drug A				Drug B			
	Time (hours)				Time (hours)			
Subject	1	2	3	6	1	2	3	6
1	1.08	1.99	1.46	1.21	1.48	2.50	2.62	1.95
2	1.19	2.10	1.21	0.96	0.62	0.88	0.68	0.48
3	1.22	1.91	1.36	0.90	0.65	1.52	1.32	0.95
4	0.60	1.10	1.03	0.61	0.32	2.12	1.48	1.09
5	0.55	1.00	0.82	0.52	1.48	0.90	0.75	0.44

394. Blood glucose levels

Crowder, M.J. and Hand, D.J. (1990) *Analysis of repeated measures.* London: Chapman and Hall, 14.

Blood glucose levels were recorded for six volunteers before and after they had eaten a test meal. Recordings were made at times –15, 0, 30, 60, 90, 120, 180, 240, 300, and 360 minutes after feeding time. This whole process was repeated six times, with the meal taken at various times of the day and night.

The researcher was primarily interested in time of day effect, and some kind of repeated measures analysis is appropriate. A score of –1.00 signifies a missing value.

<table>
<tr><th colspan="11">Time (minutes)</th></tr>
<tr><th></th><th>-15</th><th>0</th><th>30</th><th>60</th><th>90</th><th>120</th><th>180</th><th>240</th><th>300</th><th>360</th></tr>
<tr><th>Subject</th><th colspan="10">10am meal</th></tr>
<tr><td>1</td><td>4.90</td><td>4.50</td><td>7.84</td><td>5.46</td><td>5.08</td><td>4.32</td><td>3.91</td><td>3.99</td><td>4.15</td><td>4.41</td></tr>
<tr><td>2</td><td>4.61</td><td>4.65</td><td>7.90</td><td>6.13</td><td>4.45</td><td>4.17</td><td>4.96</td><td>4.36</td><td>4.26</td><td>4.13</td></tr>
<tr><td>3</td><td>5.37</td><td>5.35</td><td>7.94</td><td>5.64</td><td>5.06</td><td>5.49</td><td>4.77</td><td>4.48</td><td>4.39</td><td>4.45</td></tr>
<tr><td>4</td><td>5.10</td><td>5.22</td><td>7.20</td><td>4.95</td><td>4.45</td><td>3.88</td><td>3.65</td><td>4.21</td><td>-1.00</td><td>4.44</td></tr>
<tr><td>5</td><td>5.34</td><td>4.91</td><td>5.69</td><td>8.21</td><td>2.97</td><td>4.30</td><td>4.18</td><td>4.93</td><td>5.16</td><td>5.54</td></tr>
<tr><td>6</td><td>6.24</td><td>5.04</td><td>8.72</td><td>4.85</td><td>5.57</td><td>6.33</td><td>4.81</td><td>4.55</td><td>4.48</td><td>5.15</td></tr>
<tr><th></th><th colspan="10">2pm meal</th></tr>
<tr><td>1</td><td>4.91</td><td>4.18</td><td>9.00</td><td>9.74</td><td>6.95</td><td>6.92</td><td>4.66</td><td>3.45</td><td>4.20</td><td>4.63</td></tr>
<tr><td>2</td><td>4.16</td><td>3.42</td><td>7.09</td><td>6.98</td><td>6.13</td><td>5.36</td><td>6.13</td><td>3.67</td><td>4.37</td><td>4.31</td></tr>
<tr><td>3</td><td>4.95</td><td>4.40</td><td>7.00</td><td>7.80</td><td>7.78</td><td>7.30</td><td>5.82</td><td>5.14</td><td>3.59</td><td>4.00</td></tr>
<tr><td>4</td><td>3.82</td><td>4.00</td><td>6.56</td><td>6.48</td><td>5.66</td><td>7.74</td><td>4.45</td><td>4.07</td><td>3.73</td><td>3.58</td></tr>
<tr><td>5</td><td>3.76</td><td>4.70</td><td>6.76</td><td>4.98</td><td>5.02</td><td>5.95</td><td>4.90</td><td>4.79</td><td>5.25</td><td>5.42</td></tr>
<tr><td>6</td><td>4.13</td><td>3.95</td><td>5.53</td><td>8.55</td><td>7.09</td><td>5.34</td><td>5.56</td><td>4.23</td><td>3.95</td><td>4.29</td></tr>
<tr><th></th><th colspan="10">6am meal</th></tr>
<tr><td>1</td><td>4.22</td><td>4.92</td><td>8.09</td><td>6.74</td><td>4.30</td><td>4.28</td><td>4.59</td><td>4.49</td><td>5.29</td><td>4.95</td></tr>
<tr><td>2</td><td>4.52</td><td>4.22</td><td>8.46</td><td>9.12</td><td>7.50</td><td>6.02</td><td>4.66</td><td>4.69</td><td>4.26</td><td>4.29</td></tr>
<tr><td>3</td><td>4.47</td><td>4.47</td><td>7.95</td><td>7.21</td><td>6.35</td><td>5.58</td><td>4.57</td><td>3.90</td><td>3.44</td><td>4.18</td></tr>
<tr><td>4</td><td>4.27</td><td>4.33</td><td>6.61</td><td>6.89</td><td>5.64</td><td>4.85</td><td>4.82</td><td>3.82</td><td>4.31</td><td>3.81</td></tr>
<tr><td>5</td><td>4.81</td><td>4.85</td><td>6.08</td><td>8.28</td><td>5.73</td><td>5.68</td><td>4.66</td><td>4.62</td><td>4.85</td><td>4.69</td></tr>
<tr><td>6</td><td>4.61</td><td>4.68</td><td>6.01</td><td>7.35</td><td>6.38</td><td>6.16</td><td>4.41</td><td>4.96</td><td>4.33</td><td>4.54</td></tr>
</table>

6pm meal

1	4.05	3.78	8.71	7.12	6.17	4.22	4.31	3.15	3.64	3.88
2	3.94	4.14	7.82	8.68	6.22	5.10	5.16	4.38	4.22	4.27
3	4.19	4.22	7.45	8.07	6.84	6.86	4.79	3.87	3.60	4.92
4	4.31	4.45	7.34	6.75	7.55	6.42	5.75	4.56	4.30	3.92
5	4.30	4.71	7.44	7.08	6.30	6.50	4.50	4.36	4.83	4.50
6	4.45	4.12	7.14	5.68	6.07	5.96	5.20	4.83	4.50	4.71

2am meal

1	5.03	4.99	9.10	10.03	9.20	8.31	7.92	4.86	4.63	3.52
2	4.51	4.50	8.74	8.80	7.10	8.20	7.42	5.79	4.85	4.94
3	4.87	5.12	6.32	9.48	9.88	6.28	5.58	5.26	4.10	4.25
4	4.55	4.44	5.56	8.39	7.85	7.40	6.23	4.59	4.31	3.96
5	4.79	4.82	9.29	8.99	8.15	5.71	5.24	4.95	5.06	5.24
6	4.33	4.48	8.06	8.49	4.50	7.15	5.91	4.27	4.78	-1.00

10pm meal

1	4.60	4.72	9.53	10.02	10.25	9.29	5.45	4.82	4.09	3.52
2	4.33	4.10	4.36	6.92	9.06	8.11	5.69	5.91	5.65	4.58
3	4.42	4.07	5.48	9.05	8.04	7.19	4.87	5.40	4.35	4.51
4	4.38	4.54	8.86	10.01	10.47	9.91	6.11	4.37	3.38	4.02
5	5.06	5.04	8.86	9.97	8.45	6.58	4.74	4.28	4.04	4.34
6	4.43	4.75	6.95	6.64	7.72	7.03	6.38	5.17	4.71	5.14

395. Rat body weights with diet supplements

Crowder, M.J. and Hand, D.J. (1990) *Analysis of repeated measures*. London: Chapman and Hall, 19.

Three groups of rats were put on different diets and after a settling in period their body weights (in grams) were recorded weekly over nine weeks, with an extra observation taken on day 44 as shown in the table. A treatment was given during the sixth week. Research questions which might be addressed are whether the diets alone cause differences between the groups, whether the treatment has an effect, and whether the treatment has a differential effect according to group.

Rat Group 1	Day 1	8	15	22	29	36	43	44	50	57	64
1	240	250	255	260	262	258	266	266	265	272	278
2	225	230	230	232	240	240	243	244	238	247	245
3	245	250	250	255	262	265	267	267	264	268	269
4	260	255	255	265	265	268	270	272	274	273	275
5	255	260	255	270	270	273	274	273	276	278	280

6	260	265	270	275	275	277	278	278	284	279	281
7	275	275	260	270	273	274	276	271	282	281	284
8	245	255	260	268	270	265	265	267	273	274	278

Group 2

9	410	415	425	428	438	443	442	446	456	468	478
10	405	420	430	440	448	460	458	464	475	484	496
11	445	445	450	452	455	455	451	450	462	466	472
12	555	560	565	580	590	597	595	595	612	618	628

Group 3

13	470	465	475	485	487	493	493	504	507	518	525
14	535	525	530	533	535	540	525	530	543	544	559
15	520	525	530	540	543	546	538	544	553	555	548
16	510	510	520	515	530	538	535	542	550	553	569

396. Diet supplements

Crowder, M.J. and Hand, D.J. (1990) *Analysis of repeated measures.* London: Chapman and Hall, 28.

Fifteen guinea pigs were given a growth inhibiting substance and body weight measurements (in grams) were recorded at the ends of weeks 1, 3, 4, 5, 6, and 7. At the beginning of week 5 vitamin E therapy was started, the guinea pigs being divided into three groups of five to receive zero, low, or high doses of vitamin E. The primary research question was whether the growth patterns differed between the groups.

			Weeks			
Animal	**1**	**3**	**4**	**5**	**6**	**7**
Group 1						
1	455	460	510	504	436	466
2	467	565	610	596	542	587
3	445	530	580	597	582	619
4	485	542	594	583	611	612
5	480	500	550	528	562	576
Group 2						
6	514	560	565	524	552	597
7	440	480	536	484	567	569
8	495	570	569	585	576	677
9	520	590	610	637	671	702
10	503	555	591	605	649	675
Group 3						
11	496	560	622	622	632	670
12	498	540	589	557	568	609
13	478	510	568	555	576	605
14	545	565	580	601	633	649
15	472	498	540	524	532	583

397. Visual acuity and lens strength

Crowder, M.J. and Hand, D.J. (1990) *Analysis of repeated measures*. London: Chapman and Hall, 30.

Seven subjects had their response times measured when a light was flashed into each eye through lenses of powers 6/6, 6/18, 6/36, and 6/60. Measurements are in milliseconds, and the question was whether response time varied with lens strength. (A lens of power **a/b** means that the eye will perceive as being at **a** feet an object which is actually positioned at **b** feet.)

Subject	Left eye				Right eye			
	6/6	**6/18**	**6/36**	**6/60**	**6/6**	**6/18**	**6/36**	**6/60**
1	116	119	116	124	120	117	114	122
2	110	110	114	115	106	112	110	110
3	117	118	120	120	120	120	120	124
4	112	116	115	113	115	116	116	119
5	113	114	114	118	114	117	116	112
6	119	115	94	116	100	99	94	97
7	110	110	105	118	105	105	115	115

398. Plasma ascorbic acid levels in hospital patients

Crowder, M.J. and Hand, D.J. (1990) *Analysis of repeated measures*. London: Chapman and Hall, 32.

Twelve hospital patients were given a special diet. Measurements of plasma ascorbic acid were taken twice before treatment, three times during, and twice after, at week numbers 1, 2, 6, 10, 14, 15, 16. A question of interest is whether there is any treatment effect.

Patient	Week						
	1	**2**	**6**	**10**	**14**	**15**	**16**
1	0.22	0.00	1.03	0.67	0.75	0.65	0.59
2	0.18	0.00	0.96	0.96	0.98	1.03	0.70
3	0.73	0.37	1.18	0.76	1.07	0.80	1.10
4	0.30	0.25	0.74	1.10	1.48	0.39	0.36
5	0.54	0.42	1.33	1.32	1.30	0.74	0.56
6	0.16	0.30	1.27	1.06	1.39	0.63	0.40
7	0.30	1.09	1.17	0.90	1.17	0.75	0.88
8	0.70	1.30	1.80	1.80	1.60	1.23	0.41
9	0.31	0.54	1.24	0.56	0.77	0.28	0.40
10	1.40	1.40	1.64	1.28	1.12	0.66	0.77
11	0.60	0.80	1.02	1.28	1.16	1.01	0.67
12	0.73	0.50	1.08	1.26	1.17	0.91	0.87

399. Salsolinol excretion rates

Crowder, M.J. and Hand, D.J. (1990) *Analysis of repeated measures*. London: Chapman and Hall, 57.

Two groups of subjects, one with moderate and the other with severe dependence on alcohol, had their salsolinol excretion levels measured (in mmol) on four consecutive days. Primary interest focussed on whether the groups evolved differently over time. The raw data are skewed, so some kind of transformation will be needed.

	Day 1	Day 2	Day 3	Day 4
Group 1	0.33	0.70	2.33	3.20
	5.30	0.90	1.80	0.70
	2.50	2.10	1.12	1.01
	0.98	0.32	3.91	0.66
	0.39	0.69	0.73	2.45
	0.31	6.34	0.63	3.86
Group 2	0.64	0.70	1.00	1.40
	0.73	1.85	3.60	2.60
	0.70	4.20	7.30	5.40
	0.40	1.60	1.40	7.10
	2.60	1.30	0.70	0.70
	7.80	1.20	2.60	1.80
	1.90	1.30	4.40	2.80
	0.50	0.40	1.10	8.10

400. Response profiles to three drugs

Grizzle, J.E., Starmer, C.F. and Koch, G.G. (1969) Analysis of categorical data by linear models. *Biometrics*, **25**, 137–156.

Forty-six Subjects were each given three drugs, to each of which the response was either favourable (F) or unfavourable (U). Interest is in whether the three drugs cause similar responses.

Drug 1	Drug 2	Drug 3	Frequency
F	F	F	6
F	F	U	16
F	U	F	2
F	U	U	4
U	F	F	2
U	F	U	4
U	U	F	6
U	U	U	6

401. Patterns of psychotropic drug consumption

Murray, J.D., Dunn, G., Williams, P. and Tarnopolsky, A. (1981) Factors affecting the consumption of psychotropic drugs. *Psychological Medicine*, **11**, 551–60.

Data from a survey of West London was analysed to explore the factors influencing the pattern of consumption of psychotropic drugs. An extract from the data is given below. Interest lies in whether sex (coded as 0=male, 1=female), age, and score on the General Health Questionnaire (GHQ, 0=low, 1=high) influenced the probability of consumption and whether there was any relationship between these factors.

Sex	Age group	Mean age	GHQ	No. taking drugs	Total number
0	16-29	23.2	0	9	531
0	30-44	36.5	0	16	500
0	45-64	54.3	0	38	644
0	65-74	69.2	0	26	275
0	>74	79.5	0	9	90
0	16-29	23.2	1	12	171
0	30-44	36.5	1	16	125
0	45-64	54.3	1	31	121
0	65-74	69.2	1	16	56
0	>74	79.5	1	10	26
1	16-29	23.2	0	12	568
1	30-44	36.5	0	42	596
1	45-64	54.3	0	96	765
1	65-74	69.2	0	52	327
1	>74	79.5	0	30	179
1	16-29	23.2	1	33	210
1	30-44	36.5	1	47	189
1	45-64	54.3	1	71	242
1	65-74	69.2	1	45	98
1	>74	79.2	1	21	60

402. Wool data

Box, G.E.P. and Cox, D.R. (1964) An analysis of transformations (with discussion). *Journal of the Royal Statistical Society, Series A*, **143**, 383–430.

The data show the number of cycles to failure of samples of worsted yarn under cycles of repeated loading. There are three experimental conditions arranged in a $3 \times 3 \times 3$ factorial design:

x_1: length of test specimen (250, 300, 350 mm)
x_2: amplitude of loading cycle (8, 9, 10 mm)
x_3: load (40, 45, 50 g)

In the table these levels are denoted by −1, 0, +1 respectively.

These data can be used to illustrate analysis of variance and transformation of data. Box and Cox (1964) discuss log transformations of all variables, but since the the ranges of the predictors are not large one might consider only transforming the dependent variable.

Factor levels			Cycles to failure
x_1	x_2	x_3	
−1	−1	−1	674
−1	−1	0	370
−1	−1	+1	292
−1	0	−1	338
−1	0	0	266
−1	0	+1	210
−1	+1	−1	170
−1	+1	0	118
−1	+1	+1	90
0	−1	−1	1414
0	−1	0	1198
0	−1	+1	634
0	0	−1	1022
0	0	0	620
0	0	+1	438
0	+1	−1	442
0	+1	0	332
0	+1	+1	220
+1	−1	−1	3636
+1	−1	0	3184
+1	−1	+1	2000
+1	0	−1	1568
+1	0	0	1070
+1	0	+1	566
+1	+1	−1	1140
+1	+1	0	884
+1	+1	+1	360

403. Survival times of animals

Box, G.E.P. and Cox, D.R. (1964) An analysis of transformations (with discussion). *Journal of the Royal Statistical Society, Series A*, **143**, 383–430.

The data show survival times (in 10 hour units) of animals arranged in a 3 × 4 factorial experiment. The factors are poison (3 levels) and treatment (4 levels) and there are four replications.

These data can be used to illustrate analysis of variance. Box and Cox first transform the data, showing that a reciprocal transformation is reasonable.

		Treatment		
Poison	A	B	C	D
I	0.31	0.82	0.43	0.45
	0.45	1.10	0.45	0.71
	0.46	0.88	0.63	0.66
	0.43	0.72	0.76	0.62
II	0.36	0.92	0.44	0.56
	0.29	0.61	0.35	1.02
	0.40	0.49	0.31	0.71
	0.23	1.24	0.40	0.38
III	0.22	0.30	0.23	0.30
	0.21	0.37	0.25	0.36
	0.18	0.38	0.24	0.31
	0.23	0.29	0.22	0.33

404. Effectiveness of slimming clinics

Hand, D.J. and Taylor, C.C. (1987) *Multivariate analysis of variance and repeated measures*. London: Chapman and Hall, 150.

Slimming clinics aim to encourage people to lose weight by offering encouragement and support about dieting through regular meetings. The data here are extracted from a larger data set collected in a study to explore the effectiveness of such groups. Of particular interest was the question of whether adding a technical manual giving advice based on psychological behaviourist theory to the support offered would help the clients to control their diet. A comparison between two **conditions** was of particular interest, where condition 1 was the 'experimental' group (those given the manual) and condition 2 was the 'control' group (those without the manual). It was also thought important to distinguish between clients who had already been trying to slim and those who had not, so the subjects were selected in two **status** groups, group 1 being those who had been trying to slim for more than one year and group 2 being those who had been trying for not more than three weeks. The design is thus an unbalanced 2 × 2 factorial design.

The response variable was (weight at three months — ideal weight)/(Initial weight — ideal weight), expressed as a percentage.

These data can be used to illustrate analysis of variance.

Condition	Status	Response	Condition	Status	Response
1	1	−14.67	2	1	−3.39
1	1	−1.85	2	1	−4.00
1	1	−8.55	2	1	−2.31
1	1	−23.03	2	1	−3.60
1	1	11.61	2	1	−7.69
1	2	0.81	2	1	−13.92
1	2	2.38	2	1	−7.64
1	2	2.74	2	1	−7.59
1	2	3.36	2	1	−1.62
1	2	2.10	2	1	−12.21
1	2	−0.83	2	1	−8.85
1	2	−3.05	2	2	5.84
1	2	−5.98	2	2	1.71
1	2	−3.64	2	2	−4.10
1	2	−7.38	2	2	−5.19
1	2	−3.60	2	2	0.00
1	2	−0.94	2	2	−2.80

405. Prevalence of vertebral fractures

Cooper, C., Shah, S., Hand, D.J., Adams, J., Compston, J., Davie, M. and Woolf, A. (1991) Screening for vertebral osteoporosis using individual risk factors. *Osteoporosis International*, **2**, 48–53.

Osteoporosis is a major cause of ill health in postmenopausal women. In this study a consultant radiologist assessed 1005 women for evidence of vertebral fracture. Fractures were defined as a 20% or greater reduction in vertebral height and, separately, as a 15% or greater reduction in vertebral height. The results are shown in the table, classified by age.

Interest concerns the relationship between age and propensity to fracture.

Age	Total number	Number with vertebral fracture 20% criterion	15% criterion
<50	1	0	0
50-54	17	1	4
55-59	282	12	49
60-64	244	17	42
65-69	218	23	53
70-74	120	9	30
75-79	105	11	22
>79	18	5	7

406. Agreement between two examination markers

Dunn, G. (1989) *Design and analysis of reliability studies.* London: Edward Arnold, 23, Table 2.2.

These data show the grades awarded by two examiners to each of 29 candidates. The grades are on an ordinal scale from 0=very poor to 4=excellent.

Interest lies in the reliability of the scores.

Candidate	Examiner A	Examiner B
1	1	2
2	0	0
3	0	0
4	2	2
5	0	0
6	4	3
7	0	0
8	0	0
9	0	0
10	2	3
11	1	2
12	2	3
13	0	1
14	4	3
15	4	3
16	1	2
17	0	2
18	1	2
19	2	3
20	0	0
21	2	3
22	4	4
23	0	0
24	0	0
25	4	3
26	0	2
27	1	2
28	3	4
29	2	3

407. Subjective estimates of lengths of pieces of string

Dunn, G. (1989) *Design and analysis of reliability studies.* London: Edward Arnold, 6, Table 1.2.

These data show subjective estimates of the lengths (to the nearest 0.1 inch) of 15 pieces of string as assessed by three raters. Also given is the measured length of the string.

Interest lies in the reliability of these subjective assessments.

String	Measured length	Graham	Brian	David
1	6.3	5.0	4.8	6.0
2	4.1	3.2	3.1	3.5
3	5.1	3.6	3.8	4.5
4	5.0	4.5	4.1	4.3
5	5.7	4.0	5.2	5.0
6	3.3	2.5	2.8	2.6
7	1.3	1.7	1.4	1.6
8	5.8	4.8	4.2	5.5
9	2.8	2.4	2.0	2.1
10	6.7	5.2	5.3	6.0
11	1.5	1.2	1.1	1.2
12	2.1	1.8	1.6	1.8
13	4.6	3.4	4.1	3.9
14	7.6	6.0	6.3	6.5
15	2.5	2.2	1.6	2.0

408. Comparison of two kitchen scales

Dunn, G. (1989) *Design and analysis of reliability studies.* London: Edward Arnold, 112, Table 5.26.

These data show the weights (in grammes) of 19 items as measured on two kitchen scales. Interest lay in assessing the reliability of the two scales.

Item	Scale A	Scale B
1	300	320
2	190	190
3	80	90
4	20	50
5	200	220
6	550	550
7	400	410
8	610	600
9	740	760
10	1040	1080
11	920	940
12	1160	1180
13	1330	1330
14	1490	1510
15	1620	1590
16	1360	1330
17	1150	1140
18	1000	980
19	580	550

409. Performance in mathematical degrees by sex in England and Scotland

Cohen, G. and Fraser, E.J.P. (1992) Female participation in mathematical degrees at English and Scottish universities. *Journal of the Royal Statistical Society, Series A,* **155**, 241–258, Table 7.

The table shows the number of honours graduates and the percentage obtaining different classes of degree classified by sex and country.

Interest lay in whether females are put off studying mathematical subjects and whether the two countries differed in this regard.

Country	Gender	Number	I	Percentage with IIi	IIii	III
England	Male	4343	18	32	31	19
	Female	2143	16	30	37	18
Scotland	Male	368	27	28	30	14
	Female	261	18	29	34	19

410. Chemical process

Bissell, A.F. (1992) Lines through the origin — is NO INT the answer? *Journal of Applied Statistics,* **19**, 193–210, Table 4.

In a chemical process batches of liquid are passed through a bed containing an ingredient which is absorbed by the liquid. Normally about 6% to 6.5% (by weight) of the ingredient is absorbed but to be sure that there is enough the bed is supplied with about 7.5%. Since excess ingredient cannot be re-used there is concern that wastage and variation should be minimized. Interest lies in the relationship between quantity and percentage absorption. In the source paper the author discusses various models to fit these data.

In a simple bivariate plot of take-up against liquid there is evidence of heteroscedasticity.

Liquid (kg)	Take-up (kg)	Take-up (%)	Liquid (kg)	Take-up (kg)	Take-up (%)	Liquid (kg)	Take-up (kg)	Take-up (%)
310	14.0	4.52	330	17.1	5.18	370	21.3	5.76
490	27.2	5.55	400	20.4	5.10	450	27.4	6.09
580	31.9	5.50	560	32.5	5.80	520	28.4	5.46
650	34.1	5.25	650	39.8	6.12	650	38.5	5.92
810	50.4	6.22	800	43.8	5.48	760	50.4	6.63
1020	71.3	6.99	1020	64.3	6.30	910	53.5	5.88
1230	78.5	6.38	1200	80.8	6.73	1160	79.6	6.86
1490	98.6	6.62	1460	105.6	7.23	1380	98.9	7.17

411. Carcinoma of the uterine cervix

Agresti, A., Lipsitz, S., and Lang, B. (1992) Comparing marginal distributions of large, sparse contingency tables. *Computational Statistics and Data Analysis*, **14**, 55–73, Table 1.

Seven pathologists rating 118 slides made classifications on a 5-level ordinal scale regarding carcinoma in situ of the uterine cervix. The responses are:

1 = negative
2 = atypical squamous hyperplasia
3 = carcinoma in situ
4 = squamous carcinoma with early stromal invasion
5 = invasive carcinoma

The data comprise a large sparse contingency table. Interest lies in how one can model such a structure and in particular on testing marginal homogeneity.

Pathologist ratings A B C D E F G	Count	Pathologist ratings A B C D E F G	Count	Pathologist ratings A B C D E F G	Count
1 1 1 1 1 1 1	10	2 3 2 2 2 1 2	1	3 3 3 4 3 2 3	1
1 1 1 1 2 1 1	8	2 3 2 2 3 1 3	1	3 3 3 4 3 2 4	1
1 1 2 1 1 1 1	2	2 3 2 2 3 2 2	1	4 2 3 2 3 2 3	1
1 1 2 1 2 1 1	2	2 3 2 2 3 2 3	1	4 3 1 1 2 1 2	1
1 2 1 1 1 1 1	1	2 3 2 2 4 1 2	1	4 3 1 3 3 2 3	1
1 2 2 1 2 1 2	1	2 3 2 2 4 1 3	1	4 3 3 2 3 2 3	1
1 3 2 1 2 1 1	1	3 2 2 2 2 1 1	1	4 3 3 3 3 2 3	2
1 3 2 2 2 1 2	1	3 2 2 2 2 1 2	1	4 3 3 3 3 3 3	3
2 1 1 1 2 1 1	1	3 3 2 1 3 2 2	1	4 3 3 3 3 5 3	1
2 1 2 2 1 1 1	1	3 3 2 2 2 1 1	1	4 3 3 4 3 3 3	2
2 1 2 1 1 1 1	1	3 3 2 2 2 2 3	1	4 3 3 4 4 3 3	1
2 1 2 1 2 1 1	1	3 3 2 2 3 1 3	4	4 3 4 2 3 3 3	1
2 1 2 2 2 1 2	2	3 3 2 2 3 2 2	1	4 3 4 2 4 1 3	1
2 2 1 1 2 1 1	1	3 3 2 2 3 2 3	2	4 4 3 2 4 1 3	2
2 2 1 1 2 1 2	1	3 3 2 2 3 3 3	1	4 4 3 3 4 3 3	1
2 2 1 2 2 1 2	1	3 3 2 2 4 2 3	1	4 4 3 4 4 3 4	1
2 2 2 1 1 1 2	2	3 3 2 3 2 2 3	1	4 4 4 2 4 3 3	1
2 2 2 1 2 2 2	1	3 3 2 3 3 1 3	1	4 4 4 2 5 1 3	1
2 2 2 2 3 1 2	2	3 3 2 3 3 3 3	2	4 4 4 3 3 3 3	1
2 3 1 1 2 1 1	1	3 3 3 2 3 1 3	2	5 3 3 2 3 2 3	1
2 3 1 1 2 1 2	1	3 3 3 2 3 2 3	3	5 3 3 3 4 1 3	1
2 3 1 1 3 1 1	1	3 3 3 2 3 3 3	1	5 3 4 2 3 4 3	1
2 3 1 2 3 1 3	1	3 3 3 2 4 2 3	2	5 5 1 4 5 5 4	1
2 3 2 1 3 2 2	1	3 3 3 2 4 3 3	1	5 5 5 4 5 5 5	1
2 3 2 2 2 1 3	1	3 3 3 3 3 2 3	5	5 5 5 5 5 5 5	1
2 3 2 2 2 2 2	1	3 3 3 3 3 3 3	4		

412 Rainfall in Minneapolis/St Paul

Hinkley, D. (1977) On quick choice of power transformation. *Applied Statistics*, **26**, 67–69.

The table shows thirty successive values of March precipitation (in inches) for Minneapolis/St Paul.

They were used to illustrate the application of power transformations to symmetrize the data. The data are to be read across rows.

0.77	1.74	0.81	1.20	1.95	1.20	0.47	1.43
3.37	2.20	3.00	3.09	1.51	2.10	0.52	1.62
1.31	0.32	0.59	0.81	2.81	1.87	1.18	1.35
4.75	2.48	0.96	1.89	0.90	2.05		

413. Book condition by strength of paper

Simonoff, J.S. and Tsai, C-L. (1991) Higher order effects in log-linear models and log-non-linear models for contingency tables with ordered categories. *Applied Statistics*, **40**, 449–458.

The data arose during a study by the New York Public Library on book deterioration in 1983. They show a cross-classification of condition of book by strength of book, the latter being measured by the number of folds that the paper of the book can stand before breaking.

Interest lies in finding an appropriate model to fit the data. Is, for example, an interaction term necessary?

Strength	Intact	Slightly deteriorated	Moderately deteriorated	Extremely deteriorated
1 fold	181	14	18	43
2-4 folds	140	6	1	15
5-15 folds	44	2	0	0
>15 folds	369	7	0	0

414. Oxford and Cambridge boat race crews

The Independent, 31st March, 1992

The figures below give the weights of the members of the crews of the Oxford and Cambridge boat race in 1992, in stones and pounds. The Cambridge crew have the heavier mean, but how do the distributions compare overall? Note that the cox in each crew is only about half the weight of the other crew members.

Cambridge		Oxford	
13st	6.5lb	13	4
13	1	13	2.5
13	12.5	14	8
13	3	13	2.5
15	4	13	13.5
14	7.5	14	6.5
13	4	12	6
12	10.5	13	1
7	11	7	11.5

415. Comparisons between 48 major cities in 1991

Prices and earnings around the globe (1991) Economic Research Department, Union Bank of Switzerland, Zurich.

Working hours = weighted average of 12 different occupations.

Price level = cost of a basket of 112 goods and services, weighted by consumer habits, excluding rent (Zurich=100).

Salary level = levels of hourly earnings in 12 different occupations, weighted according to occupational distributions, net after deducting tax and social insurance contributions (Zurich=100).

City	Working hours	Price level	Salary level
Amsterdam	1714	65.6	49.0
Athens	1792	53.8	30.4
Bogota	2152	37.9	11.5
Bombay	2052	30.3	5.3
Brussels	1708	73.8	50.5

Buenos Aires	1971	56.1	12.5
Cairo	—	37.1	—
Caracas	2041	61.0	10.9
Chicago	1924	73.9	61.9
Copenhagen	1717	91.3	62.9
Dublin	1759	76.0	41.4
Dusseldorf	1693	78.5	60.2
Frankfurt	1650	74.5	60.4
Geneva	1880	95.9	90.3
Helsinki	1667	113.6	66.6
Hong Kong	2375	63.8	27.8
Houston	1978	71.9	46.3
Jakarta	—	43.6	—
Johannesburg	1945	51.1	24.0
Kuala Lumpur	2167	43.5	9.9
Lagos	1786	45.2	2.7
Lisbon	1742	56.2	18.8
London	1737	84.2	46.2
Los Angeles	2068	79.8	65.2
Luxembourg	1768	71.1	71.1
Madrid	1710	93.8	50.0
Manila	2268	40.0	4.0
Mexico City	1944	49.8	5.7
Milan	1773	82.0	53.3
Montreal	1827	72.7	56.3
Nairobi	1958	45.0	5.8
New York	1942	83.3	65.8
Nicosia	1825	47.9	28.3
Oslo	1583	115.5	63.7
Panama	2078	49.2	13.8
Paris	1744	81.6	45.9
Rio de Janeiro	1749	46.3	10.5
Sao Paulo	1856	48.9	11.1
Seoul	1842	58.3	32.7
Singapore	2042	64.4	16.1
Stockholm	1805	111.3	39.2
Sydney	1668	70.8	52.1
Taipei	2145	84.3	34.5
Tel Aviv	2015	67.3	27.0
Tokyo	1880	115.0	68.0
Toronto	1888	70.2	58.2
Vienna	1780	78.0	51.3
Zurich	1868	100.0	100.0

416. Height and resting pulse measurements for a sample of hospital patients

The data show the heights (in cm) and resting pulses (beats per minute) for a sample of hospital patients. One question of possible interest is whether or not there is a relationship between height and pulse rate.

Height	Pulse	Height	Pulse	Height	Pulse
160	68	170	80	168	90
167	80	148	82	178	80
162	84	175	76	182	76
175	80	160	84	167	80
185	80	153	70	170	84
162	80	185	80	160	80
173	92	165	82	182	80
167	92	165	84	168	80
170	80	172	116	155	80
170	80	185	80	175	104
163	80	163	95	168	80
158	80	177	80	180	68
157	80	165	76	175	84
160	78	182	100	145	64
170	90	162	88	170	84
177	80	172	90	175	72
166	72	177	90		

417. Coal miners' breathing difficulties

Kullback, S. and Fisher, M. (1973) Partitioning second-order interaction in three-way contingency tables. *Applied Statistics*, **22**, 172–184, Table 1.

The table shows coal miners without radiological pneumoconiosis classified according to whether or not they had breathlessness and wheeze and by age group. Particular interest here focussed on whether or not age could be partitioned into two subgroups, in each of which simplifying models could be fitted.

	Breathlessness			
	yes		**no**	
	Wheeze		**Wheeze**	
Age	**yes**	**no**	**yes**	**no**
20-24	9	7	95	1841
25-29	23	9	105	1654
30-34	4	19	177	863
35-39	121	48	257	2357
40-44	169	54	273	1778
45-49	269	88	324	1712
50-54	404	117	245	1324
55-59	406	152	225	967
60-64	372	106	132	526

418. Toxicity of potassium cyanate on trout eggs

O'Hara Hines, R.J. and Carter, E.M. (1993) Improved added variable and partial residual plots for the detection of influential observations in generalized linear models. *Applied Statistics*, **42**, 3–20, Table 2.

The toxicant potassium cyanate was applied in six different concentrations to vials of trout fish eggs. For half of the vials toxicant was applied immediately after fertilization, while for the other half the eggs were allowed to water-harden for several hours after fertilization and before the toxicant was applied. The number of eggs in each vial and the number of dead eggs at day 19 was counted. Particular interest lay in modelling the relationship between mortality and toxicant concentration.

Conc. (mg/l)	Water hardening		No water hardening	
	Total number	Number dead	Total number	Number dead
90	111	8	130	7
	97	10	179	25
	108	10	126	5
	122	9	129	3
180	68	4	114	12
	109	6	149	4
	109	11	121	4
	118	6	105	0
360	98	6	102	4
	110	5	145	21
	129	9	61	1
	103	17	118	3
720	83	2	99	29
	87	3	109	53
	118	16	99	40
	100	9	70	0
1440	140	60	100	14
	114	47	127	10
	103	49	132	8
	110	20	113	3
2880	143	79	145	113
	131	85	103	84
	111	78	143	105
	111	74	102	78

419. Asthma death, corticosteroid use, and Fenoterol use

Walker, A.M. and Lanes, S.F. (1991) Misclassification of covariates. *Statistics in Medicine*, **10**, 1181–1196, Table 1.

The table shows death/hospitalization cross-classified by Fenoterol use divided according to the use/no use of corticosteroids. These data can be used to illustrate log-linear models and odds ratio calculations. The authors of the source paper were in fact interested in whether the overall association between death and Fenoterol use was induced by confounding with severity of the asthma. Steroid use was used as an indicator of severity and their interest was whether the heterogeneity between steroid strata could have arisen because steroid use is an imperfect measure of the risk of death in asthma.

	No corticosteroids		Corticosteroids	
	Fenoterol	No Fenoterol	Fenoterol	No Fenoterol
Deaths	34	50	26	7
Hospitalized	151	213	38	66

420. Seed germination

Smith, P.J. and Heitjan, D.F. (1993) Testing and adjusting for departures from nominal dispersion in generalized linear models. *Applied Statistics*, **42**, 31–41, Table 1.

The table shows the number of seeds germinating in a factorial experiment in which two types of seed and two root extracts (bean and cucumber) were compared. Particular interest here was on modelling overdispersion.

O. aegyptiaca 73				*O. aegypticiaca* 75			
Bean		Cucumber		Bean		Cucumber	
y	m	y	m	y	m	y	m
10	39	5	6	8	16	3	12
23	62	53	74	10	30	22	41
23	81	55	72	8	28	15	30
26	51	32	51	23	45	32	51
17	39	46	79	0	4	3	7
		10	13				

421. Depression and friendship amongst children

Goodyer, I., Germany, E., Gowrusankur, J. and Altham, P. (1991) Social influences on the course of anxious and depressive disorders in school-age children. *The British Journal of Psychiatry*, **158**, 676–684, Table 4.

These data are taken from a study of the influence of social factors on childhood anxiety and depression. Below are shown the proportion of children with good or moderate/poor friendships diagnosed as anxious or depressed at the time of presentation of disorder and at the follow-up rated as recovered or not recovered. Interest lies in whether the presence of good friendships is protective.

	Diagnosis at presentation			
	Anxious		Depressed	
	Friendships			
	Poor/ mod.	Good	Poor/ mod.	Good
Children's report				
recovered	0	18	1	11
not recovered	0	3	7	9
Mother's report				
recovered	0	13	0	15
not recovered	0	8	8	5
Psychiatrist's report				
recovered	0	9	1	13
not recovered	0	12	7	7

422. Social class of fathers and sons

Payne, A.C. (1992) *Confounding variables and selection effects in a follow-up study.* Southampton University MSc dissertation, Table 5.9, Faculty of Mathematical Studies.

The table shows the social class of a sample of 845 Hertfordshire men, cross-classified by the social class of their father. In this table 1 is 'low' social class and 9 is 'high'. Interest lies in change in social class over the generations.

		Son's social class								
		1	2	3	4	5	6	7	8	9
Father's	1	1	0	3	7	2	1	0	0	0
social	2	4	20	9	22	16	2	0	0	0
class	3	4	12	5	7	5	2	0	0	0
	4	12	54	32	143	50	10	2	0	2
	5	5	29	34	116	67	22	4	0	2
	6	0	8	8	50	18	12	0	0	0
	7	0	2	0	1	1	1	1	0	0
	8	3	1	0	3	3	1	0	0	0
	9	1	1	6	10	7	2	1	0	0

423. Survival data for patients with chronic active hepatitis

Pocock, S.J. (1986) *Clinical trials: a practical approach.* John Wiley & Sons, Table 14.6.

Original source: Kirk, A.P., Jain, S., Pocock, S. *et al.* (1980) Late results of the Royal Free Hospital prospective controlled trial of prednisolone therapy in hepatitis B surface antigen negative chronic active hepatitis, *Gut,* **21**, 78–83.

These data show survival times (in months) for patients suffering from chronic active hepatitis. Times are given for a control group and a group being treated with prednisolone. D=Dead, A=still alive (censored).

Control survival times	Prednisolone survival times
2 D	2 D
3 D	6 D
4 D	12 D
7 D	54 D
10 D	56 A
22 D	68 D
28 D	89 D
29 D	96 D
32 D	96 D
37 D	125 A
40 D	128 A
41 D	131 A
54 D	140 A
61 D	141 A
63 D	143 D
71 D	145 A
127 A	146 D
140 A	148 A
146 A	162 A
158 A	168 D
167 A	173 A
182 A	181 A

424. Survival times of leukaemia patients

Feigl, P. and Zelen, M. (1965) Estimation of exponential survival probabilities with concomitant information. *Biometrics,* **21**, 826–38.

The table shows the time to death (in weeks) and the white blood count (WBC) for two groups of leukaemia patients. Interest lies in comparing the groups taking into account differences arising from the explanatory variable. Note also that the groups have not been created by random allocation but have arisen naturally.

AG positive		AG negative	
WBC	Time to death	WBC	Time to death
2300	65	4400	56
750	156	3000	65
4300	100	4000	17
2600	134	1500	7
6000	16	9000	16
10500	108	5300	22
10000	121	10000	3
17000	4	19000	4
5400	39	27000	2
7000	143	28000	3
9400	56	31000	8
32000	26	26000	4
35000	22	21000	3
100000	1	79000	30
100000	1	100000	4
52000	5	100000	43
100000	65		

425. Dissimilarity ratings of Second World War politicians

Everitt, B.S. (1987) *Introduction to optimization methods and their application in statistics*, London: Chapman and Hall, Table 6.7.

Two subjects assessed the degree of dissimilarity between 12 World War II politicians. The lower triangle below contains the assessments of one subject and the upper triangle that of the other. Interest lies in exploring the structure underlying the dissimilarities — for example, can they be explained by a small number of underlying factors?

		1	2	3	4	5	6	7	8	9	10	11	12
1	Hitler	*	2	7	8	5	9	2	6	8	8	8	9
2	Mussolini	3	*	8	8	8	9	1	7	9	9	9	9
3	Churchill	4	6	*	3	5	8	7	2	8	3	5	6
4	Eisenhower	7	8	4	*	8	7	7	3	8	2	3	8
5	Stalin	3	5	6	8	*	7	7	5	6	7	9	5
6	Attlee	8	9	3	9	8	*	9	7	7	4	7	5
7	Franco	3	2	5	7	6	7	*	5	9	8	8	9
8	De Gaulle	4	4	3	5	6	5	4	*	6	5	6	5
9	Mao Tse Tung	8	9	8	9	6	9	8	7	*	8	8	6
10	Truman	9	9	5	4	7	8	8	4	4	*	4	6
11	Chamberlain	4	5	5	4	7	2	2	5	9	5	*	8
12	Tito	7	8	2	4	7	8	3	2	4	5	7	*

426. Rate of formation of a chemical impurity

Box, G.E.P., Hunter, W.G. and Hunter, J.S. (1978) *Statistics for experimenters*, New York: John Wiley & Sons, Table 14.5.

The mean value of the initial rate of formation of a particular chemical impurity which caused discoloration was thought to be linearly related to two other variables: the concentration of monomer and the concentration of dimer. The rate was thought to be zero when both the concentrations were zero. The table below shows the results of measurements from 6 experiments.

Monomer concentration	Dimer concentration	Initial rate of formation of impurity
0.34	0.73	5.75
0.34	0.73	4.79
0.58	0.69	5.44
1.26	0.97	9.09
1.26	0.97	8.59
1.82	0.46	5.09

427. Road distances between major UK towns

Chapman & Hall 1993 diary.

These data show the distances, in miles, between 23 major UK towns. They can be used as data for a two-dimensional multidimensional scaling exercise to see how closely the geographic positions of the towns can be reconstructed.

The key to the towns is: A=Aberdeen, B=Birmingham, C=Brighton, D=Bristol, E=Cardiff, F=Carlisle, G=Dover, H=Edinburgh, I=Fort William, J=Glasgow, K=Holyhead, L=Hull, M=Inverness, N=Leeds, O=Liverpool, P=London, Q=Manchester, R=Newcastle, S=Norwich, T=Nottingham, U=Penzance, V=Plymouth, W=Sheffield.

427. Road distances between major UK towns

	A	B	C	D	E	F	G	H	I	J	K	L	M	N	O	P	Q	R	S	T	U	V
B	431																					
C	611	185																				
D	515	88	170																			
E	535	108	205	47																		
F	232	198	378	282	302																	
G	595	206	78	210	245	398																
H	126	298	478	381	402	99	466															
I	159	407	587	491	511	209	608	133														
J	146	297	477	381	401	98	497	46	102													
K	461	153	333	237	205	228	353	328	438	327												
L	360	135	283	233	253	173	264	231	382	272	223												
M	106	456	636	540	560	258	657	158	65	172	488	431										
N	335	115	264	220	240	123	271	206	333	222	168	61	382									
O	358	102	282	185	205	125	302	225	335	224	105	130	383	75								
P	546	120	60	120	155	313	79	413	523	412	268	190	571	199	216							
Q	352	89	269	172	193	119	289	219	329	218	125	99	378	44	35	204						
R	237	202	350	300	320	57	352	108	241	152	268	142	268	93	175	285	144					
S	497	176	169	233	268	285	171	368	494	384	308	151	543	173	242	115	185	254				
T	402	55	196	146	166	190	217	273	400	289	177	93	449	73	109	131	70	160	120			
U	701	274	287	196	234	468	364	568	678	567	423	419	726	406	370	312	359	486	425	331		
V	630	203	216	125	162	397	292	496	607	496	352	348	655	335	300	241	287	414	354	260	78	
W	376	86	234	184	204	164	255	247	374	263	159	67	422	35	79	169	37	134	148	44	370	299

428. Penicillin manufacture

Box, G.E.P., Hunter, W.G. and Hunter, J.S. (1978) *Statistics for experimenters*, New York: John Wiley & Sons, Table 7.1.

Four methods, denoted A, B, C, and D, for manufacturing penicillin were compared in a randomized block experiment. The blocks were 'blends' containing sufficient material for four runs. Interest lies in seeing if there is a difference between the methods and, if so, which is the best. The yields are shown in the first column of values under each treatment and the order in the block is given in the second column of values.

| | | Treatment | | |
	A	B	C	D
Block				
1	89 1	88 3	97 2	94 4
2	84 4	77 2	92 3	79 1
3	81 2	87 1	87 4	85 3
4	87 1	92 3	89 2	84 4
5	79 3	81 4	80 1	88 2

429. Early detection of autism

Baron-Cohen, S., Allen, J. and Gillberg, C. (1992) Can autism be detected at 18 months? *British Journal of Psychiatry*, **161**, 839–843.

Currently autism is only detected at about three years of age and these data arose from a study to see if detection at 18 months was possible. As part of the study two groups of subjects were tested.

Group 1 consisted of 50 randomly selected 18 month old children attending a London health centre for their routine 18 month check-up. Group 2 consisted of 41 younger siblings of children with autism.

Both groups were tested using the 'Checklist for Autism in Toddlers'. This has two sections, section A assesses nine areas of development and section B compares aspects of the child's actual play behaviour with that reported by the parent. The table shows the percentages of each group 'passing' each item on the checklist. Interest, of course, lies in whether one can distinguish between the groups using these data.

	Group 1 (n=50)	Group 2 (n=41)
Section A questions		
1	90	92.7
2	94	97.5
3	100	95.0
4	100	95.1
5	86	82.9
6	98	87.8
7	92	87.8
8	100	100.0
9	94	92.7
Section B items		
i	100	96.8
ii	98	90.3
iii	82	74.2
iv	88	80.6

430. Adolescent attempted suicide rates

Hawton, K. and Fagg, J. (1992) Deliberate self-poisoning and self-injury in adolescents. *British Journal of Psychiatry*, **161**, 816–823, Table 1.

These data arose in a study of trends in attempted suicide referrals to the general hospital in Oxford of 10-19 year old pateints during 1976-1989. Interest lies in trends over time and differences between the two sexes.

	Females		**Males**	
Age (years)	**Persons**	**Episodes**	**Persons**	**Episodes**
10	2	2	0	0
11	4	4	2	2
12	21	22	3	3
13	61	64	11	11
14	158	174	26	27
15	232	254	47	52
16	271	351	75	90
17	299	375	119	141
18	304	383	147	173
19	310	386	190	227

431. Lung cancer cases and person-years experience by age and period of diagnosis

Holford, T.R. (1992) Analysing the temporal effects of age, period, and cohort. *Statistical Methods in Medical Research*, **1**, 317–337, Table 1.

These data were presented in a study discussing the problems of interpreting results from statistical models that incorporate time effects. Table (a) shows the number of lung cancer cases classified by age and period of diagnosis in Connecticut. Table (b) shows corresponding person-years experience. The author of this study points out that the categorization introduces some ambiguity: 'For example, someone aged 20-29 when diagnosed in 1935 will have been born some time from 1905 to 1914; and those aged 20-29 in 1944 were born from 1915 to 1924. Hence, the incidence rate in the first row and first column of the table refers to individuals born during the 20-year span from 1905 to 1924, although the length of time someone might contribute to this rate would vary from ten years if born in the middle of the decade, to less than a year if born early or late in the decade.'

The study used these data to discuss construction of age-period-cohort models.

(a) Number of cases

Males

Age	1935-44	1945-54	1955-64	1965-74	1975-84
20-9	1	3	4	6	7
30-9	10	20	28	31	40
40-9	70	115	195	289	281
50-9	247	543	885	1300	1418
60-9	395	1057	1992	2780	3769
70-9	209	790	2001	3017	4354
80-9	60	231	673	1453	2270

Females

Age	1935-44	1945-54	1955-64	1965-74	1975-84
20-9	4	2	2	6	7
30-9	6	9	14	21	28
40-9	28	32	89	154	241
50-9	52	75	189	512	799
60-9	69	118	223	706	1764
70-9	68	142	289	587	1832
80-9	32	73	151	309	868

(b) Person-years experience

Males

Age	1935-44	1945-54	1955-64	1965-74	1975-84
20-9	1537781	1380360	1555934	2322128	2769374
30-9	1406807	1615355	1632000	1863489	2343684
40-9	1258708	1493910	1800315	1727315	1800233
50-9	1143763	1232189	1514848	1789483	1660060
60-9	770224	980496	1095932	1336181	1561113
70-9	437017	567892	723242	756609	941025
80-9	145147	207148	273417	345493	389562

Females

Age	1935-44	1945-54	1955-64	1965-74	1975-84
20-9	1562465	1411502	1618906	2412148	2757255
30-9	1458565	1704171	1699896	1935928	2439026
40-9	1288044	1536560	1879937	1813764	1906093
50-9	1108337	1256742	1559464	1895399	1769078
60-9	769207	995569	1184060	1464623	1735186
70-9	487210	641265	865618	1004712	1232792
80-9	183561	273824	383557	555759	709808

432. Yields of wheat

Graybill, F. (1954) Variance heterogeneity in a randomised block experiment. *Biometrics*, **10**, 516–520.

Reproduced in Mudholkar, G.S. and Sarkar, I.C. (1992) Testing homoscedasticity in a two-way table. *Biometrics*, **48**, 883–888, Table 1.

These data were collected at the Oklahoma Agricultural Experimental Station and constitute a randomized block experiment with 13 blocks (the rows of the table) and 4 varieties of wheat (the columns of the table). Particular interest lay in exploring homoscedasticity between varieties.

Yields of wheat (cwt/acre)

Location	Variety			
	1	2	3	4
1	43.60	24.05	19.47	19.41
2	40.40	21.76	16.61	23.84
3	18.08	14.19	16.60	16.08

4	19.57	18.61	17.78	18.29
5	45.20	29.33	20.19	30.08
6	25.87	25.60	23.31	27.04
7	55.20	38.77	21.15	39.95
8	55.32	34.19	18.56	25.12
9	19.79	21.65	23.31	22.45
10	46.24	31.52	22.48	29.28
11	14.88	15.68	19.79	22.56
12	7.52	4.69	20.53	22.08
13	41.17	32.59	29.25	43.95

433. Estimating numbers of unobserved species

Heltshe, J.F. and Forrester, N.E. (1983) Estimating species richness using the jackknife procedure. *Biometrics*, **39**, 1–11, Table 5.

One approach to estimating the number of distinct species in a region is to divide the region up into quadrats and count the species occurrences within each quadrat. The table shows the species occurrence in 10 quadrats in a subtidal marsh creek in the Pettaquamscutt River in southern Rhode Island, and was collected by Jeffrey Hyland of the Graduate School of Oceanography of the University of Rhode Island in April 1973. The data are also discussed in Mingoti S.A. and Meeden G. (1992) Estimating the total number of distinct species using presence and absence data. *Biometrics*, **48**, 863–875.

| | | | | Quadrat number | | | | | |
Species	1	2	3	4	5	6	7	8	9	10
Streblospio benedict		13	21	14	5	22	13	4	4	27
Nereis succines	2	2	4	4	1	1	1		1	6
Polydora ligni		1						1		
Scoloplos robustus	1		1	2		6			1	2
Eteone heterpoda			1	2			1			1
Heteromastus filiformis	1	1	2	1		1			1	5
Capitella capitata	1									
Scolecolepides viridis	2									
Hypaniola grayi		1								
Branis clavata			1							
Macoma balthica			3							2
Ampelisca abdita			5	1		2				3
Neopanope texana								1		
Tubifocodies sp.	8	36	14	19	3	22	6	8	5	41

434. Distance and direction travelled by blue periwinkles

Fisher, N.I. and Lee, A.J. (1992) Regression models for an angular response. *Biometrics*, **48**, 665–677, Table 1.

The table shows measurements of distance (cm) and direction (angular degrees) travelled by 31 blue periwinkles after they had been transplanted downshore from the height at which they normally lived. Interest in the study focussed particularly on constructing a regression model for the dependence of the mean and dispersion of direction moved on distance travelled.

Case	Distance	Direction	Case	Distance	Direction
1	107	67	17	21	165
2	46	66	18	1	133
3	33	74	19	71	101
4	67	61	20	60	105
5	122	58	21	71	71
6	69	60	22	71	84
7	43	100	23	57	75
8	30	89	24	53	98
9	12	171	25	38	83
10	25	166	26	70	71
11	37	98	27	7	74
12	69	60	28	48	91
13	5	197	29	7	38
14	83	98	30	21	200
15	68	86	31	27	56
16	38	123			

435. Birth season and psychosis

Fombonne, E. (1989) Season of birth and childhood psychosis. *British Journal of Psychiatry*, **155**, 655–661, Table 1.

The study in which these data are described was concerned with investigating the hypothesis that the season of birth influences the probability of childhood psychosis. For the study all children admitted to a particular child psychiatry facility in Paris with the diagnosis of childhood psychosis were included as 'cases'. A sample of controls was selected from the records of children who had attended the out-patient clinic during the same period. These controls were matched to the cases by sex and year of birth and were within two years of admission of the cases. Figures for the monthly birth distribution for France are also given in the paper. Interest centres on whether the risk of child psychosis varies between seasons.

	Cases	General population	Controls
January	13	0.84	83
February	12	0.78	71
March	16	0.87	88
April	18	0.86	114
May	21	0.91	86
June	18	0.85	93
July	15	0.87	87
August	14	0.83	70
September	13	0.81	83
October	19	0.81	80
November	21	0.76	97
December	28	0.80	88

436. Psychotropic drug use by sex, age, and physical illness

Vazquez-Barquero, J.L., Manrique, J.F.D., Pena, C., Gonzalez, A.A., Cuesta, M.J. and Artal, J.A. (1989) Patterns of psychotropic drug use in a Spanish rural community. *British Journal of Psychiatry*, **155**, 633–641, Table VI.

The table shows the cross classification of the respondents to a survey according to psychotropic drug use, age, sex, and whether or not they had a physical illness. The primary question is whether there is a relationship between psychotropic drug use and physical illness. Note that the numbers taking psychotropic drugs are very small.

There is an ambiguity in the tables given in the paper: we believe that the < and > symbols should include equality.

	Physically ill		Not physically ill	
	Total number	Number taking drugs	Total number	Number taking drugs
Males				
<34	27	1	152	1
35-54	47	1	138	2
>55	127	9	90	1
Females				
<34	42	4	140	1
35-54	116	26	109	1
>55	164	32	71	5

437. Mouse lymphoma mutation

Snee, R.D. (1986) An alternative approach to fitting models when re-expression of the response is useful. *Journal of Quality Technology*, **18**, 211–225.

These data are also analysed in Hinkley, D.V. (1989) Modified profile likelihood in transformed linear models. *Applied Statistics*, **38**, 495–506.

The table shows results from a bioassay arranged as a 2 × 6 design, with two replicates in each cell. The first factor is simply the trial group, the second factor is dose (in micrograms per millilitre), and the response is mutation frequency per million survivors. The objective is to find a good model for these data.

| | Response | |
Dose	Group 1	Group 2
0	35, 38	21, 18
12	52, 45	42, 23
25	55, 52	58, 31
50	120, 77	57, 93
100	183, 158	160, 150
200	395, 299	170, 272

438. Stopping distance

Snee, R.D. (1986) An alternative approach to fitting models when re-expression of the response is useful. *Journal of Quality Technology*, **18**, 211–225.

The table shows the stopping distance (feet) for cars travelling at the indicated speeds (miles per hour). There are replicate values for some speeds. Interest lies in finding a model for these data.

Speed	Stopping distance	Speed	Stopping distance
4	4	21	55,39,42
5	2,8,8,4	22	35
7	6,7	24	56
8	9,8,13,11	25	33,59,48,56
9	5,5,13	26	39,41
10	8,17,14	27	78,57
12	11,21,19	28	64,84
13	18,27,15	29	68,54
14	14,16	30	60,101,67
15	16	31	77
16	19,14,34	35	85,107
17	29,22	36	79
18	47,29,34	39	138
19	30	40	110,134
20	48		

439. Decontaminants for *M. bovis*

Trajstman, A.C. (1989) Indices for comparing decontaminants when data come from dose-response survival and contamination experiments. *Applied Statistics*, **38**, 481–494, Table 1.

The aim of this experiment was to compare the effectiveness of two decontaminants used in the primary isolation of *M. bovis*. Each of the two decontaminants (1-hexadecylpyridium chloride, HPC for short, and oxalic acid) was tested at several doses, and at each dose 10 plates were used. Control data, with no decontaminant, were also collected on 20 plates. These data are needed to estimate the initial numbers of colony forming units set down on each plate. The numbers in the body of the table give counts of colonies on each plate at stationarity.

Dose (% weight/volume)	Number of colonies									
HPC										
0.75	2	4	8	9	10	1	0	5	14	7
0.375	11	12	13	12	11	13	17	16	21	2
0.1875	16	6	20	23	23	39	18	23	33	21
0.09375	33	46	42	18	35	20	19	29	41	36
0.075	30	30	27	53	51	39	31	36	38	22
0.0075	53	62	38	54	54	38	46	58	54	57
0.00075	3	42	45	49	32	39	40	34	45	51
Oxalic acid										
5	14	15	6	13	4	1	9	6	12	13
0.5	27	33	31	30	26	41	33	40	31	20
0.05	33	26	32	24	30	52	28	28	26	22
0.005	36	-	54	31	37	50	73	44	50	37
Control experiments										
	52	80	55	50	58	50	43	50	53	54
	44	51	34	37	46	56	64	51	67	40

440. Breast cancer and ambulatory status over time

De Stavola, B.L. (1988) Testing departures from time homogeneity in multistate markov processes. *Applied Statistics*, **37**, 242–250, Table 1.

The table shows ambulatory status before treatment and at 3, 6, 12, 24, and 60 months after treatment of 37 breast cancer patients treated for spinal metastases at the London Hospital. The codes in the table are 0=dead, 1=inability to walk, 2=ability to walk, and NK=alive but status unknown. Interest lies in producing a model for change over time.

Patient	Initial status	Follow up times (months)					
		0	3	6	12	24	60
1	2	2	2	2	2	2	0
2	2	2	2	2	0		
3	1	1	0				
4	1	2	1	2	2	2	0
5	2	2	2	2	2	1	0
6	2	2	2	2	0		
7	2	2	1	2	2	0	
8	1	2	1	0			
9	2	2	2	2	2		
10	1	1	0				
11	1	1	0				
12	2	2	2	2	0		
13	1	1	0				
14	1	2	2	2	0		
15	1	1					
16	1	1	0				
17	2	2	2	2	0		
18	1	1	1	0			
19	2	2	0				
20	2	2	1	1	1	0	
21	1	1	0				
22	1	2	2	2	2	2	
23	1	1	0				
24	2	2	1	NK	0		
25	2	2	2	2	2	0	
26	1	1	1	1	1	0	
27	1	2	1	1	1	0	
28	1	2	2	2	2	0	
29	1	2	1	1	0		
30	1	2	2	1	0		
31	1	2	1	1	1		
32	2	2	2	1	1	0	
33	1	2	1	1	0		
34	2	2	2	2	0		
35	2	2	2	0			
36	1	1	0				
37	1	1	0				

441. Numbers of icebergs sighted

Shaw, N. (1942) *Manual of meteorology*, Vol. 2, London: Cambridge University Press, 7.

Reprinted in Mosteller, F. and Tukey, J.W. (1977) *Data analysis and regression.* Reading, Massachusetts: Addison-Wesley. Exhibit 1, 519.

The table shows the number of icebergs sighted monthly south of Newfoundland and south of the Grand Banks in 1920.

	Month											
	J	**F**	**M**	**A**	**M**	**J**	**J**	**A**	**S**	**O**	**N**	**D**
Newfoundland	3	10	36	83	130	68	25	13	9	4	3	2
Grand Banks	0	1	4	9	18	13	3	2	1	0	0	0

442. Ratings of synchronized swimming

Fligner, M.A. and Verducci, J.S. (1988) A nonparametric test for judges' bias in an athletic competition. *Applied Statistics*, **37**, 101–110, Table 1.

The table shows the total scores assigned by each of 5 judges to each of 40 competitors in a synchronized swimming event at the 1986 National Olympic Festival in Houston, Texas. Interest lies in assessing the reliability of the judges' ratings.

	Judge				
Contestant	**1**	**2**	**3**	**4**	**5**
1	33.1	32.0	31.2	31.2	31.4
2	26.2	29.2	28.4	27.3	25.3
3	31.2	30.1	30.1	31.2	29.2
4	27.0	27.9	27.3	24.7	28.1
5	28.4	25.3	25.6	26.7	26.2
6	28.1	28.1	28.1	32.0	28.4
7	27.0	28.1	28.1	28.1	27.0
8	25.1	27.3	26.2	27.5	27.3
9	31.2	29.2	31.2	32.0	30.1
10	30.1	30.1	28.1	28.6	30.1
11	29.0	28.1	29.2	29.0	27.0
12	27.0	27.0	27.3	26.4	25.3
13	31.2	33.1	31.2	30.3	29.2

14	32.3	31.2	32.3	31.2	31.2
15	29.5	28.4	30.3	30.3	28.4
16	29.2	29.2	29.2	30.9	28.1
17	32.3	31.2	29.2	29.5	31.2
18	27.3	30.1	29.2	29.2	29.2
19	26.4	27.3	27.3	28.1	26.4
20	27.3	26.7	26.4	26.4	26.4
21	27.3	28.1	28.4	27.5	26.4
22	29.5	28.1	27.3	28.4	26.4
23	28.4	29.5	28.4	28.6	27.5
24	31.2	29.5	29.2	31.2	27.3
25	30.1	31.2	28.1	31.2	29.2
26	31.2	31.2	31.2	31.2	30.3
27	26.2	28.1	26.2	25.9	26.2
28	27.3	27.3	27.0	28.1	28.1
29	29.2	26.4	27.3	27.3	27.3
30	29.5	27.3	29.2	28.4	28.1
31	28.1	27.3	29.2	28.1	29.2
32	31.2	31.2	31.2	31.2	28.4
33	28.1	27.3	27.3	28.4	28.4
34	24.0	28.1	26.4	25.1	25.3
35	27.0	29.0	27.3	26.4	28.1
36	27.5	27.5	24.5	25.6	25.3
37	27.3	29.5	26.2	27.5	28.1
38	31.2	30.1	27.3	30.1	29.2
39	27.0	27.5	27.3	27.0	27.3
40	31.2	29.5	30.1	28.4	28.4

443. Numbers of revertant colonies of TA98 Salmonella

Margolin, B.H., Kaplan, N. and Zeiger, E. (1981) Inference sensitivity for Poisson mixtures. *Biometrika,* **65**, 591–602.

These data are also analysed in Breslow, N.E. (1984) Extra-Poisson variation in log-linear models. *Applied Statistics,* **33**, 38–44. For each of three plates at each of six doses of quinoline (in microgrammes per plate) they show the number of revertant colonies of TA98 Salmonella.

Dose of quinoline					
0	**10**	**33**	**100**	**333**	**1000**
15	16	16	27	33	20
21	18	26	41	38	27
29	21	33	60	41	42

444. Landsat multi-spectral scanner data

Campbell, N.A. and Kiiveri, H.T. (1988) Spectral-temporal indices for discrimination. *Applied Statistics*, **37**, 51–62, Table 1.

The aim of the study in which these data were collected was to discriminate between three classes of vegetation: crop, pasture, and bush. Samples of pixels of each were taken, the respective sample sizes being 88, 63, and 420. Measurements on four spectral bands (green, red, and two near infrared) for three overpasses were available. The table shows the means, standard deviations, and correlations.

The aim is to see if these data permit discrimination between the three vegetation classes.

b4	b5	b6	b7	b4	b5	b6	b7	b4	b5	b6	b7
Crop means											
26.9	32.9	45.2	42.6	27.4	31.0	58.0	57.9	31.2	34.8	84.0	86.8
Pasture means											
30.8	42.3	50.1	43.4	32.5	44.8	59.3	53.2	39.7	56.4	93.9	87.7
Bush means											
18.0	16.4	19.0	18.9	20.2	20.3	24.0	24.1	25.1	26.8	33.7	33.8
Standard deviations											
1.52	2.84	2.88	2.80	1.47	2.48	4.14	3.10	2.17	3.94	4.39	3.88
Correlations											
1.00	0.70	0.57	0.48	0.27	0.30	0.24	0.17	0.27	0.25	0.25	0.17
	1.00	0.64	0.49	0.32	0.39	0.29	0.28	0.25	0.26	0.32	0.25
		1.00	0.68	0.27	0.38	0.40	0.36	0.19	0.22	0.31	0.28
			1.00	0.19	0.33	0.46	0.44	0.19	0.22	0.31	0.30
				1.00	0.50	0.34	0.29	0.36	0.31	0.30	0.24
					1.00	0.49	0.46	0.41	0.33	0.43	0.36
						1.00	0.69	0.25	0.25	0.36	0.34
							1.00	0.14	0.24	0.38	0.36
								1.00	0.80	0.60	0.50
									1.00	0.69	0.54
										1.00	0.76
											1.00

445. Effectiveness of insecticides

Giltinan, D.M., Capizzi, T.P., and Malani, H. (1988) Diagnostic test for similar action of two compounds. *Applied Statistics*, **37**, 39–50, Table 2.

These data arose from an experiment to investigate the joint activity of two insecticides, A and B, against the tobacco budworm *Heliothis virescens*. Various mixture ratios were tested, with various concentrations of each insecticide. The response variable was the proportion of insects killed after 96 hours.

Although the original paper used these data to explore deviations from similar action of the two drugs, they can be used for other teaching purposes.

Mixture	Amount of A (ppm)	Amount of B (ppm)	Number of dead insects	Number of insects tested
B	0	30.00	26	30
B	0	15.00	19	30
B	0	7.50	7	30
B	0	3.75	5	30
A25:B75	6.50	19.50	23	30
A25:B75	3.25	9.75	11	30
A25:B75	1.625	4.875	3	30
A25:B75	0.813	2.438	0	30
A50:B50	13.00	13.00	15	30
A50:B50	6.50	6.50	5	30
A50:B50	3.25	3.25	4	29
A50:B50	1.625	1.625	0	29
A75:B25	19.50	6.50	20	30
A75:B25	9.75	3.25	13	30
A75:B25	4.875	1.625	6	29
A75:B25	2.438	0.813	0	30
A	30.00	0	23	30
A	15.00	0	21	30
A	7.50	0	13	30
A	3.75	0	5	30

446. Strengths of glass fibres

Smith, R.L. and Naylor, J.C. (1987) A comparison of maximum likelihood and Bayesian estimators for the three-parameter Weibull distribution. *Applied Statistics*, **36**, 358–369, Table 1.

The table shows the strengths of 1.5 cm glass fibres, measured at the National Physical Laboratory, England. Unfortunately, the units of measurement are not given in the paper. Interest lay in fitting an appropriate distribution to the data.

0.55	0.74	0.77	0.81	0.84
0.93	1.04	1.11	1.13	1.24
1.25	1.27	1.28	1.29	1.30
1.36	1.39	1.42	1.48	1.48
1.49	1.49	1.50	1.50	1.51
1.52	1.53	1.54	1.55	1.55
1.58	1.59	1.60	1.61	1.61
1.61	1.61	1.62	1.62	1.63
1.64	1.66	1.66	1.66	1.67
1.68	1.68	1.69	1.70	1.70
1.73	1.76	1.76	1.77	1.78
1.81	1.82	1.84	1.84	1.89
2.00	2.01	2.24		

447. Psychoactive drug use

Everitt, S. and Dunn, G. (1991) *Applied Multivariate Data Analysis*, London: Edward Arnold, Table 4.1.

(Huba, *et al* (1981) A comparison of two latent variable causal models for adolescent drug use. *Journal of Personality and Social Psychology*, **40**, 180–193), collected data from 1634 American students in Los Angeles on the extent of use of 13 psychoactive drugs. Usage was coded as 1=never tried, 2=only once, 3=a few times, 4=many times, 5=regularly. The 13 drugs were d1=cigarettes, d2=beer, d3=wine, d4=liquor, d5=cocaine, d6=tranquillizers, d7=drug store medications used to get high, d8=heroin and other opiates, d9=marijuana, d10=hashish, d11=inhalents (glue, gasoline, etc.), d12=hallucinogenics (LSD, mescaline, etc.), d13=amphetamine stimulants.

The correlation matrix of these data is shown below. One question of interest is whether this matrix can be explained in terms of fewer latent variables — using principal components analysis or factor analysis.

d1	d2	d3	d4	d5	d6	d7	d8	d9	d10	d11	d12	d13
1												
.447	1											
.422	.619	1										
.435	.604	.583	1									
.114	.068	.053	.115	1								
.203	.146	.139	.258	.349	1							
.091	.103	.110	.122	.209	.221	1						
.082	.063	.066	.097	.321	.355	.201	1					
.513	.445	.365	.482	.186	.315	.150	.154	1				
.304	.318	.240	.368	.303	.377	.163	.219	.534	1			
.245	.203	.183	.255	.272	.323	.310	.288	.301	.302	1		
.101	.088	.074	.139	.279	.367	.232	.320	.204	.368	.340	1	
.245	.199	.184	.293	.278	.545	.232	.314	.394	.467	.392	.511	1

448. Estimated abdominal disease diagnoses

Habbema, J.D.F., Hilden, J. and Bjerregaard, B. (1978) The measurement of performance in probabilistic diagnosis. *Methods of Information in Medicine*, **17**, 217–226, Table 3.

A modified version of the classification rule which assumes independence between the variables was used to assign 50 patients with acute abdominal pain to one of three classes: 1=nonspecific abdominal pain, 2=acute appendicitis, 3=other diseases. Also known are the true classes. Interest here centres on defining suitable rules to calculate the reliability and discriminability of the classification rule.

True class	Prob. d1.	Prob. d2.	Prob. d3.
1	0.43	0.30	0.54
1	0.86	0.10	0.13
1	0.63	0.00	0.37
1	0.17	0.03	0.79
1	0.52	0.07	0.42
1	0.28	0.01	0.71
1	0.68	0.11	0.20
1	0.78	0.01	0.21
1	0.45	0.04	0.51
1	0.82	0.05	0.13
1	0.83	0.06	0.11
1	0.35	0.04	0.61
1	0.64	0.12	0.24
2	0.15	0.77	0.08
2	0.08	0.84	0.09
2	0.06	0.89	0.05
2	0.74	0.10	0.15
2	0.01	0.98	0.01
2	0.04	0.91	0.05
2	0.04	0.91	0.05
2	0.02	0.89	0.09
3	0.36	0.01	0.64
3	0.01	0.00	0.99
3	0.17	0.00	0.83
3	0.33	0.00	0.67
3	0.22	0.04	0.75
3	0.25	0.01	0.74
3	0.25	0.33	0.43
3	0.23	0.52	0.26
3	0.07	0.01	0.92
3	0.06	0.00	0.94
3	0.57	0.01	0.42
3	0.00	0.13	0.87
3	0.28	0.01	0.71
3	0.05	0.38	0.57
3	0.10	0.11	0.80
3	0.11	0.02	0.87
3	0.04	0.00	0.96
3	0.20	0.00	0.80
3	0.06	0.00	0.94
3	0.50	0.36	0.14
3	0.31	0.08	0.62
3	0.05	0.01	0.94
3	0.09	0.00	0.91
3	0.12	0.00	0.87
3	0.05	0.00	0.95
3	0.20	0.00	0.80
3	0.75	0.04	0.20
3	0.35	0.02	0.64
3	0.11	0.03	0.86

449. Statistics examination results

D.J. Hand.

These data show the results of 10 students sitting 14 examination papers for a degree in statistics. Each mark is a percentage, and the issue is how best to combine them to produce an overall rating of the students. For example, should the papers be standardised in some way before combining them, and if so, how?

Student							Paper							
	1	2	3	4	5	6	7	8	9	10	11	12	13	14
1	48	65	68	60	35	61	70	64	68	78	64	66	80	80
2	69	80	73	72	55	58	64	75	88	78	75	71	82	96
3	25	65	63	60	37	9	69	62	64	71	62	62	71	42
4	32	60	65	42	53	49	58	66	80	82	66	60	75	70
5	48	55	68	52	35	76	68	71	75	93	71	55	77	76
6	33	65	55	42	42	59	56	58	83	81	58	60	70	64
7	89	65	75	70	32	85	62	85	78	73	85	84	75	84
8	84	85	70	56	40	84	57	82	70	81	82	76	74	80
9	38	85	65	54	45	50	59	60	75	74	60	41	78	50
10	26	70	65	66	35	51	61	60	77	69	60	44	75	66

450. Eight variables measured on female psychiatric patients

Conrad, S. (1989) *Assignments in Applied Statistics*. Chichester: Wiley, 126.

Eight variables were scored on each of 118 female psychiatric patients. The data presented below are a subset of these.

The 8 variables are age, IQ, anxiety (1=none, 2=mild, 3=moderate, 4=severe), depression (1-4 as for anxiety), can you sleep normally? (1=yes, 2=no), have you lost interest in sex? (1=no, 2=yes), have you thought recently about ending your life? (1=no, 2=yes), weight change over last six months (in lb). *=missing (the 3 for sleep in the 3rd patient is clearly a misprint).

Age	IQ	Anxiety	Depression	Sleep	Sex	Life	Weight
39	94	2	2	2	2	2	4.9
41	89	2	2	2	2	2	2.2
42	83	3	3	3	2	2	4.0

30	99	2	2	2	2	2	-2.6
35	94	2	1	1	2	1	-0.3
44	90	*	1	2	1	1	0.9
31	94	2	2	*	2	2	-1.5
39	87	3	2	2	2	1	3.5
35	*	3	2	2	2	2	-1.2
33	92	2	2	2	2	2	0.8
38	92	2	1	1	1	1	-1.9
31	94	2	2	2	*	1	5.5
40	91	3	2	2	2	1	2.7
44	86	2	2	2	2	2	4.4
43	90	3	2	2	2	2	3.2
32	*	1	1	1	2	1	-1.5
32	91	1	2	2	*	1	-1.9
43	82	4	3	2	2	2	8.3
46	86	3	2	2	2	2	3.6
30	88	2	2	2	2	1	1.4
34	97	3	3	*	2	2	*
37	96	3	2	2	2	1	*
35	95	2	1	2	2	1	-1.0
45	87	2	2	2	2	2	6.5
35	103	2	2	2	2	1	-2.1
31	*	2	2	2	2	1	-0.4
32	91	2	2	2	2	1	-1.9
44	87	2	2	2	2	2	3.7
40	91	3	3	2	2	2	4.5
42	89	3	3	2	2	2	4.2

451. Suicide thoughts

Goldberg, D. (1972) *The detection of psychiatric illness by questionnaire.* London: Oxford University Press, 126, paragraph G.19.

The table gives the counts of responses of people from three groups (normals, mild psychiatric illness, and severe psychiatric illness) to the question 'Have you recently found that the idea of taking your own life kept coming into your mind?'

	Definitely not	I don't think so	Has crossed my mind	Definitely has
Normal	90	5	3	1
Mild	43	18	21	15
Severe	34	8	21	36

452. Similarities between eight offences

Everitt, S. and Dunn, G. (1991) *Applied Multivariate Data Analysis*. London: Edward Arnold, Table 5.15.

Raters were asked to assess eight offences and state how unlike the others each one was, in terms of seriousness. The percentages of raters judging offences as very dissimilar are given in the table. Interest lies in trying to summarise the data in a convenient way, perhaps using multidimensional scaling techniques.

The offences are 1=assault and battery, 2=rape, 3=embezzlement, 4=perjury, 5=libel, 6=burglary, 7=prostitution, 8=receiving stolen goods.

Offence							
1	2	3	4	5	6	7	8
0							
21.1	0						
71.2	54.1	0					
36.4	36.4	36.4	0				
52.1	54.1	52.1	0.7	0			
89.9	75.2	36.4	54.1	53.0	0		
53.0	73.0	75.2	52.1	36.4	88.3	0	
90.1	93.2	71.2	63.4	52.1	36.4	73.0	0

453. Language abilities

Everitt, S. and Dunn, G. (1991) *Applied Multivariate Data Analysis*. London: Edward Arnold, Table 5.9.

The table shows the percentages of persons claiming to speak a foreign language at a level 'enough to make yourself understood'. These data can be regarded as an asymmetric proximity matrix. Interest lies in displaying them, for example via multidimensional scaling and unfolding methods.

Key: a=German, b=Italian, c=French, d=Dutch, e=Flemish, f=English, g=Portugese, h=Swedish, i=Danish, j=Norwegian, k=Finnish, l=Spanish.

	Language											
Country	a	b	c	d	e	f	g	h	i	j	k	l
West Germany	100	2	10	2	1	21	0	0	0	0	0	1
Italy	3	100	11	0	0	5	0	0	0	0	0	1

France	7	12	100	1	1	10	1	2	3	0	0	7
Netherlands	47	2	16	100	100	41	0	0	0	0	0	2
Belgium	15	2	44	0	59	14	0	0	0	0	0	1
Great Britain	7	3	15	0	0	100	0	0	0	0	0	2
Portugal	0	1	10	0	0	9	100	0	0	0	0	2
Sweden	25	1	6	0	0	43	0	100	10	11	5	1
Denmark	36	3	10	1	1	38	0	22	100	20	0	1
Norway	19	1	4	0	0	34	1	25	19	100	0	0
Finland	11	1	2	0	0	12	0	23	0	0	100	0
Spain	1	2	11	0	0	5	0	0	0	0	0	100

454. Heat evolved by setting cement

Draper, N. and Smith, H. (1966) *Applied Regression Analysis*, New York: John Wiley & Sons, 366.

These data were originally described in Woods, H., Steinour, H.H. and Starke, H.R. (1932) Effects of composition of Portland Cement on heat evolved during hardening, *Industrial and Engineering Chemistry*, **24**, 1207–1214. They show the heat (in calories per gram of cement) evolved while samples of cement set. The percentages by weight of four constituents of each sample were measured, and the objective is to formulate a rule to predict the heat evolved from the composition.

The four constituents were:

a = tricalcium aluminate
b = tricalcium silicate
c = tetracalcium alumino ferrite
d = dicalcium silicate

a	b	c	d	heat
7	26	6	60	78.5
1	29	15	52	74.3
11	56	8	20	104.3
11	31	8	47	87.6
7	52	6	33	95.9
11	55	9	22	109.2
3	71	17	6	102.7
1	31	22	44	72.5
2	54	18	22	93.1
21	47	4	26	115.9
1	40	23	34	83.8
11	66	9	12	113.3
10	68	8	12	109.4

455. Quality control in bread baking

Chau, A. K-M. (1992) A comparison of methods of analysis for a quality improvement study. MSc Thesis, University of Southampton, 74, and description, 23–24, Tables 1.1, 1.2.

In addition to the major ingredient of wheat, six minor ingredients, here labelled A, B, C, D, E, and F, can be added to bread making flour. In addition, during production, there are six uncontrollable factors, here called 'production factors': yeast level, proof time, water addition level, moulding pressure, degree of mixing, and oven bake.

Market research has shown that in general people prefer light fluffy loaves — corresponding to a high specific volume. The response variable below is specific volume.

In this study the effects of A to F were studied. Quoting from Chau page 23:

'The lowest level of each ingredient is zero. Previous experiments in the baking industry indicated that a second-order (quadratic) model may be necessary. Therefore a 1/4 replicate of a 2^6 factorial design was used together with 12 axial points, chosen to give orthogonality, and a single centre point. The 2-factor interactions AD, AF, BD, and CE were considered to be important. So in this central composite design, the defining contrast was I=ABCD=BCEF=ADEF. There are a total of 29 flour formulations. The six production factors were grouped together to form three noise factors, each having two levels labelled H and L. The production factors are grouped in such a way that the effects of the factors in the same group are acting in the same direction. For example, it is known from previous studies that high specific volume is related to high yeast level and long proof time, and low specific volume is related to low yeast level and short proof time.

A 1/2 replicate of the 2^3 factorial design with defining contrast I=PQR was used in the experiment. This experiment lasted for 4 consecutive days, and each day a different combination of noise factors was employed. Unfortunately it is not possible to estimate the effect of individual noise factors because the noise factors are confounded with the days. Every day a single dough was made up and baked from each of the 29 flour formulations.'

In the table, a=1.6644

Noise Factor	Level	
	H	**L**
P	High yeast level Long proof time	Low yeast level Short proof time
Q	Over mixing High water addition level Light moulding pressure	Under mixing Low water addition level Heavy moulding pressure
R	Over bake	Under bake

Ingedient value						Average specific volume Day			
A	B	C	D	E	F	1	2	3	4
-1	-1	-1	-1	-1	-1	519	446	337	415
-1	-1	-1	-1	1	1	503	468	343	418
-1	-1	1	1	-1	1	567	471	355	424
-1	-1	1	1	1	-1	552	489	361	425
-1	1	-1	1	-1	1	534	466	356	431
-1	1	-1	1	1	-1	549	461	354	427
-1	1	1	-1	-1	-1	560	480	345	437
-1	1	1	-1	1	1	535	477	363	418
1	-1	-1	1	-1	-1	558	483	376	418
1	-1	-1	1	1	1	551	472	349	426
1	-1	1	-1	-1	1	576	487	358	434
1	-1	1	-1	1	-1	569	494	357	444
1	1	-1	-1	-1	1	562	474	358	404
1	1	-1	-1	1	-1	569	494	348	400
1	1	1	1	-1	-1	568	478	367	463
1	1	1	1	1	1	551	500	373	462
a	0	0	0	0	0	567	471	358	445
-a	0	0	0	0	0	530	467	362	431
0	a	0	0	0	0	575	492	359	428
0	-a	0	0	0	0	565	467	356	437
0	0	a	0	0	0	581	466	355	442
0	0	-a	0	0	0	564	487	346	420
0	0	0	a	0	0	552	477	359	437
0	0	0	-a	0	0	530	468	370	431
0	0	0	0	a	0	572	488	360	436
0	0	0	0	-a	0	554	482	347	461
0	0	0	0	0	a	536	492	360	436
0	0	0	0	0	-a	525	481	363	434
0	0	0	0	0	0	562	495	359	428

Noise factor combinations used on each day:

	Noise factor level		
Day	P	Q	R
1	H	H	L
2	H	L	H
3	L	L	L
4	L	H	H

456. Death rates from heart disease among doctors

Breslow, N. (1985) Cohort analysis in epidemiology. In *Celebration of Statistics,* ed. A.C. Atkinson and S.E. Fienberg. New York: Springer-Verlag, 109–143.

The table shows the death rates from coronary heart disease among British male doctors, classified according to whether or not they smoked. Interest lies in comparing the two groups.

| | Person-years | | Coronary deaths | |
Age	Non-smokers	Smokers	Non-smokers	Smokers
35-44	18790	52407	2	32
45-54	10673	43248	12	104
55-64	5710	28612	28	206
65-74	2585	12663	28	186
75-84	1462	5317	31	102

457. Renal transplant data

Henderson, R. and Milner, A. (1991) Aalen plots under proportional hazards. *Applied Statistics*, **40**, 401–409, Table 1.

The table shows the graft survival times (*t*) in months of 148 renal transplant patients. Also given is (*x*) the total number of HLA-B or DR antigen mismatches between donor and recipient. Particular interest here lay in whether the effect of the latter variable, the covariate, was time dependent.

Those indicated * are observed failure times. The remainder are censored.

t	x	t	x	t	x	t	x	t	x
0.035*	3	3.803	3	12.213*	1	19.508	2	32.672	1
0.068*	0	4.311	1	12.508*	3	19.574	3	32.705	2
0.100*	0	4.867	0	12.533	2	19.733	0	33.148	1
0.101*	1	5.180*	1	13.467	0	20.148	2	33.567	1
0.167	4	6.233	2	13.800	2	20.180	0	33.770	1
0.168*	2	6.367	2	14.267	0	20.900*	2	33.869	2
0.197	1	6.600	1	14.475	4	21.167	0	34.836	0
0.213*	1	6.600	0	14.500	1	21.233	0	34.869	1
0.233	1	7.180*	3	15.213	1	21.600	3	34.934	2
0.234*	2	7.667	1	15.333	0	22.100	1	35.738	0
0.508*	0	7.733*	1	15.525	1	22.148	2	36.180	1
0.508	2	7.800	2	15.533	2	22.180	0	36.213	1

0.533*	3	7.933	1	15.541	1	22.180	0	39.410	1
0.633	0	7.967	1	15.934	0	22.267	0	39.433	0
0.767*	3	8.016*	2	16.200	1	22.300	2	39.672	0
0.768*	4	8.300*	1	16.300	0	22.500	1	40.001	0
0.770	0	8.410	0	16.344	1	22.533	1	41.733	2
1.066*	4	8.607	1	16.600	0	22.867	1	41.734	0
1.267	2	8.667*	1	16.700	1	23.738	1	42.311	2
1.300*	3	8.800	1	16.933	3	24.082	1	42.869	0
1.600*	1	9.100	0	17.033	3	24.180	0	43.180	0
1.639	2	9.233*	1	17.067	0	24.705	0	43.279	1
1.803	2	10.541	2	17.475	1	25.705	2	43.902	2
1.867*	4	10.607	3	17.667	1	25.213	1	44.267	2
2.180*	3	10.633	1	17.700	1	29.705	3	44.475	1
2.667*	4	10.667*	2	17.967	1	30.443	1	44.900	1
2.967	1	10.869	3	18.115	2	31.667	0	45.148	1
3.328	2	11.067*	2	18.115	1	31.934	2	46.451	0
3.393*	3	11.180	0	18.933	0	32.180	1		
3.700*	4	11.443	0	18.934	1	32.367	0		

458. Strengths of welds in high density polyethylene

Buxton, J.R. (1991) Some comments on the use of response variable transformations in empirical modelling. *Applied Statistics*, **40**, 391–400, Tables 1 and 2.

Dumb-bell shapes of high density polyethylene were made using injection moulding, cut into two pieces, and hot plate welded back together. The quality of the weld was then measured by the ratio of the yield stress of the welded bar to the mean yield stress of unwelded bars. This ratio is the 'weld factor', *WF*.

Four control variables were involved: hot plate temperature (*pt*, in degrees Centigrade), heating time (*ht*, in seconds), welding time (*wt*, in seconds), and pressure on the weld (*wp*, in bars). The levels of these variables are given in Table (a). The design and the results are shown in Table (b).

The study describing the analysis of these data pointed out that care has to be taken when applying empirical transformations.

(a)

Code	*pt*	*ht*	*wt*	*wp*
-2	220	10	10	1.5
-1	245	20	15	2.0
0	270	30	20	2.5
1	295	40	25	3.0
2	320	50	30	3.5

(b)

pt	ht	wt	wp	WF	pt	ht	wt	wp	WF
1	1	1	1	0.82	0	2	0	0	0.88
1	1	1	-1	0.87	0	-2	0	0	0.66
1	1	-1	1	0.83	0	0	2	0	0.84
1	1	-1	-1	0.86	0	0	-2	0	0.81
1	-1	1	1	0.82	0	0	0	2	0.88
1	-1	1	-1	0.80	0	0	0	-2	0.81
1	-1	-1	1	0.77	0	0	0	0	0.82
1	-1	-1	-1	0.58	0	0	0	0	0.86
-1	1	1	1	0.89	0	0	0	0	0.80
-1	1	1	-1	0.86	0	0	0	0	0.84
-1	1	-1	1	0.84	0	0	0	0	0.86
-1	1	-1	-1	0.82	0	0	0	0	0.83
-1	-1	1	1	0.67	0	0	0	0	0.81
-1	-1	1	-1	0.77	0	0	0	0	0.79
-1	-1	-1	1	0.74	0	0	0	0	0.80
-1	-1	-1	-1	0.40	0	0	0	0	0.82
2	0	0	0	0.83	0	0	0	0	0.86

459. Subjective health assessment in 5 regions

Turrall, K. (1992) *An analysis of 5 health and lifestyle surveys.* MSc dissertation, Southampton University, Faculty of Mathematics, Table 3.52, 43.

The table shows subjective assessments of personal health by respondents in 5 separate health and lifestyle surveys. Interest lies in whether or not the regions differ in this regard.

		Health	
Region	Good	Fairly good	Not good
Southampton	954	444	78
Swindon	985	504	87
Jersey	459	175	43
Guernsey	377	176	35
West Dorset	926	503	109

460. AIDS incidence data for the USA

Rosenberg, P.S. and Gail, M.H. (1991) Backcalculation of flexible linear models of the Human Immunodeficiency Virus infection curve. *Applied Statistics*, **40**, 269–282, Table 1.

The table shows AIDS incidence data for the USA adjusted for reporting delays. Interest in the source paper lay in backcalculating numbers and times of previous infections, but here they can be used to fit models for forecasting and extrapolation.

Quarterly calendar period	Observed number of cases
1977:1-1984:4	374
1982:1	185
1982:2	200
1982:3	293
1982:4	374
1983:1	554
1983:2	713
1983:3	763
1983:4	857
1984:1	1147
1984:2	1369
1984:3	1563
1984:4	1726
1985:1	2142
1985:2	2525
1985:3	2951
1985:4	3160
1986:1	3819
1986:2	4321
1986:3	4863
1986:4	5192
1987:1	6155
1987:2	6816
1987:3	7491
1987:4	7726
1988:1	8483

461. Stock recruitment of Skeena River sockeye salmon

Kettl, S. (1991) Accounting for heteroscedasticity in the transform both sides regression model. *Applied Statistics*, **40**, 261–268, Table 1.

The table shows stock recruitment of the Skeena River sockeye salmon, by year from 1940 to 1967. Two variables are given: the number of spawners (mature fish) and recruits into the fishery. Observations in years 1951 and 1955 are known to be extreme points because of a rockslide in 1951.

The aim is to fit a model to these data.

Year	Spawners	Recruits	Year	Spawners	Recruits
1940	963	2215	1954	511	1393
1941	572	1334	1955	87	363
1942	305	800	1956	370	368
1943	272	438	1957	448	2067
1944	824	3071	1958	819	644
1945	940	957	1959	799	1747
1946	486	934	1960	273	744
1947	307	971	1961	936	1087
1948	1066	2257	1962	558	1335
1949	480	1451	1963	597	1981
1950	393	686	1964	848	627
1951	176	127	1965	619	1099
1952	237	700	1966	397	1532
1953	700	1381	1967	616	2086

462. Tumour response to chemotherapy

Holtbrugge, W. and Schumacher, M. (1991) A comparison of regression models for the analysis of ordered categorical data. *Applied Statistics*, **40**, 249–259, Table 1.

The table shows the response of patients suffering from a particular type of lung cancer to two different kinds of chemotherapy (sequential therapy and alternating therapy). The patients were randomly assigned to the two treatment groups using a stratified procedure to balance for sex.

Interest lies in finding a good model for the ordered categorial data.

		Response			
Therapy	Sex	Progressive disease	No change	Partial remission	Complete remission
Sequential	Male	28	45	29	26
	Female	4	12	5	2
Alternative	Male	41	44	20	20
	Female	12	7	3	1

463. Cervical cancer deaths in four European countries

Whittemore, A.S. and Gong, G. (1991) Poisson regression with misclassified counts: application to cervical cancer mortality rates. *Applied Statistics*, **40**, 81–93, Table 1.

The table shows (a) numbers of cervical cancer deaths and (b) woman-years at risk for four European countries in the period 1969-73, classified by age group. Although the cited paper was concerned with problems arising from misclassification, these data can be used for simple models of how the dates depend on age.

(a) Number of deaths

Country	Age group 25-34	35-44	45-54	55-64
England and Wales	192	860	2762	3035
Belgium*	8	81	242	268
France	96	477	998	1117
Italy	45	255	621	839

(* Belgian data only given for 1969-1972)

(b) Woman-years at risk

Country	Age group 25-34	35-44	45-54	55-64
England and Wales	15399	14268	15450	15142
Belgium*	2328	2557	2268	2253
France	15324	16186	14432	13201
Italy	19115	18811	16234	15246

(* Belgian data only given for 1969-1972)

464. Performance of a credit scoring instrument

Chandler, G.G. and Johnson, R.W. (1992) The benefit to consumers from generic scoring models based on credit reports. *IMA Journal of Mathematics Applied in Business and Industry*, **4**, 61–72, Table 1.

Financial credit granting agencies rate applicants for credit using a score card. For a sample of over 800000 new accounts booked and scored during 1988 and 1989 their scores, the bankruptcy rate for applicants with that score, the charge-off rate, and the major delinquency rate are given. Account status was determined 12 months after booking. Interest lies in modelling the relationship between score and risk.

Score intervals	Percent of accounts	Bankruptcy rate %	Charge-off rate %	Major delinquency rate %
<1	11.42	0.01	0.05	0.11
1-50	5.95	0.03	0.07	0.22
51-100	6.34	0.05	0.07	0.25
101-150	9.14	0.07	0.22	0.53
151-200	8.30	0.07	0.17	0.57
201-250	8.43	0.11	0.25	0.72
251-300	9.48	0.18	0.51	1.08
301-350	8.06	0.18	0.53	1.29
351-400	6.99	0.29	0.61	1.52
401-450	5.91	0.38	0.73	1.80
451-500	4.79	0.52	0.82	2.12
501-550	3.84	0.59	0.98	2.27
551-600	3.06	0.62	1.17	2.62
601-650	2.40	0.85	1.33	3.28
651-700	1.75	1.16	1.47	3.37
701-750	1.33	1.64	1.51	4.10
751-800	0.92	1.85	2.06	4.48
801-850	0.65	2.20	2.86	5.01
851-900	0.44	2.76	3.60	6.15
901-950	0.30	2.40	3.77	6.86
951-1000	0.20	3.80	5.12	7.02
>1000	0.30	6.35	5.82	9.24

465. Weights of mice

Rissanen, J. (1989) *Stochastic complexity in statistical enquiry*, Singapore: World Scientific, 90, Table 1.

The table shows the weights of 13 mice weighed every 3 days after birth. The aim of the study is to predict the weights in the last column from those in the preceding six.

Day

0.109	0.388	0.621	0.823	1.078	1.132	1.191
0.218	0.393	0.568	0.729	0.839	0.852	1.004
0.211	0.394	0.549	0.700	0.783	0.870	0.925
0.209	0.419	0.645	0.850	1.001	1.026	1.069
0.193	0.362	0.520	0.530	0.641	0.640	0.751
0.201	0.361	0.502	0.530	0.657	0.762	0.888
0.202	0.370	0.498	0.650	0.795	0.858	0.910
0.190	0.350	0.510	0.666	0.819	0.879	0.929
0.219	0.399	0.578	0.699	0.709	0.822	0.953
0.255	0.400	0.545	0.690	0.796	0.825	0.836
0.224	0.381	0.577	0.756	0.869	0.929	0.999
0.187	0.329	0.441	0.525	0.589	0.621	0.796
0.278	0.471	0.606	0.770	0.888	1.001	1.105

466. Poisson random variate generation

Kemp, C.D. and Kemp, W. (1991) Poisson random variate generation. *Applied Statistics*, **40**, 143–158, Table 1.

In statistical simulation studies one has to generate random numbers from known distributions and, given that large numbers of values will typically need to be generated, it is important that the programs should be as fast as possible. In these data the distribution was the Poisson distribution. Two situations were studied, corresponding to the Poisson parameter varying (left hand side of the table) and fixed (right hand side of the table). In the former, adjustment to the times has been made to account for the extra computation needed to generate the value of the parameter.

Three different generators were compared, KEMPOIS, KPOISS, and PTPE. Interest lies in deciding which is the fastest. This will involve consideration of issues such as the ranges over which each method is best, comparison of variances, etc.

Each entry in the table gives the time (in seconds) needed to generate 100000 values.

	Varying			Fixed		
Parameter	KEMPOIS	KPOISS	PTPE	KEMPOIS	KPOISS	PTPE
10	27	52	60	18	48	50
15	27	50	59	19	46	50
30	28	49	52	20	45	43
50	30	48	51	22	44	42
100	32	47	47	24	43	37
200	35	47	45	27	42	36
500	42	46	45	34	42	35
600	44	46	45	35	42	35
700	45	46	45	37	42	35
1000	49	46	45	41	42	34

467. Survival times of gastric cancer patients

Gamerman, D. (1991) Dynamic Bayesian models for survival data. *Applied Statistics*, **40**, 63–79, Table 1.

The table shows the survival times of two groups of 45 patients suffering from gastric cancer. Group 1 received chemotherapy and radiation. Group 2 just received chemotherapy. An asterisk indicates censoring.

Interest, of course, lies in comparing the survival times of the two groups.

Group 1				Group 2		
17	185	542		1	383	778
42	193	567		63	383	786
44	195	577		105	388	797
48	197	580		125	394	955
60	208	795		182	408	968
72	234	855		216	460	977
74	235	1174*		250	489	1245
95	254	1214		262	523	1271
103	307	1232*		301	524	1420
108	315	1366		301	535	1460*
122	401	1455*		342	562	1516*
144	445	1585*		354	569	1551
167	464	1622*		356	675	1690*
170	484	1626*		358	676	1694
183	528	1736*		380	748	

468. Time for assays to register HIV positivity

Makuch, R.W., Escobar, M. and Merill, S. (1991) A two sample test for incomplete multivariate data. *Applied Statistics*, **40**, 202–212, Table 1.

The table shows the number of days before HIV positivity was registered in an assay. Each patient was measured on four occasions, at monthly intervals. There are two groups of patients (drug 1 and drug 2). The asterisk signifies a censored value.

Interest lies in testing to see if there is a difference between the two groups. Things are complicated as the data are multivariate and censored.

Group	Month 1	Month 2	Month 3	Month 4
1	8	0*	25*	21
1	6	4	5	5
1	6	5	28*	18
1	14*	35*	23	19*
1	7	0*	13	0*
1	5	4	27	8
1	5	21*	6	14
1	6	10	14	18
1	7	4	15	8
1	6	5	5	5
1	4	5	6	3
1	5	4	7	5

1	21*	5	0*	6
1	13	27*	21*	8
1	4	27*	7	6
1	6	3	7	8
1	6	0*	5	5
1	6	0*	4	6
1	7	9	6	7
1	8	15	8	0*
1	18*	27*	18*	9
1	16	14	14	6
1	15	9	12	12
2	4	5	4	3
2	8	22	25*	0*
2	6	6	8	5
2	7	10	10	18
2	5	14	17*	6
2	3	5	8	6
2	6	11	6	13
2	6	0*	15	7
2	6	12	19	8
2	6	25*	0*	22*
2	4	7	5	7
2	5	7	4	6
2	3	9	7	6
2	9	17	0*	21*
2	6	4	8	14
2	5	5	7	16*
2	12	18*	14	0*
2	9	11	15	18*
2	6	5	9	0*
2	18*	8	10	13
2	4	4	5	10
2	3	10	0*	21*
2	8	7	10	12
2	3	6	7	9

469. Mean daily SIDS deaths by temperatures

Fung, Po-Lin (1992) *A study of the relationship between sudden infant death and environmental temperature in England and Wales analyzed using time series regression for counts*. MSc Dissertation, Faculty of Social Sciences, University of Southampton, 22, Table 3.1.

Sudden Infant Death Syndrome (SIDS), or 'Cot Deaths', account for over 1500 deaths per year in the United Kingdom. The cause of these deaths is still unknown. One hypothesized link is to environmental temperature. This table shows the mean daily numbers of SIDS deaths on days at the given temperatures. Column 1 gives the temperature (in °C), column 2 gives the total number of SIDS deaths at the indicated temperatures, column 3 gives the mean number of SIDS deaths on days with the indicated temperatures for infants aged less than 3 months, and column 4 gives the same for infants aged between 3 months and 1 year. The period of data collection was 8 January 1979 to 31 November 1985.

Temp	Total number of SIDS deaths	<3 months	3-12 months
16.0-16.9	171	1.070	1.105
17.0-17.9	153	1.033	0.967
18.0-18.9	102	1.020	1.206
19.0-19.9	79	1.114	0.962
20.0-20.9	56	0.857	0.804
21.0-21.9	52	0.673	0.846
22.0-22.9	30	0.733	0.833
23.0-23.9	17	1.118	0.471
24.0-24.9	5	1.200	1.200
>=25.0	3	3.667	0.333

470. Number of SIDS deaths per day

Fung, Po-Lin (1992) *A study of the relationship between sudden infant death and environmental temperature in England and Wales analyzed using time series regression four counts.* MSc Dissertation, Faculty of Social Sciences, University of Southampton, 63.

Sudden Infant Death Syndrome (SIDS), or 'Cot Deaths' account for over 1500 deaths per year in the United Kingdom. The cause of these deaths is still unknown. One hypothesised link is to environmental temperature. The table shows the mean temperatures of days between 8 January 1979 and 31 November 1985 in which this temperature was above 19°C and the number of SIDS deaths which occurred on the following day.

0 19.7 0 20.9 0 19.1 0 19.8 1 20.4 1 22.9 0 21.5 2 20.0 0 19.4 1 21.5 1 21.1 1 19.9 1 21.9 0 19.7 0 20.1 0 20.5 0 20.5 2 19.5 1 20.6 1 21.1 0 19.1 1 19.4 2 19.0 4 19.0 0 19.6 0 20.7 1 19.6 0 21.4 0 22.4 1 19.1 0 19.0 0 20.1 1 20.4 0 19.6 1 22.3 3 21.2 1 19.6 0 21.9 1 24.8 1 19.1

0 20.4 1 20.5 0 21.6 3 23.1 1 19.3 1 20.5 0 22.6 0 21.9 0 21.6 5 19.0 1 20.0 1 19.5 0 20.5 1 20.0 1 19.9 3 19.3 0 22.2 1 22.3 2 20.7 1 22.6 0 23.1 1 22.9 1 21.4 1 21.0 0 19.0 0 19.6 2 22.3 2 23.0 0 19.6 3 21.4 0 20.8 1 19.5 0 20.2 0 19.1 2 19.8 0 20.7 1 21.5 2 22.4 3 22.6 0 23.1

0 23.1 0 20.5 0 21.1 1 19.0 1 20.8 0 21.3 0 20.9 0 20.0 1 20.6 1 19.7 2 19.0 1 21.0 1 19.9 3 20.1 1 22.4 3 19.3

2 21.1 1 19.6 2 19.0 0 20.4 0 22.2 0 22.3 1 21.8 3 21.2

3 23.6 0 22.0 0 21.3 2 22.8 3 24.1 1 25.5 4 23.9 4 25.6 1 27.4 0 23.4 0 23.1 1 20.6 0 20.3 2 21.0 0 19.3 0 21.4

1 22.8 0 22.8 2 21.1 1 24.1 0 23.8 2 23.6 1 19.4 0 20.4

2 21.6 0 20.5 1 19.7 0 21.6 0 22.0 1 20.2 1 20.8 0 21.6 1 23.3 1 20.9 1 20.0 1 21.3 2 21.2 2 19.0 1 20.0 1 21.0 0 19.9 1 20.9 0 20.1 3 19.5 2 19.5 2 20.1 0 19.6 4 19.7 0 19.5 0 20.6 1 21.1 2 21.3 0 23.4 1 19.7 2 20.0 0 19.6

0 20.6 0 20.9 0 22.1 2 23.3 1 19.0 0 20.1 3 19.0 0 19.3 0 20.9 0 21.8 2 24.3 3 21.1 0 23.7 0 20.8 0 19.1 2 21.0 1 21.6 2 20.4 0 19.4 0 19.3 0 22.8 0 22.0 2 22.8 0 21.8

2 21.7 0 21.3 2 20.6 0 19.4 0 21.9 0 21.2 0 20.6 1 21.5 1 21.4 0 19.9 2 19.9 1 20.8 1 19.1 1.19.2 2 19.1 2 20.4

1 23.3 0 20.7 1 19.2 1 19.4 0 20.5 1 21.8 0 19.4 0 20.8 2 22.6 0 19.3 2 19.1 0 21.0 0 21.4 1 19.9 1 20.7 5 19.5 2 19.5 1 19.9 4 19.4 3 22.4 0 19.1 1 19.1 3 19.3 1 22.9

1 21.7 0 19.7 0 20.4 1 22.7 0 22.8 1 22.2 1 21.6 0 21.1

1 22.4 0 21.4 2 19.8 2 21.2 0 24.3 1 23.9 0 19.8 0 19.5

0 19.3 1 19.3

471. Predicting benefit from group psychotherapy

Pearson, M.J. and Girling, A.J. (1990) The value of the Claybury Selection Battery in predicting benefit from group psychotherapy. *British Journal of Psychiatry*, **157**, 384–388, Table 1.

Ratings of the effectiveness of group therapy were carried out at 3 months and 12 months on patients who continued in therapy (the 'stayers' in the table) and patients who left therapy (the 'leavers'). The table shows the numbers falling into each of 5 rating categories, with 1 being the best outcome and 5 the worst. The question is to see if there is any difference between the two groups. This is complicated by the zero entries in the table.

Rating	3 months		12 months	
	Stayers	Leavers	Stayers	Leavers
1	8	0	12	0
2	6	0	1	0
3	9	0	5	4
4	2	0	1	1
5	4	3	3	5

472. African statisticians in 1977

Ntozi, J.P.M. (1992) Training of African statisticians at a professional level. *Journal of Official Statistics*, **8**, 467–479, Table 1.

The table gives the numbers of statisticians in different African countries in 1977, classified according to the type of employer. Countries not included are Egypt, Namibia, Reunion, South Africa, Zimbabwe, and Western Sahara. A - signifies not stated, but probably means none.

The data can be used to illustrate cluster analysis, correspondence analysis, multidimensional scaling, or related techniques.

Country	CSO	Other gov't	Other sectors
North Africa			
Algeria	17	3	19
Libya Arab Jamahiriya	11	8	3
Morocco	50	52	87
Sudan	58	10	10
Tunisia	42	8	34
West Africa			
Benin	8	1	9
Cape Verde	2	-	-
Gambia	3	-	3
Ghana	23	5	12
Guinea	3	-	6
Guinea Bissau	-	-	-
Ivory Coast	9	12	25
Liberia	16	5	1
Mali	20	22	32
Mauritania	7	-	2
Niger	8	1	-
Nigeria	86	32	20
Senegal	14	8	19
Sierra Leone	14	-	2
Togo	24	3	9
Upper Volta	7	1	8
Central Africa			
Angola	3	-	5
Burundi	3	1	-
Cameroon Un. Rep.	31	21	24
Central African Republic	5	2	2
Chad	3	-	-
Congo	7	5	4
Equatorial Guinea	-	-	-
Gabon	7	1	2
Rwanda	1	4	-
Sao Tome and Principe	-	-	-
Zaire	34	14	1

East Africa			
Botswana	11	-	-
Comoros	-	-	-
Djibouti	1	2	-
Ethiopia	24	26	12
Kenya	23	-	6
Lesotho	12	1	-
Madagascar	28	11	26
Malawi	16	3	-
Mauritius	9	6	7
Mozambique	4	-	5
Seychelles	3	-	-
Somalia	9	3	1
Swaziland	7	1	2
Tanzania	30	4	4
Uganda	17	16	34
Zambia	23	8	5

473. Railway accidents on British Rail, 1970-83

Chatfield, C. (1988) *Problem solving: a statistician's guide.* London: Chapman and Hall, Table A.1, 86.

The table lists different kinds of railway accident on British Rail for the years 1970 to 1983. The question raised in the source is whether the data provide evidence of a deterioration in safety standards.

		Collisions		Derailments		
Year	Number of train accidents	Between passenger trains	Between passenger and freight trains	Passenger	Freight	Train miles (millions)
70	1493	3	7	20	331	281
71	1330	6	8	17	235	276
72	1297	4	11	24	241	268
73	1274	7	12	15	235	269
74	1334	6	6	31	207	281
75	1310	2	8	30	185	271
76	1122	2	11	33	152	265
77	1056	4	11	18	158	264
78	1044	1	6	21	152	267
79	1035	7	10	25	150	265
80	930	3	9	21	107	267
81	1014	5	12	25	109	260
82	879	6	9	23	106	231
83	1069	1	16	25	107	249

474. Australian higher education students, 1981-88

Selected Higher Education Statistics (1989), Canberra: Australian Government Publishing Service, Table T12.

The table shows the numbers of students in Australian higher education over the years 1981 to 1988, cross-classified by sex and age group. Questions which these data can be used to address include such things as modelling change over time of a 2 × 4 contingency table.

The table excludes those who did not state their age.

	<20		20-24		25-29		>29	
Year	M	F	M	F	M	F	M	F
1981	46687	46460	59698	43007	31227	20159	44558	41296
1982	46977	48064	59629	43801	30972	20418	46435	44060
1983	47220	49062	61120	45009	30644	20530	47753	45894
1984	48301	50591	61252	46256	31034	20828	49858	48317
1985	49617	54223	61377	47956	30669	21999	51970	51478
1986	52165	59198	61914	51123	30869	23443	54464	56083
1987	56099	65741	60759	52558	29251	23088	50022	55733
1988	61068	72985	62896	56899	29327	24768	52333	60249

475. Contents of books on multivariate analysis

Gifi, A. (1920) *Nonlinear multivariate analysis*, Chichester: John Wiley & Sons, Table 1.1.

The table contains the numbers of pages in each of 20 books on multivariate analysis devoted to the following topics:

M =   mathematics other than statistics, i.e. linear algebra, matrices, transformation groups, sets, relations.

C =   correlation and regression, including path analysis, linear structural and functional equations.

F =   factor analysis and principal components analysis.

CA =   canonical correlation analysis.

D =   discriminant analysis, classification, cluster analysis.

S =   statistics, including distributional theory, hypothesis testing, and estimation; also statistical analysis of categorical data.

MA =   MANOVA and the general multivariate linear model.

The introduction to the source presents a correspondence analysis of these data to produce low dimensional representations of the books and content.

	M	C	F	CA	D	S	MA
Roy (1957)	31	0	0	0	0	164	11
Kendall (1957)	0	16	54	18	27	13	14
Kendall (1975)	0	40	32	10	42	60	0
Anderson (1958)	19	0	35	19	28	163	52
Cooley and Lohnes (1962)	14	7	35	22	17	0	56
Cooley and Lohnes (1971)	20	69	72	33	55	0	32
Morrison (1967)	74	0	86	14	0	84	48
Morrison (1976)	78	0	80	5	17	105	60
Van de Geer (1967)	74	19	33	12	26	0	0
Van de Geer (1971)	80	68	67	15	29	0	0
Dempster (1969)	108	48	4	10	46	108	0
Tatsuoka (1971)	109	13	5	17	39	32	46
Harris (1975)	16	35	69	24	0	26	41
Dagnelie (1975)	26	86	60	6	48	48	28
Green and Carroll (1976)	290	10	6	0	8	0	2
Cailliez and Pages (1976)	184	48	82	42	134	0	0
Giri (1977)	29	0	0	0	41	211	32
Gnanadesikan (1977)	0	19	56	0	39	75	0
Kshirsagar (1978)	0	22	45	42	60	230	59
Thorndike (1978)	30	128	90	28	48	0	0

476. Age of menarche in Polish girls

Morgan, B.J.T. (1992) *Analysis of quantal response data*, London: Chapman & Hall, 7, Table 1.4.

For 3918 Warsaw girls, the table shows their age group and whether or not menstruation had started. Interest lies in trying to fit a model. Note that the data set is large, so accurate fitting is possible.

Mean age of group (years)	Number having menstruated	Number of girls
9.21	0	376
10.21	0	200
10.58	0	93
10.83	2	120
11.08	2	90
11.33	5	88
11.58	10	105
11.83	17	111
12.08	16	100

12.33	29	93
12.58	39	100
12.83	51	108
13.08	47	99
13.33	67	106
13.58	81	105
13.83	88	117
14.08	79	98
14.33	90	97
14.58	113	120
14.83	95	102
15.08	117	122
15.33	107	111
15.58	92	94
15.83	112	114
17.58	1049	1049

477. Level of Lake Victoria Nyanza and numbers of sunspots

Shaw, N. (1942) *Manual of meteorology*, Vol. 1, London: Cambridge Univesity Press, 284.

Reprinted in Mosteller, F. and Tukey, J.W. (1977) *Data analysis and regression.* Reading, Massachusetts: Addison-Wesley. Exhibit 1, 518.

The table shows the mean annual level of Lake Victoria Nyanza for the years 1902 to 1921 relative to a fixed standard (units not given in source) and the number of sunspots in the same years.

These data can be used for simple regression analysis.

Year	Level	Number of sunspots	Year	Level	Number of sunspots
1902	-10	5	1912	-11	4
1903	13	24	1913	-3	1
1904	18	42	1914	-2	10
1905	15	63	1915	4	47
1906	29	54	1916	15	57
1907	21	62	1917	35	104
1908	10	49	1918	27	81
1909	8	44	1919	8	64
1910	1	19	1920	3	38
1911	-7	6	1921	-5	25

478. Surgical deaths

Mosteller, F. and Tukey, J.W. (1977) *Data analysis and regression.* Reading, Massachusetts: Addison-Wesley. Exhibit 1, 515.

For two areas in the United States the table shows the numbers of patients, classified by sex and age, who underwent surgery and the numbers who died in the hospital following surgery.

The source uses these data as an exercise on standardization, standardizing the two areas to permit a fair comparison. One can also do logistic regression.

Area 1

Age	Total undergoing surgery		Number dying	
	Males	Females	Males	Females
0-4	2104	1952	34	22
5-14	4272	3911	9	11
15-24	2835	2989	23	5
25-34	2785	2606	19	8
35-44	1930	1886	16	15
45-54	1497	1524	59	40
55-64	960	1013	101	52
65-75	652	855	185	118
76-83	186	287	97	108
>83	69	125	68	103

Area 2

Age	Total undergoing surgery		Number dying	
	Males	Females	Males	Females
0-4	703	689	12	3
5-14	1739	1758	5	2
15-24	1233	1244	14	1
25-34	989	1004	8	3
35-44	897	922	9	13
45-54	921	961	28	15
55-64	686	739	68	37
65-75	611	784	159	73
76-83	189	290	86	88
>83	52	124	70	119

479. Mean monthly temperature at various altitudes

Report of the Royal Society, IGY Antarctic Expedition to Halley Bay.

Extract printed in Mosteller, F. and Tukey, J.W. (1977) *Data analysis and regression*, Reading, Massachusetts: Addison-Wesley, Exhibit 1, 497.

The table shows the mean monthly temperature at noon GMT at heights where the air pressure had fallen to the given values; the measurements were taken during balloon ascents. All values are negative and are given in tenths of degrees. Interest lies in finding the relationship between pressure, month, and temperature.

Pressure (mb)	J	F	M	A	M	J	J	A	S	O	N	D
30	343	423	529	687	787	870	917	885	827	651	394	326
40	354	425	530	665	779	849	901	891	852	704	443	348
50	368	425	517	640	778	837	888	886	850	720	464	350
60	375	428	509	636	761	822	870	877	839	728	483	371
80	385	431	498	610	719	789	843	851	929	731	520	392
100	396	435	496	585	696	761	811	831	820	740	597	403
150	412	435	472	557	644	717	768	792	789	740	601	446
200	423	435	468	551	630	701	750	762	759	720	626	461
250	453	456	492	584	652	666	701	719	708	677	628	540
300	502	496	548	571	614	616	639	654	644	616	586	556
400	430	427	463	472	509	500	521	525	512	499	465	459
500	334	329	361	373	411	396	417	417	399	385	354	355
700	179	186	212	227	270	254	257	272	248	225	204	209
850	93	123	139	177	208	206	207	218	211	154	128	128

480. Aircraft air-conditioning systems failures

Proschan, F. (1963) Theoretical explanation of observed decreasing failure rate, *Technometrics*, **5**, 375–383.

The time intervals between failures of 12 type 720 aircraft are given below. Interest lies in finding a suitable model to predict time to failure.

Aircraft

1	2	3	4	5	6	7	8	9	10	11	12
194	413	90	74	55	23	97	50	359	50	487	102
15	14	10	57	320	261	51	44	9	254	18	209
41	58	60	48	56	87	11	102	12	5	100	14
29	37	186	29	104	7	4	72	270	283	7	57
33	100	61	502	220	120	141	22	603	35	98	54
181	65	49	12	239	14	18	39	3	12	5	32
	9	14	70	47	62	142	3	104		85	67
	169	24	21	246	47	68	15	2		91	59
	447	56	29	176	225	77	197	438		43	134
	184	20	386	182	71	80	188			230	152
	36	79	59	33	246	1	79			3	27
	201	84	27		21	16	88			130	14
	118	44			42	106	46				230
		59			20	206	5				66
		29			5	82	5				61
		118			12	54	36				34
		25			120	31	22				
		156			11	216	139				
		310			3	46	210				
		76			14	111	97				
		26			71	39	30				
		44			11	63	23				
		23			14	18	13				
		62			11	191	14				
					16	18					
					90	163					
					1	24					
					16						
					52						
					95						

481. Female age distribution in Mexico in 1960

Mosteller, F. and Tukey, J.W. (1977) *Data analysis and regression*, Reading, Massachusetts: Addison-Wesley, Exhibit 1, 477.

The table shows the reported age distribution of females in Mexico in 1960. Mosteller and Tukey (1977, page 476) remark on the tendency people have to report their ages rounded to the nearest multiple of 5 years. Can such an effect be detected here?

Age	Number	Age	Number	Age	Number
0	558	26	243	52	42
1	513	27	220	53	61
2	582	28	283	54	66
3	604	29	182	55	148
4	584	30	412	56	69
5	566	31	113	57	46
6	562	32	208	58	83
7	524	33	163	59	48
8	529	34	146	60	245
9	430	35	310	61	20
10	497	36	168	62	43
11	369	37	130	63	32
12	455	38	224	64	32
13	398	39	129	65	103
14	404	40	331	66	29
15	382	41	51	67	23
16	366	42	136	68	40
17	346	43	91	69	16
18	403	44	77	70	109
19	300	45	231	71	9
20	409	46	90	72	25
21	226	47	77	73	15
22	325	48	148	74	14
23	294	49	78	75	48
24	289	50	281		
25	380	51	42		

482. Volume of hens' eggs

Dempster, A.P. (1969) *Elements of continuous multivariate analysis.* Reading Massachusetts: Addison-Wesley, 151.

The table shows L, the log of the length of the longest diameter, W, the log of the diameter of the largest circular cross-section, and V, the log of the volume, of 12 hens' eggs. Interest lies in constructing an expression to relate these three variables.

L	W	V	L	W	V
.7659	.6360	1.750	.7747	.6156	1.726
.7353	.6198	1.701	.7718	.6239	1.714
.7416	.6280	1.714	.7889	.6114	1.714
.7600	.6280	1.738	.7659	.6072	1.714
.7861	.6239	1.750	.7689	.6156	1.714
.7539	.6156	1.675	.7478	.6239	1.726

483. Smoking and health

Best, E.W.R. and Walker, C.B. (1964) A Canadian study of smoking and health. *Canadian Journal of Public Health*, **55**, 1.

Also given in Mosteller, F. and Tukey, J.W. (1977) *Data analysis and regression*, Reading, Massachusetts: Addison-Wesley, Exhibit 1, 559.

The table shows the male populations and the numbers of deaths in each of several age groups, classified according to whether or not they were smokers and, if so, what sort of smoker. Particular interest lies in comparing the groups.

Age	Nonsmokers		Cigar and pipe only		Cigarette and other		Cigarette only	
	Pop.	Deaths	Pop.	Deaths	Pop.	Deaths	Pop.	Deaths
40-44	656	18	145	2	4531	149	3410	124
45-49	359	22	104	4	3030	169	2239	140
50-54	249	19	98	3	2267	193	1851	187
55-59	632	55	372	38	4682	576	3270	514
60-64	1067	117	846	113	6052	1001	3791	778
65-69	897	170	949	173	3880	901	2421	689
70-74	668	179	824	212	2033	613	1195	432
75-80	361	120	667	243	871	337	436	214
>80	274	120	537	253	345	189	113	63

484. Herniorrhaphy

Mosteller, F. and Tukey, J.W. (1977) *Data analysis and regression*, Reading, Massachusetts: Addison-Wesley, Exhibit 8, 567–568.

The table shows the experience of 32 patients undergoing an elective herniorrhaphy. These data can be used to relate post-operative and pre-operative measures. Appropriate methods are regression and canonical correlations analysis or other multivariate techniques.

The outcome measures are:

LEAVE: condition upon leaving the operating room
   1 = routine recovery
   2 = went to intensive care for observation overnight
   3 = went to intensive care unit; moderate care required
   4 = went to intensive care unit; intensive care required

NURSE: level of nursing required one week after operation
   1 = intense   3 = moderate
   2 = heavy   4 = light

LOS: length of stay in hospital after operation (days)

Variables describing the preoperative condition are:

PSTAT: physical status, discounting that associated with the operation, on a scale of 1-5, 1 being perfect health and 5 being very poor health.

BUILD: body build

1 = emaciated	4 = fat
2 = thin	5 = obese
3 = average	

CARDIAC or RESP: preoperative complications

1 = none	3 = moderate
2 = mild	4 = severe

Patient	Age (yrs)	Sex	PSTAT	BUILD	CARDIAC	RESP	LEAVE	LOS	NURSE
1	78	m	2	3	1	1	2	9	3
2	60	m	2	3	2	2	2	4	-
3	68	m	2	3	1	1	1	7	4
4	62	m	3	5	3	1	1	35	3
5	76	m	3	4	3	2	2	9	4
6	76	m	1	3	1	1	1	7	-
7	64	m	1	2	1	2	1	5	-
8	74	f	2	3	2	2	1	16	3
9	68	m	3	4	2	1	1	7	-
10	79	f	2	2	1	1	2	11	3
11	80	f	3	4	4	1	1	4	-
12	48	m	1	3	1	1	1	9	3
13	35	f	1	4	1	2	1	2	-
14	58	m	1	3	1	2	1	4	-
15	40	m	1	4	1	1	1	3	-
16	19	m	1	3	1	1	1	4	-
17	79	m	3	2	3	3	3	3	-
18	51	m	1	3	1	1	1	5	-
19	57	m	2	3	2	1	1	8	3
20	51	m	3	3	3	2	1	8	4
21	48	m	1	3	1	1	1	3	-
22	48	m	1	3	1	1	1	5	-
23	66	m	1	3	1	1	1	8	4
24	71	m	2	3	2	2	2	2	-
25	75	f	3	1	3	1	2	7	-
26	2	f	1	3	1	1	1	0	-
27	65	f	2	3	1	1	2	16	3
28	42	f	2	3	1	1	2	3	-
29	54	m	2	2	2	2	2	2	-
30	43	m	1	2	1	1	1	3	-
31	4	m	2	2	2	1	1	3	-
32	52	m	1	3	1	1	1	8	3

485. Birthweight by gestational age

Bland, J.M., Peacock, J.L., Andeson, H.R., and Brooke, O.G. (1990) The adjustment of birthweight for very early gestational ages: two related problems in statistical analysis. *Applied Statistics*, **39**, 229–239.

The table shows the mean birthweight (in grams) for infants born to Caucasian mothers in St George's Hospital London between August 1982 and March 1984. Also given are the gestational ages (weeks) and standard deviations. The distribution of gestational ages is markedly left-skewed, so that some points will be influential. The analysis should take this into account and should also take into account the accuracies of the estimates at each gestational age.

Gestational age (weeks)	No. of births	Mean birthweight (g)	s.d.
22	1	520	-
23	1	700	-
24	0	-	-
25	1	1000	-
26	0	-	-
27	1	1170	-
28	6	1198	121
29	1	1480	-
30	3	1617	589
31	6	1693	319
32	7	1720	438
33	7	2340	313
34	7	2516	572
35	29	2796	448
36	43	2804	444
37	114	3108	344
38	222	3204	444
39	353	3353	427
40	401	3478	408
41	247	3587	440
42	53	3612	371
43	9	3390	408
44	1	3740	-

486. Underdispersed word count data

Bailey, B.J.R. (1990) A model for function word counts. *Applied Statistics*, **39**, 107–114, Table 1.

In studies aimed at characterising an author's style, samples of n words are taken and the number of function words in each sample counted. Often binomial or Poisson distributions are assumed to hold for the proportions of function words. The table below shows the combined frequencies (x) of the articles 'the', 'a', and 'an' in samples from Macauley's 'Essay on Milton', taken from the Oxford edition of Macauley's (1923) literary essays. Nonoverlapping samples were drawn from the opening words of two randomly chosen lines from each of the 50 pages of printed text, 10 word samples being simply extensions of 5 word samples. The data show clear evidence of underdispersion.

x	5-word frequency	x	10-word frequency
0	45	0	27
1	49	1	44
2	6	2	26
>2	0	3	3
		>3	0

487. Red core disease in strawberries

Jansen, J. (1990) On the statistical analysis of ordinal data when extravariation is present. *Applied Statistics*, **39**, 75–84, Table 1.

The disease red core in strawberries is caused by the fungus *Phytophtora fragariae*. Here 12 populations of strawberries were arranged in a randomized block experiment with four blocks. Observations on 9 or 10 plants were taken in each plot. The final score for each plant was one of three ordered categories in order of increasing damage. The 12 populations arose as a factorial structure: 3 genotypes (male parent) by 4 genotypes (female parent).

One of the problems with this set of data is that the response is only measured on an ordinal scale. Moreover, analysis shows that between-plot variation should be incorporated into the model.

Male parent	Female parent	Block 1			Block 2			Block 3			Block 4		
		Disease category											
		1	2	3	1	2	3	1	2	3	1	2	3
1	1	0	3	6	2	2	6	2	3	5	2	5	3
1	2	2	3	5	0	3	7	4	6	0	2	3	5
1	3	3	4	3	7	2	1	1	1	7	2	3	5
1	4	0	5	5	5	4	1	2	8	0	1	4	5
2	1	1	4	4	2	2	6	1	2	7	1	5	4
2	2	1	4	5	3	4	2	1	6	3	4	2	4
2	3	4	3	3	5	1	4	3	3	4	4	2	2
2	4	1	4	5	1	2	6	8	2	0	2	5	3
3	1	0	0	9	3	5	2	2	5	3	0	0	0
3	2	5	3	2	3	2	5	3	6	1	2	1	7
3	3	0	3	6	2	5	3	1	3	6	0	3	7
3	4	3	0	7	5	2	3	7	3	0	3	4	3

488. Failure times for a piece of electronic equipment

Juran, J. and Gryna, F. (1980) *Quality planning and analysis*, New York: McGraw-Hill, 204.

These data are also analysed in Schneider, H., Lin, B-S. and O'Cinneide, C. (1990) Comparison of nonparametric estimators for the renewal function. *Applied Statistics*, **39**, 55–61.

The table shows the failures times (hours) of 107 units of a piece of electronic equipment.   Schneider *et al.* suggest that there is an early failure period of approximately 20 hours and a wear-out period of about 100 hours, the period between being fairly constant. Their interest was in the expected number of failures at the end of the early period and at the beginning of the wear-out period.

1.0	6.4	19.2	54.2	88.4	114.8
1.2	6.8	28.1	55.6	89.9	115.1
1.3	6.9	28.2	56.4	90.8	117.4
2.0	7.2	29.0	58.3	91.1	118.3
2.4	7.9	29.9	60.2	91.5	119.7
2.9	8.3	30.6	63.7	92.1	120.6
3.0	8.7	32.4	64.6	97.9	121.0
3.1	9.2	33.0	65.3	100.8	122.9
3.3	9.8	35.3	66.2	102.6	123.3
3.5	10.2	36.1	70.1	103.2	124.5
3.8	10.4	40.1	71.0	104.0	125.8
4.3	11.9	42.8	75.1	104.3	126.6
4.6	13.8	43.7	75.6	105.0	127.7
4.7	14.4	44.5	78.4	105.8	128.4
4.8	15.6	50.4	79.2	106.5	129.2
5.2	16.2	51.2	84.1	110.7	129.5
5.4	17.0	52.0	86.0	112.6	129.9
5.9	17.5	53.3	87.9	113.5	

489. Survival time of patients with lymphocytic non-Hodgkins lymphoma

Dinse, G.E. (1982) Nonparametric estimation for partially complete time and type of failure data. *Biometrics*, **38**, 417–431.

These data are also analysed in Kimber, A.C. (1990) Exploratory data analysis for possibly censored data from skewed distributions. *Applied Statistics*, **39**, 21–30. They show the survival times (in weeks) of patients with lymphocytic non-Hodgkins lymphoma, classified into two groups: asymptomatic and symptomatic. Interest lies in comparing the survival times of the two groups. An asterisk indicates a censored observation.

Asymptomatic			Symptomatic
50	257	349*	49
58	262	354*	58
96	292	359	75
139	294	360*	110
152	300*	365*	112
159	301	378*	132
189	306*	381*	151
225	329*	388*	276
239	342*	281	
242	346*	362*	

490. Prestige, income, education, and suicide rates for 36 occupations

Labovitz, S. (1970) The assignment of numbers to rank order categories. *American Sociological Review*, **35**, 515–524, Table 1.

The table shows 36 occupations in the United States with, for each of them, (1) the NORC occupational prestige rating, (2) the male suicide rate (amongst males aged 20-64), (3) the median income (by 1949 in dollars), and (4) the median number of school years completed (in 1950). The source paper was concerned with the effect of treating ordinal scales as intervals, but clearly this data set could be used for other teaching purposes, such as regression.

Occupation	(1)	(2)	(3)	(4)
Accountants and auditors	82	23.8	3977	14.4
Architects	90	37.5	5509	16+
Authors, editors, and reporters	76	37.0	4303	15.6
Chemists	90	20.7	4091	16+
Clergymen	87	10.6	2410	16+
College presidents, professors, and instructors (n.e.c.)	93	14.2	4366	16+
Dentists	90	45.6	6448	16+
Engineers, civil	88	31.9	4590	16+
Lawyers and judges	89	24.3	6284	16+
Physicians and surgeons	97	31.9	8302	16+
Social welfare, recreation, and group workers	59	16.0	3176	15.8
Teachers (n.e.c.)	73	16.8	3456	16+
Managers, officials, and proprietors (n.e.c.) - self-employed - manufacturing	81	64.8	4700	12.2
Managers, officials, and proprietors (n.e.c.) - self-employed - wholesale and retail trade	45	47.3	3806	11.6
Bookkeepers	39	21.9	2828	12.7
Mail-carriers	34	16.5	3480	12.2

Insurance agents and brokers	41	32.4	3771	12.7
Salesmen and sales clerks (n.e.c.), retail trade	16	24.1	2543	12.1
Carpenters	33	32.7	2450	8.7
Electricians	53	30.8	3447	11.1
Locomotive engineers	67	34.2	4648	8.8
Machinists and job setters, metal	57	34.5	3303	9.6
Mechanics and repairmen, automobile	26	24.4	2693	9.4
Plumbers and pipe fitters	29	29.4	3353	9.3
Attendants, auto service and parking	10	14.4	1898	10.3
Mine operators and labourers (n.e.c.)	15	41.7	2410	8.2
Motormen, street, subway, and elevated railway	19	19.2	3424	9.2
Taxicab-drivers and chauffeurs	10	24.9	2213	8.9
Truck and tractor drivers, deliverymen and routemen	13	17.9	2590	9.6
Operatives and kindred workers (n.e.c.), machinery, except electrical	24	15.7	2915	9.6
Barbers, beauticians, and manicurists	20	36.0	2357	8.8
Waiters, bartenders, and counter and fountain workers	7	24.4	1942	9.8
Cooks, except private household	16	42.2	2249	8.7
Guards and watchmen	11	38.2	2551	8.5
Janitors, sextons, and porters	8	20.3	1866	8.2
Policemen, detectives, sheriffs, bailiffs, marshals, and constables	41	47.6	2866	10.6

491. Natural scientists and engineers in the US

Bayer, A.E. (1968) The effect of international interchange of high-level manpower on the United States. *Social Forces*, **46**, 465–477, Table 3.

The table shows the numbers of domestic graduates in natural science and engineering and the number of immigrant natural scientists and engineers in the United States between 1949 and 1964. Interest lies in modelling the change over time of the proportion of natural scientists and engineers who are immigrants.

Year	Domestic	Immigrant
1949-1950	209179	2753
1951-1952	166409	4675
1953-1954	118717	5918
1955-1956	119600	6652
1957-1958	151271	11013
1959-1960	175917	9407
1961-1962	183439	8027
1963-1964	196697	11181

492. Correlation between nine statements about pain

Skevington, S.M. (1990) A standardised scale to measure beliefs about controlling pain. (B.P.C.Q.): a preliminary study. *Psychological Health*, **4**, 221–232.

The subset of data presented here are also analysed in Arnold, G.M. and Collins, A.J. (1993) Interpretation of transformed axes in multivariate analysis. *Applied Statistics*, **42**, 381–400, Table 3. They show the correlations between nine statements about pain by 123 people suffering from extreme pain. Each statement was scored on a scale of 1 to 6 ranging from disagreement to agreement. These data are suitable for multivariate techniques such as factor analysis and principal components analysis.

1 = whether or not I am in pain in the future depends on the skills of the doctors.

2 = whenever I am in pain, it is usually because of something I have done or not done.

3 = whether or not I am in pain depends on what the doctors do for me.

4 = I cannot get any help for my pain unless I go to seek medical advice.

5 = when I am in pain I know that it is because I have not been taking proper exercise or eating the right food.

6 = people's pain results from their own carelessness.

7 = I am directly responsible for my pain.

8 = relief from pain is chiefly controlled by the doctors.

9 = people who are never in pain are just plain lucky.

	Statement							
	1	2	3	4	5	6	7	8
2	-.0385							
3	.6066	-.0693						
4	.4507	-.1167	.5916					
5	.0320	.4881	.0317	-.0802				
6	-.2877	.4271	-.1336	-.2073	.4731			
7	-.2974	.3045	-.2404	-.1850	.4138	.6346		
8	.4526	-.3090	.5886	.6286	-.1397	-.1329	-.2599	
9	.2952	-.1704	.3165	.3680	-.2367	-.1541	-.2893	.4047

493. Manufacture of truck leaf springs

Pignatiello, J.J. and Ramberg, J.S. (1985) Discussion on Kacker's paper. *Journal of Quality Technology*, **17**, 198–209.

These data are also analysed in Kirmani, S.N.U.A. and Shyamal Das Peddada (1993) Stochastic ordering approach to off-line quality control. *Applied Statistics*, **42**, 271–281, Table 3. They show the design and results of an experiment in the manufacture of leaf springs for trucks. The response variable is the free height of the spring in the unloaded condition, which has a target value of 8 inches. There are five factors, shown below, and each factor combination has three observations. Interest lies in determining which factors and factor combinations are important, both in terms of the mean result and in terms of its variance.

	Low (-)	High (+)
B, high heat temperature, °F	1840	1880
C, heating time, seconds	25	23
D, transfer time, seconds	12	10
E, hold-down time, seconds	2	3
O, quench oil temperature, °F	130-150	150-170

B	C	D	O	E	Sample mean	Sample variance
-	-	-	-	-	7.79	0.0003
+	-	-	-	+	8.07	0.0273
-	+	-	-	+	7.52	0.0012
+	+	-	-	-	7.63	0.0104
-	-	+	-	+	7.94	0.0036
+	-	+	-	-	7.95	0.0496
-	+	+	-	-	7.54	0.0084
+	+	+	-	+	7.69	0.0156
-	-	-	+	-	7.29	0.0373
+	-	-	+	+	7.73	0.0645
-	+	-	+	+	7.52	0.0012
+	+	-	+	-	7.65	0.0092
-	-	+	+	+	7.40	0.0048
+	-	+	+	-	7.62	0.0042
-	+	+	+	-	7.20	0.0016
+	+	+	+	+	7.63	0.0254

494. Housing satisfaction around Montevideo, Minnesota

Brier, S.S. (1980) Analysis of contingency tables under cluster sampling. *Biometrika*, **67**, 591–596.

These data are also presented in Wilson, J.R. (1989) Chi-square tests for overdispersion with multiparameter estimates. *Applied Statistics*, **38**, 441–453, Table 1. Households around Montevideo, Minnesota were classed as either metropolitan or non-metropolitan and a random sample of 20 neighbourhoods was taken from each class. Five households were randomly selected from each of the sampled neighbourhoods and each household gave a single response (unsatisfied (US), satisfied (S), or very satisfied (VS)) about their satisfaction with their home. The table shows the responses of the neighbourhoods which gave a complete five responses. The issue of interest, of course, is whether there is a difference between the metropolitan and non-metropolitan areas.

Non-metropolitan			Metropolitan		
US	S	VS	US	S	VS
3	2	0	0	4	1
3	2	0	0	5	0
0	5	0	0	3	2
3	2	0	3	2	0
0	5	0	2	3	0
4	1	0	1	3	1
3	2	0	4	1	0
2	3	0	4	0	1
4	0	1	0	3	2
0	4	1	1	2	2
2	3	0	0	5	0
4	1	0	3	2	0
4	1	0	2	3	0
1	2	2	2	2	1
4	1	0	4	0	1
1	3	1	0	4	1
4	1	0	4	1	0
5	0	0			

495. Car changing patterns

Harshman, R.A., Green, P.E., Wind, Y., and Lundy, M.E. (1982) A model for the analysis of asymmetric data in marketing research. *Marketing Science, 1*, 205–242.

These data are also analysed in van der Heijden, P.G.M., de Falguerolles, A., and de Leeuw, J. (1989) A combined approach to contingency table analysis using correspondence analysis and log-linear analysis. *Applied Statistics, 38*, 249–292. They show a survey of recent new car buyers in 1979, cross-classified according to the characteristics of their old and new cars. The rows denote cars disposed of and the columns new cars. The letters mean:

a = subcompact/domestic
b = subcompact/captive imports
c = subcompact/imports
d = small speciality/domestic
e = small speciality/captive imports
f = small speciality/imports
g = low price compacts
h = medium price compacts
i = import compact
j = midsize domestic
k = midsize imports
l = midsize speciality
m = low price standard
n = medium price standard
o = luxury domestic
p = luxury import

The data can be used to demonstrate the analysis of contingency tables and correspondence analysis.

	a	b	c	d	e	f	g	h
a	23272	1487	10501	18994	49	2319	12349	4061
b	3254	1114	3014	2656	23	551	959	894
c	11344	1214	25986	9803	47	5400	3262	1353
d	11740	1192	11149	38434	69	4880	6047	2335
e	47	6	0	117	4	0	0	49
f	1172	217	3622	3453	16	5249	1113	313
g	18441	1866	12154	15237	65	1626	27137	6182
h	10359	693	5841	6368	40	610	6223	7469
i	2613	481	6981	1853	10	1023	1305	632
j	33012	2323	22029	29623	110	4193	20997	12155
k	1293	114	2844	1242	5	772	1507	452
l	12981	981	8271	18908	97	3444	3693	1748
m	27816	1890	12980	15993	34	1323	18928	5836
n	17293	1291	11243	11457	41	1862	7731	6178
o	3733	430	4647	5913	6	622	1652	1044
p	105	40	997	603	0	341	75	55

	i	j	k	l	m	n	o	p
a	545	12622	481	16329	4253	2370	949	127
b	223	1672	223	2012	926	540	246	37
c	2257	5195	1307	8347	2308	1611	1071	288
d	931	8503	1177	23898	3238	4422	4114	410
e	0	110	0	10	0	0	0	0
f	738	1631	1070	4937	338	901	1310	459
g	835	20909	566	15342	9728	3610	910	170
h	564	9620	453	9731	3601	5498	764	85
i	1536	2738	1005	990	454	991	543	127
j	2533	53002	2140	61350	28006	33913	9808	706
l	565	3820	3059	2357	589	1052	871	595
l	935	11551	1314	56025	10959	18688	12541	578
m	1182	28324	938	37380	67964	28881	6585	300
n	1288	20942	1048	30189	15318	81808	21974	548
o	476	3068	829	8571	2964	9187	63509	1585
p	176	151	589	758	158	756	1234	3124

496. *In vitro* chromosome aberration assays

Williams, D.A. (1988) Tests for differences between several small proportions. *Applied Statistics,* **37**, 421–434, Table 1.

A chromosome aberration assay determines whether or not a substance induces structural changes in chromosomes. These data show the results of assays of two substances at four doses (the first being a zero dose control). At each dose a number of cells is sampled and the number aberrant determined by microscopic examination. In this case the aim is to see if there is a difference between the two substances.

Substance	Dose (mg/ml)	Number of cells sampled	Number of cells aberrant
A	0	400	3
	20	200	5
	100	200	14
	200	200	4
B	0	400	5
	62.5	200	2
	125	200	2
	250	200	4
	500	200	7

497. Visits to dentist

Sikkel, D. and Jelierse, G. (1988) Renewal theory and retrospective questions. *Applied Statistics,* **37**, 412–420, Table 1.

The table shows the reported number of contacts with the dentist for a sample of people in the Netherlands. A simple model might be the negative binomial distribution, but this turns out not to fit well — for example, the empirical distribution is bimodal. According to Sikkel and Jelierse, this reflects the fact that there is a group which regularly visit the dentist (and do so twice a year) and a group which rarely visit the dentist. Interest lies in finding a suitable model.

Number of contacts in 11.5 months	Count
0	4244
1	1719
2	3337
3	461
4	190
>4	267

498. Myocardial infarction and obesity, smoking, and vasectomy

Walker, A.M. (1982) Efficient assessment of confound effects in matched follow-up studies. *Applied Statistics*, **31**, 293–297, Table 1.

Thirty-six pairs of men from a cohort of 4830 men were matched on the basis of year of birth and calendar time of follow-up, with one man in each pair being vasectomized and one not. Shown below are the occurrences of non-fatal myocardial infarction. There were no pairs for which both men suffered a myocardial infarction (MI). The table shows, for each pair, which of the two men (vasectomized, V; or non-vasectomized, N) had the infarct, and gives details of obesity (- for not present, + for present: see source for details) and smoking history (- for non-smoking, + for smoking: see source for full details) for each man. Interest, of course, lies in the effect of the obesity, vasectomy, and smoking factors on the probability of an infarct.

Pair	Which suffered MI	Vasectomized		Not-vasectomized	
		Obese	Smoker	Obese	Smoker
1	N	-	-	+	-
2	V	+	-	-	+
3	V	-	+	-	-
4	N	-	+	-	+
5	N	+	-	+	+
6	V	-	-	-	+
7	V	+	-	-	+
8	V	-	-	-	-
9	N	-	-	-	+
10	V	-	+	-	+
11	N	-	+	-	+
12	N	-	-	-	-
13	V	+	+	-	+
14	N	-	-	-	+
15	V	-	+	-	-
16	V	+	+	+	+
17	V	-	-	-	-
18	N	+	+	-	+
19	V	+	+	-	-
20	N	-	-	-	+
21	N	-	-	-	+
22	N	-	+	-	-
23	V	-	+	-	+
24	V	-	+	-	-
25	V	+	+	-	-
26	N	-	-	-	-
27	V	-	+	-	+
28	V	+	-	-	-
29	N	-	-	-	+
30	V	-	-	-	-
31	N	-	+	+	+
32	V	-	+	-	-
33	N	+	-	-	-
34	N	+	-	-	+
35	V	-	-	-	-
36	V	-	-	-	-

499. Specific heat of water at various temperatures

Mosteller, F. and Tukey, J.W. (1977) *Data analysis and regression*. Reading, Massachusetts: Addison-Wesley. Exhibit 1, 503.

Six experimenters measured the specific heat of water at various temperatures. Interest lies in the reliability of the measurements and, of course, in obtaining an accurate estimate.

			Temperature (°C)			
Investigator	5	10	15	20	25	30
Liidin	1.0027	1.0010	1.0000	0.9994	0.9993	0.9996
Dieterici	1.0050	1.0021	1.0000	0.9987	0.9983	0.9984
Bonsfreld	1.0039	1.0016	1.0000	0.9991	0.9989	0.9990
Ronland	1.0054	1.0019	1.0000	0.9979	0.9972	0.9969
Bartollis	1.0041	1.0017	1.0000	0.9994	1.0000	1.0016
Janke	1.0040	1.0016	1.0000	0.9991	0.9987	0.9988

500. Numbers of packets of cereal purchased over 13 weeks

Barnett, V. and Lewis, T. (1984) *Outliers in statistical data*. Chichester: John Wiley & Sons, Table 1.1.

The table shows the frequency distribution for numbers of packets of cereal purchased over 13 weeks by 2000 customers. Questions of interest concern whether there are outliers and how they should be treated.

0	1149	14	8	29	1	43	0
1	199	15	2	30	1	44	0
2	129	16	7	31	0	45	0
3	87	18	3	32	0	46	0
4	71	19	1	33	0	47	0
5	43	20	2	34	0	48	0
6	49	21	0	35	0	49	0
7	46	22	0	36	0	50	0
8	44	23	1	37	0	51	0
9	24	24	0	38	0	52	1
10	45	25	1	39	1	>52	0
11	22	26	3	40	0		
12	23	27	2	41	0		
13	33	28	0	42	0		

501. Accelerated life tests on electrical insulation

Schmee, J. and Hahn, G.J. (1979) A simple method for regression analysis with censored data. *Technometrics,* **21**, 417–432.

These data are also analysed in Naylor, J.C. and Smith, A.F.M. (1982) Applications of a method for the efficient computation of posterior distributions. *Applied Statistics,* **31**, 214–255. They show the results (in hours) of temperature accelerated life tests on electrical insulation in 40 motorettes. Ten motorettes were tested at each of four temperatures. Interest lies in modelling the distribution of failure times and comparing the results at the different temperatures. Censoring is indicated by *.

Temperature (°C)

150	170	190	220
8064*	1764	408	408
8064*	2772	408	408
8064*	3444	1344	504
8064*	3542	1344	504
8064*	3780	1440	504
8064*	4860	1680*	528*
8064*	5196	1680*	528*
8064*	5448*	1680*	528*
8064*	5448*	1680*	528*
8064*	5448*	1680*	528*

502. Cardiac oxygen consumption and mechanical function

Sweeting, T.J. (1982) A Bayesian analysis of some pharmacological data using a random coefficient regression model. *Applied Statistics,* **31**, 205–213, Table 1.

The table shows values of cardiac oxygen consumption (MVO), mean left ventricular pressure (LVP), stroke volume (SV), heart rate (HR), aortic pressure (AP), and tension-time index (TTI) for a dog both with and without a beta-blocking agent (atenolol). Interest lies in modelling the relationship between the mechanical heart function measures and oxygen consumption, and in particular whether there is any difference between the consumption in the two groups over and above that which can be attributed to the mechanical function.

	LVP	SV	HR	AP	MVO
Control	32	19	50	102	78
condition	33	15	70	110	92
	45	8	110	115	116
	30	14	70	96	90
	38	7	110	100	106
	24	13	70	85	78
	44	8	110	110	99

Atenolol	24	11	70	75	74
	30	6	110	75	89
	20	9	70	60	53
	26	6	110	60	79

503. Coffee consumption and coronary heart disease

Table (a) is from Paul, O. (1968) Stimulants and coronaries. *Postgraduate Medicine*, **44**, 196–199.

Table (b) is from Paul, O., MacMillan, A., McKean, H. and Park, H. (1968) Sucrose intake and coronary heart disease. *Lancet*, **ii**, 1049–1051.

The data are also analysed in Greenland, S. and Mickey, R.M. (1988) Closed form and dually consistent methods for inference on strict collapsibility in $2 \times 2 \times K$ and $2 \times J \times K$ tables. *Applied Statistics*, **37**, 335–343.

Table (a) shows a cross-classification of coronary heart disease (CHD) by coffee drinking for a cohort of 1718 men aged 40-55. Table (b) shows a cross-classification of coronary heart disease by coffee drinking and smoking in a case control study of 66 CHD cases and 85 unmatched controls from the same population as Table (a).

These data can be used to illustrate analysis of 2×2 tables and odds ratios.

(a)

	Heavy coffee drinkers (>= 100 cups/month)	Moderate and non-drinkers (<100 cups/month)
CHD cases	38	39
Non-cases	752	889

(b)

	Heavy smokers (1 pack/day or more)		Others (under 1 pack/day)	
	Coffee use		Coffee use	
	Heavy	Other	Heavy	Other
CHD cases	25	11	15	15
Non-cases	14	8	21	42

504. Strength of sheet metal

John, J.A. (1978) Outliers in factorial experiments. *Applied Statistics,* **27**, 111–119.

The data arise from a confounded 2^5 experiment to determine how the strength of sheet metal depends on five factors describing the coating process. Of particular interest here is the detection of outliers in these data. In the table T is treatment and Y is yield.

T	Y	T	Y	T	Y	T	Y
(1)	1.4	d	5.0	e	1.7	de	9.5
a	1.2	ad	9.0	ae	2.0	ade	5.9
b	3.6	bd	12.0	be	3.1	bde	12.6
ab	1.2	abd	5.4	abe	1.2	abde	6.3
c	1.5	cd	4.2	ce	1.9	cde	8.0
ac	1.4	acd	4.4	ace	1.2	acde	4.2
bc	1.5	bcd	9.3	bce	1.0	bcde	7.7
abc	1.6	abcd	2.8	abce	1.8	abcde	6.0

505. Mortality from testicular cancer in England and Wales

Osmond, C. (1985) Biplot models applied to cancer mortality rates. *Applied Statistics,* **34**, 63–70, Table 1.

For five-year periods from 1926 to 1980 the table shows the death rates per million person-years at risk classified by age at death. These data can be used to illustrate biplot techniques.

Age at death	Period of death										
	1926 -30	1932 -35	1936 -40	1941 -45	1946 -50	1951 -55	1956 -60	1961 -65	1966 -70	1971 -75	1976 -80
15-19	2	3	3	4	6	3	3	5	6	6	6
20-24	5	5	7	7	10	11	12	15	15	18	16
25-29	8	8	11	10	14	16	21	19	23	24	23
30-34	10	11	13	13	16	15	18	22	20	25	22
35-39	12	12	15	15	15	17	14	14	17	21	19
40-44	11	14	11	11	14	12	11	10	14	12	12
45-49	9	9	9	11	12	10	10	11	9	11	8
50-54	9	12	7	10	10	8	8	10	10	7	6
55-59	6	8	9	7	9	8	7	8	7	9	8
60-64	11	9	12	10	9	10	8	6	7	7	5
65-69	16	15	17	13	11	16	16	12	9	10	10
70-74	18	18	21	20	13	14	16	10	16	11	10
75-79	29	20	18	14	19	24	16	14	13	17	11

506. Fathers' and sons' occupations

Pearson, K. (1904) On the theory of contingency and its relation to association and normal correlation. Reprinted in 1948 in *Karl Pearson's Early Statistical Papers*, Cambridge: Cambridge University Press, 443–475.

This is a sparse contingency table showing the cross-classification of occupations of fathers (rows) by occupations of sons (columns). These data have been used to illustrate methods for finding the source of a significant chi-squared test of independent rows and columns and for locating outliers. See, for example, Kotze, T.J.v W. and Hawkins, D.M. (1984) The identification of outliers in two-way contingency tables using 2 × 2 subtables. *Applied Statistics*, **33**, 215–223, Table 3.

	a	b	c	d	e	f	g	h	i	j	k	l	m	n
a	28	0	4	0	0	0	1	3	3	0	3	1	5	2
b	2	51	1	1	2	0	0	1	2	0	0	0	1	1
c	6	5	7	0	9	1	3	6	4	2	1	1	2	7
d	0	12	0	6	5	0	0	1	7	1	2	0	0	10
e	5	5	2	1	54	0	0	6	9	4	12	3	1	13
f	0	2	3	0	3	0	0	1	4	1	4	2	1	5
g	17	1	4	0	14	0	6	11	4	1	3	3	17	7
h	3	5	6	0	6	0	2	18	13	1	1	1	8	5
i	0	1	1	0	4	0	0	1	4	0	2	1	1	4
j	12	16	4	1	15	0	0	5	13	11	6	1	7	15
k	0	4	2	0	1	0	0	0	3	0	20	0	5	6
l	1	3	1	0	0	0	1	0	1	1	1	6	2	1
m	5	0	2	0	3	0	1	8	1	2	2	3	23	1
n	5	3	0	2	6	0	1	3	1	0	0	1	1	9

507. Numbers of purchases of toilet tissue

Dunn, R., Reader, S. and Wrigley, N. (1983) An investigation of the assumptions of the NBD model as applied to purchasing at individual stores. *Applied Statistics*, **32**, 249–259.

The data give observed number of purchases of toilet tissue at four stores. Two of the stores (Tesco City Centre, and Leo's Superstore) are centrally located and two (Co-op Countisbury Avenue and Maelfa) are in suburban areas. One might be interested in finding a suitable model for these distributions (one might try negative binomial) and perhaps in comparing the central and suburban results.

| No. of purchases | Central stores | | Suburban stores | |
	Tesco City Centre	Leo's Superstore	Co-op Countisbury Avenue	Intern'l Maelfa
0	326	366	371	409
1	43	29	12	7
2	20	13	10	6
3-4	23	14	12	8
5-6	18	5	8	2
7-9	9	7	11	2
10-14	5	8	15	5
15-19	6	8	4	8
20+	4	4	11	7

508. Haematology of paint sprayers

Royston, J.P. (1983) Some techniques for assessing multivariate normality based on the Shapiro-Wilk W. *Applied Statistics,* **32,** 121–133.

The data in this example are a subset of data obtained in a health survey of paint sprayers in a car assembly plant. There are six variables:

HAEMO = haemoglobin concentration
PCV = Packed cell volume
WBC = white blood cell count
LYMPHO = lymphocyte count
NEUTRO = neutrophil count
LEAD = serum lead concentration

Of particular interest to the source was whether the data were normally distributed (after a log transformation). The WBC measurement for case 93 looks like a data entry error.

Case	HAEMO	PCV	WBC	LYMPHO	NEUTRO	LEAD
1	13.4	39	4100	14	25	17
2	14.6	46	5000	15	30	20
3	13.5	42	4500	19	21	18
4	15.0	46	4600	23	16	18
5	14.6	44	5100	17	31	19
6	14.0	44	4900	20	24	19
7	16.4	49	4300	21	17	18
8	14.8	44	4400	16	26	29
9	15.2	46	4100	27	13	27
10	15.5	48	8400	34	42	36
11	15.2	47	5600	26	27	22

12	16.9	50	5100	28	17	23
13	14.8	44	4700	24	20	23
14	16.2	45	5600	26	25	19
15	14.7	43	4000	23	13	17
16	14.7	42	3400	9	22	13
17	16.5	45	5400	18	32	17
18	15.4	45	6900	28	36	24
19	15.1	45	4600	17	29	17
20	14.2	46	4200	14	25	28
21	15.9	46	5200	8	34	16
22	16.0	47	4700	25	14	18
23	17.4	50	8600	37	39	17
24	14.3	43	5500	20	31	19
25	14.8	44	4200	15	24	19
26	14.9	43	4300	9	32	17
27	15.5	45	5200	16	30	20
28	14.5	43	3900	18	18	25
29	14.4	45	6000	17	37	23
30	14.6	44	4700	23	21	27
31	15.3	45	7900	43	23	23
32	14.9	45	3400	17	15	24
33	15.8	47	6000	23	32	21
34	14.4	44	7700	31	39	23
35	14.7	46	3700	11	23	23
36	14.8	43	5200	25	19	22
37	15.4	45	6000	30	25	18
38	16.2	50	8100	32	38	18
39	15.0	45	4900	17	26	24
40	15.1	47	6000	22	33	16
41	16.0	46	4600	20	22	22
42	15.3	48	5500	20	23	23
43	14.5	41	6200	20	36	21
44	14.2	41	4900	26	20	20
45	15.0	45	7200	40	25	25
46	14.2	46	5800	22	31	22
47	14.9	45	8400	61	17	17
48	16.2	48	3100	12	15	18
49	14.5	45	4000	20	18	20
50	16.4	49	6900	35	22	24
51	14.7	44	7800	38	34	16
52	17.0	52	6300	19	21	16
53	15.4	47	3400	12	19	18
54	13.8	40	4500	19	23	21
55	16.1	47	4600	17	28	20
56	14.6	45	4700	23	22	27
57	15.0	44	5800	14	39	21
58	16.2	47	4100	16	24	18
59	17.0	51	5700	26	29	20
60	14.0	44	4100	16	24	18
61	15.4	46	6200	32	25	16
62	15.6	46	4700	28	16	16
63	15.8	48	4500	24	20	23
64	13.2	38	5300	16	26	20
65	14.9	47	5000	22	25	15
66	14.9	47	3900	15	19	16

67	14.0	45	5200	23	25	17
68	16.1	47	4300	19	22	22
69	14.7	46	6800	35	25	18
70	14.8	45	8900	47	36	17
71	17.0	51	6300	42	19	15
72	15.2	45	4600	21	22	18
73	15.2	43	5600	25	28	17
74	13.8	41	6300	25	27	15
75	14.8	43	6400	36	24	18
76	16.1	47	5200	18	28	25
77	15.0	43	6300	22	34	17
78	16.2	46	6000	25	25	24
79	14.8	44	3900	9	25	14
80	17.2	44	4100	12	27	18
81	17.2	48	5000	25	19	25
82	14.6	43	5500	22	31	19
83	14.4	44	4300	20	20	15
84	15.4	48	5700	29	26	24
85	16.0	52	4100	21	15	22
86	15.0	45	5000	27	18	20
87	14.8	44	5700	29	23	23
88	15.4	43	3300	10	20	19
89	16.0	47	6100	32	23	26
90	14.8	43	5100	18	31	19
91	13.8	41	8100	52	24	17
92	14.7	43	5200	24	24	17
93	14.6	44	9899	69	28	18
94	13.6	42	6100	24	30	15
95	14.5	44	4800	14	29	15
96	14.3	39	5000	25	20	19
97	15.3	45	4000	19	19	16
98	16.4	49	6000	34	22	17
99	14.8	44	4500	22	18	25
100	16.6	48	4700	17	27	20
101	16.0	49	7000	36	28	18
102	15.5	46	6600	30	33	13
103	14.3	46	5700	26	20	21

509. Measurement of protein content in ground wheat

Fearn, T. (1983) A misuse of ridge regression in the calibration of a near infrared reflectance instrument. *Applied Statistics,* **32**, 73–79.

The table shows the results of an experiment to calibrate a near infrared reflectance instrument for the measurement of protein content of ground wheat samples. The second column shows the protein content, measured by the standard Kjeldahl method. The final six columns show measurements of the reflectance of near infrared radiation of the wheat samples at six wavelengths in the range 1680–2310. In the source paper the aim was to find a linear combination of the last six columns which could be used to predict protein content. Although the high correlations between the explanatory variables may suggest the use of ridge regression, this is not appropriate as the relevant information in the explanatory variables is associated with small eigenvalues of their correlation matrix.

Sample	Protein (%)	L1	L2	L3	L4	L5	L6
1	9.23	468	123	246	374	386	-11
2	8.01	458	112	236	368	383	-15
3	10.95	457	118	240	359	353	-16
4	11.67	450	115	236	352	340	-15
5	10.41	464	119	243	366	371	-16
6	9.51	499	147	273	404	433	5
7	8.67	463	119	242	370	377	-12
8	7.75	462	115	238	370	353	-13
9	8.05	488	134	258	393	377	-5
10	11.39	483	141	264	384	398	-2
11	9.95	463	120	243	367	378	-13
12	8.25	456	111	233	365	365	-15
13	10.57	512	161	288	415	443	12
14	10.23	518	167	293	421	450	19
15	11.87	552	197	324	448	467	32
16	8.09	497	146	271	407	451	11
17	12.55	592	229	360	484	524	51
18	8.38	501	150	274	406	407	11
19	9.64	483	137	260	385	374	-3
20	11.35	491	147	269	389	391	1
21	9.70	463	121	242	366	353	-13
22	10.75	507	159	285	410	445	13
23	10.75	474	132	255	376	383	-7
24	11.47	496	152	276	396	404	6

510. Winning times in men's running events

Chatterjee, S. and Chatterjee, S. (1982) New lamps for old: an exploratory analysis of running times in Olympic Games. *Applied Statistics*, **31**, 14–22.

The table shows the winning times (in seconds) for men's running events in the Olympic Games between 1900 and 1976 for four different distances. The altitude at which the Games took places is also given. Interest lies in finding a model which best fits the observed times.

Year	100 metres	200 metres	400 metres	800 metres	Altitude
1900	10.80	22.20	49.40	121.40	25
1904	11.00	21.60	49.20	116.00	455
1908	10.80	22.60	50.00	112.80	8
1912	10.80	21.70	48.20	111.90	46
1920	10.80	22.00	49.60	113.40	3
1924	10.60	21.60	47.60	112.40	25
1928	10.80	21.80	47.80	111.80	8
1932	10.30	21.20	46.20	109.80	340
1936	10.30	20.70	46.50	112.90	115
1948	10.30	21.10	46.20	109.20	8
1952	10.40	20.70	45.90	109.20	25
1956	10.50	20.60	46.70	107.70	3
1960	10.20	20.50	44.90	106.30	66
1964	10.00	20.30	45.10	105.10	45
1968	9.90	19.80	43.80	104.30	7349
1972	10.14	20.00	44.66	105.90	1699
1976	10.06	20.23	44.26	103.50	104

DATA STRUCTURE INDEX

Title	Data structure		Filename
1. Germinating seeds	48	3	binary, categorical, count	GERMIN.DAT
2. Guessing lengths	44	1	numeric	LENGTHS.DAT
3. Darwin's cross-fertilized and self-fertilized plants	69	1	numeric	DARWIN.DAT
	15	2	rep(2) numeric	
4. Intervals between cars on the M1 motorway	41	1	numeric	INTERVAL.DAT
5. Tearing factor for paper	20	2	numeric(2)	TEARING.DAT.
6. Abrasion loss	30	3	numeric(3)	ABRASION.DAT
7. Mortality and water hardness	61	3	binary, rate(2)	WATER.DAT
8. Tensile strength of cement	21	2	numeric(2)	CEMENT.DAT
9. Weight gain in rats	40	3	binary(2),numeric	WEIGHT.DAT
10. Weight of chickens	24	3	nominal, categorical, numeric	CHICKENS.DAT
11. Flicker frequency	27	4	nominal(3), numeric [latin square]	FLICKER.DAT
12. Effect of ammonium chloride on yield	16	5	nominal, binary(3), numeric [randomized blocks]	CHLORIDE.DAT
13. Comparing dishwashing detergents	36	3	nominal(2), count [bibd]	DISHWASH.DAT
14. Software system failures	136	1	numeric [survival]	SOFTWARE.DAT
15. Piston-ring failures	12	3	categorical, nominal, count	PISTON.DAT
16. Strength of chemical pastes	60	3	nominal(2), numeric	PASTES.DAT

Title	Data structure			Filename
17. Human age and fatness	18	3	numeric, percentage, binary	HUMAN.DAT
18. Motion sickness	49	3	binary(2), numeric [survival]	MOTION.DAT
19. Plum root cuttings	960	3	binary(3)	PLUM.DAT
20. Spectacle wearing and delinquency	16	2	binary(2)	GLASSES.DAT
21. Yield of isatin derivative	16	5	binary(4), numeric	ISATIN.DAT
22. Fecundity of fruitflies	75	2	nominal, numeric	FRUITFLY.DAT
23. Butterfat	100	3	nominal, binary, numeric	BUTTER.DAT
24. Snoring and heart disease	2484	2	categorical, binary	SNORING.DAT
25. Trees' nearest neighbours	6	6	count(6)	TREES.DAT
26. Air pollution in US cities	41	7	numeric(6), count	USAIR.DAT
27. Acacia ants	28	2	binary(2)	ACACIA.DAT
28. Vaccination	24	6	nominal(3), binary, count(2)	VACCINE.DAT
29. Peppers in glasshouses	24	6	binary(5), numeric	PEPPERS.DAT
30. A clinical trial in lymphoma	273	2	categorical, binary	CLINICAL.DAT
31. Danish do-it-yourself	1591	6	numeric, binary(4), categorical	DANISH.DAT
32. Testing cement	36	3	nominal(2), numeric	TESTING.DAT
33. Irises	150	5	numeric(4), nominal	IRISES.DAT
34. Water voles	14	13	percentage(13)	VOLES.DAT
	14 × 14		[dissimilarity matrix]	
35. Facilities in East Jerusalem	8	9	percentage(9)	FACILITY.DAT
36. Yields of winter wheat	12	4	rep(4) numeric	WHEAT.DAT
37. WISC blocks	24	3	binary, numeric(2)	BLOCKS.DAT
38. Byssinosis	5419	6	categorical, binary(4), numeric	LUNG.DAT
39. A tomato crossing experiment	1611	2	binary(2)	TOMATO.DAT
40. Coronary heart disease	1329	3	numeric(2), binary	CORONARY.DAT
41. Toxaemia of pregnancy	113384	4	categorical(2), binary(2)	TOXAEMIA.DAT
42. Starting positions in horse racing	144	1	rank	HORSE.DAT

Title	Data structure			Filename
43. Saltiness judgements	9	1	numeric	SALT.DAT
44. Oral socialization and explanations of illness	39	2	binary, categorical	ILLNESS.DAT
45. Dopamine and schizophrenia	25	2	binary, numeric	DOPAMINE.DAT
46. US cancer mortality	11	5	rep(2) count(2), numeric	USCANCER.DAT
47. Cholesterol and behaviour type	3154	2	binary, numeric	BEHAVE.DAT
48. Aflatoxin in peanuts	34	2	numeric, percentage	PEANUTS.DAT
49. Anscombe's correlation data	11	2	numeric(2)	ANSCOMBE.DAT
	11	2	numeric(2)	
	11	2	numeric(2)	
	11	2	numeric(2)	
50. Caffeine and finger tapping	30	2	numeric(2)	CAFFEINE.DAT
51. Jackal mandible lengths	20	2	binary, numeric	JACKAL.DAT
52. Birds in paramo vegetation	14	5	count, numeric(4)	PARAMO.DAT
53. Extinction of marine genera	48	2	numeric, percentage	MARINE.DAT
54. Household expenditures	40	5	binary, numeric(4)	HOUSE.DAT
55. Cork deposits	28	4	numeric(4)	CORK.DAT
56. Creatinine kinase and heart attacks	360	2	binary, numeric	HEART.DAT
57. Petrol expenditure	32	4	numeric(4)	PETROL.DAT
58. North Buckinghamshire moths	24	1	count	MOTHS.DAT
59. Crowds and threatened suicide	21	2	binary(2)	SUICIDE.DAT
60. Putting the shot	20	1	numeric	SHOT.DAT
61. Wooden toy prices	31	1	numeric	TOY.DAT
62. Food prices	10	2	rep(2) numeric	FOOD.DAT
63. UK earnings ratios	1	60	rep(12) numeric(5)	EARNINGS.DAT
64. Census data on 10 towns	10	4	percentage(4)	CENSUS.DAT
65. Expenditure on food	11	2	numeric, percentage	EXPEND.DAT
66. Anaerobic threshold	1	106	rep(53) numeric(2)	ANAEROB.DAT
67. Road casualties on Fridays	1	24	rep(24) count [time series]	FRIDAY.DAT

Title	Data structure			Filename
68. UK sprinters	10	2	maxima(2)	UKSPRINT.DAT
69. Unemployment and expenditure on motoring	11	2	percentage(2)	MOTORING.DAT
70. Cloud seeding	24	7	binary(2), numeric(5)	CLOUD.DAT
71. Calculator random digits	10	1	count	RANDOM.DAT
72. Captopril and blood pressure	15	4	rep(2) numeric(2)	BLOOD.DAT
73. Food prices and house prices	1	38	rep(19) numeric(2) [time series]	PRICES.DAT
74. UNICEF data on child mortality	14	3	rate, percentage(2)	UNICEF.DAT
	16	2	rate, percentage	
	12	2	rate, percentage	
75. Length of stay on a psychiatric observation ward	336	3	binary(2), numeric	WARD.DAT
76. Hodgkin's disease	538	2	categorical, nominal	HODGKINS.DAT
77. A Church Assembly vote	451	2	nominal(2)	CHURCH.DAT
78. A vandalized experiment	36	4	nominal, numeric(3) [latin square]	VANDAL.DAT
79. Chest, waist and hips measurements	20	3	numeric(3)	MEASURE.DAT
80. Morse code mistakes	10 × 10		[confusion matrix]	MORSE.DAT
81. Distribution of minimum temperatures	3 (≤)49		rep(49)minima [time series]	MINTEMP.DAT
82. Rat skeletal muscle	25	4	count(4)	MUSCLE.DAT
83. Ground cover under apple trees	48	4	nominal(2), numeric(2)	APPLE.DAT
84. Lung cancer and occupation	25	2	numeric(2)	CANCER.DAT
85. House insulation: Whitburn	15	2	binary, numeric	WHITBURN.DAT
86. House insulation: Bristol	20	2	binary, numeric	BRISTOL.DAT
87. Engine capacity and fuel consumption	9	2	numeric(2)	ENGINE.DAT
88. House insulation: Whiteside's data	1	113	rep(56) numeric(2), binary	INSULATE.DAT
89. 1981 coffee prices	15	1	numeric	COFFEE.DAT
90. House temperatures: Neath Hill	15	2	numeric(2)	TEMPERAT.DAT

Title	Data structure		Filename
91. Sex differences at school	749	2	SEX.DAT
	749	2	
			categorical, binary
			categorical, binary
92. Destination of school leavers	115	2	SCHOOL.DAT
93. House insulation: Pennyland	26	2	PENNY.DAT
	27	2	
	16	2	
	30	2	
94. Wave energy device mooring	18	2	WAVE.DAT
95. The effect of light on mustard root growth	31	3	MUSTARD.DAT
96. Height and weight of 11-year-old girls	30	2	HEIGHT.DAT
97. Consulting the doctor about a child	15	2	DOCTOR.DAT
98. House heating: Great Linford	10	3	HEATING.DAT
99. Births in Basel	1	12	BASEL.DAT
100. Homing in desert ants	20	2	ANTS.DAT
101. Homing in pigeons	18	2	PIGEONS.DAT
102. Breast cancer mortality and temperature	16	2	BREAST.DAT
103. Deaths from sport parachuting	5	4	SPORT.DAT
104. Cotton imports in the 18th century	1	31	COTTON.DAT
105. British exports in the 19th century	1	31	EXPORTS.DAT
106. Size of ships in 1907	25	3	SHIPSIZE.DAT
107. Car sales in Quebec	1	108	QUEBEC.DAT
108. Sales and advertising	1	72	SALES.DAT
109. Canadian lynx trappings	1	114	LYNX.DAT
110. Jet fighters	22	6	JET.DAT
111. Head size in brothers	25	4	HEADSIZE.DAT
112. Sunspots	1	100	SUNSPOTS.DAT

Data structure descriptions (in order):

- categorical, binary
- categorical, binary
- categorical, numeric
- numeric, binary
- numeric, binary
- rep(2) numeric
- numeric, binary
- numeric(2)
- binary(2), numeric
- numeric(2)
- numeric, percentage
- numeric(3)
- rep(12) count
- binary, numeric
- binary, numeric
- numeric, rate
- rep(3) count, count
- rep(31) numeric [time series]
- numeric [time series]
- numeric, count, binary
- rep(108) count [time series]
- rep(36) numeric(2) [time series]
- rep(114) count [time series]
- numeric(5), binary
- numeric(4)
- rep(100) count [time series]

Title	Data structure		Filename
113. Airline passenger numbers	1	144	rep(144) count [time series]	AIRLINE.DAT
114. Yields from a batch chemical process	1	70	rep(70) numeric [time series]	YIELDS.DAT
115. Distances in Sheffield	20	2	numeric(2)	DISTANCE.DAT
116. University of Iowa enrolments	1	29	rep(29) count [time series]	IOWA.DAT
117. Insects in dunes	33	1	count	DUNES.DAT
118. Women's heights in Bangladesh	1243	1	numeric	WOMEN.DAT
119. Size of families in California	1800	1	count	FAMILIES.DAT
120. Hutterite population structure	1236	2	count(2)	HUTTERIT.DAT
121. Whooping cranes	35	2	count(2)	CRANES.DAT
122. Pork and cirrhosis	10	3	numeric(3)	PORK.DAT
123. Pet birds and lung cancer	429	2	binary(2)	PETBIRDS.DAT
124. Water injections and whiplash injuries	40	3	binary(3)	WHIPLASH.DAT
125. Telling cancer patients about their disease	27	14	count(7), proportion(7)	QUESTION.DAT
126. Diet and neural tube defects	367	2	categorical, binary	TUBE.DAT
	176	2	categorical, binary	
127. Age at marriage in Guatemala	90	1	numeric	MARRIAGE.DAT
128. Sand-flies	521	3	binary, numeric(2)	SANDFLY.DAT
129. Tonsil size and carrier status	1398	2	binary, categorical	TONSIL.DAT
130. Duration of pregnancy	1669	2	numeric, categorical	PREGNANT.DAT
131. 20 000 die throws	6	1	count	DIETHROW.DAT
132. Iron in slag	53	2	numeric(2)	IRON.DAT
133. Lowering blood pressure during surgery	54	3	numeric(3)	SURGERY.DAT
134. Crime in the USA: 1960	47	14	numeric(13), binary	USCRIME.DAT
135. Prater's gasoline yields	32	5	numeric(4), percentage	PRATER.DAT
136. Strength of cables	108	2	nominal, numeric	CABLES.DAT
137. Control of leatherjackets	72	3	nominal(2), numeric [randomised blocks]	CONTROL.DAT
138. Spinning synthetic yarn	36	4	nominal(2), categorical, numeric [latin square]	YARN.DAT

Title	Data structure			Filename
139. Forearm lengths	140	1	numeric	FOREARM.DAT
140. Lifetime of electric lamps	300	1	numeric	LAMPS.DAT
141. Computer failures	1	128	rep(128) count [time series]	COMPFAIL.DAT
142. Replacement value of books	100	2	numeric(2)	VALUE.DAT
143. Sulphinpyrazone and heart attacks	1475	2	binary(2)	ATTACKS.DAT
144. Tibetan skulls	32	6	binary, numeric(5)	TIBETAN.DAT
145. East Midlands village dialects	25 ×	25	[similarity matrix]	DIALECT.DAT
146. Lung cancer mortality	24	4	numeric, count(2), nominal	MORTAL.DAT
147. Aphids	19 ×	19	[correlation matrix]	APHIDS.DAT
148. Pit props	13 ×	13	[correlaton matrix]	PITPROPS.DAT
149. Silver content of Byzantiné coins	27	2	nominal, percentage	SILVER.DAT
150. Shoshoni American Indians	20	1	ratio	SHOSHONI.DAT
151. Component lifetimes	20	1	numeric [survival]	LIFETIME.DAT
152. Duckweed	2	9	rep(9) count	DUCKWEED.DAT
153. Simulated polynomial regression data	21	3	numeric(3)	SIMULATE.DAT
154. Heights of elderly females	351	1	numeric	ELDERLY.DAT
155. Etruscan and Italian skulls	154	2	binary, numeric	ETRUSCAN.DAT
156. A random pattern screen	1000	1	count	PATTERN.DAT
157. Rainfall in Australia	1	47	rep(47) maxima	RAINFALL.DAT
158. Scottish soldiers	5732	1	numeric	SOLDIERS.DAT
159. Borrowing library books	114035	1	count	LIBRARY.DAT
	18854	1	count	
160. Nerve impulse times	1	799	rep(799) numeric [time series]	NERVE.DAT
161. Corneal thickness of eyes	8	2	rep(2) numeric	CORNEAL.DAT
162. Memory retention	1	26	rep(13) numeric(2)	MEMORY.DAT
163. Student's yeast cell counts	1600	2	nominal, count	YEAST.DAT
	20 ×	20	[spatial]	
164. A comparison of growing conditions	24	2	nominal, numeric	GROWING.DAT
165. Anacapa pelican eggs	65	2	numeric(2)	PELICAN.DAT

Title	Data structure			Filename
				STEEL.DAT
166. Steel ball bearings	binary, numeric	20	2	MURDER.DAT
167. Murder rates	rep(2) numeric	30	2	MODEL.DAT
168. Kulasekeva model	count	240	1	
	count	122	1	EPILEPSY.DAT
169. Epileptic seizures	count	193	1	
	count	422	1	
	count	351	1	
				REDDEER.DAT
170. Red deer	numeric(2) [truncated]	78	2	ALBINISM.DAT
171. The incidence of albinism	count(2)	209	2	QUEUES.DAT
172. Females in queues	count	100	1	STUDENT.DAT
173. Student absenteeism	rep(2) count	113	2	LEADING.DAT
174. Leading digits	categorical	305	1	FUNGUS.DAT
175. Spores of the fungus *Sordaria*	proportion	907	1	LITTERS.DAT
176. Litters of pigs	count(2)	1961	2	SHOOTS.DAT
177. Shoots of *Armeria maritima*	count	100	1	
	count	100	1	
				NFLMATCH.DAT
178. American N.F.L. matches	numeric(2)	42	2	LEUKAEM.DAT
179. Survival times in leukaemia	numeric [survival]	43	1	TENSILE.DAT
180. Tensile strength	numeric	30	1	FATIGUE.DAT
181. Fatigue-life failures	numeric [survival]	22	1	LINKED.DAT
182. Linked transmission failures	numeric(2) [survival]	15	2	BIOLOGY.DAT
183. Published research papers in biology	count	1534	1	ROACHES.DAT
184. Counting cockroaches	rep(4) count(2)	1	8	SUNFISH.DAT
185. Survival times of green sunfish	binary, numeric [survival]	40	2	PLAGUE.DAT
186. The great plague of 1665	rep(9) count(2)	1	18	LOG.DAT
187. Log-series data	count	1469	1	
	count	211	1	
	count	102	1	
	count	73	1	

Title	Data structure		Filename
188. Eye and hair colour in children	27748	3	nominal(2), binary — EYE.DAT
189. Suicide figures in prisons	1	12	rep(6) count(2) — PRISON.DAT
190. Counting beetles	16	4	categorical(2), nominal, count [latin square] — BEETLES.DAT
191. The number of words in a sentence	600	1	count — WORDS.DAT
192. The size of gangs	895	1	count — GANGS.DAT
193. Skin graft survival times	11	3	rep(2) numeric, binary [survival] — GRAFT.DAT
194. Angles of spiders' webs	10	1	numeric — WEBS.DAT
195. Finger length	3000	1	numeric — FINGER.DAT
196. The Charlier model	7640	1	count — CHARLIER.DAT
197. Disease clusters	1	62	rep(62) count [time series] — CLUSTERS.DAT
198. Multiple sclerosis	43	4	binary(2), numeric(2) — MS.DAT
199. The number of pigs in a litter	378	1	count — PIGS.DAT
200. The lengths of scallops	222	1	numeric — SCALLOPS.DAT
201. Smoking habits in children	1	37	rep(4) count(9), binary [transition matrices] — SMOKING.DAT
202. The genesis of the t-test	10	2	numeric(2) — TTEST.DAT
203. Dimensions of cuckoos' eggs	243	1	numeric — CUCKOOS.DAT
	243	1	numeric
204. Time intervals between coal mining disasters	1	190	rep(190) count [time series] — COAL.DAT
205. Ticks on sheep	82	1	count — TICKS.DAT
206. Brownlee's stack loss data	1	84	rep(21) numeric(4) [time series] — BROWNLEE.DAT
207. Heine-Euler extensions	100	1	count — HEINE.DAT
	1096	1	count
	103	1	count

Title	Data structure			Filename
227. 'Bliss' data sets	150	1	count	BLISS.DAT
	400	1	count	
	400	1	count	
	414	1	count	
	240	1	count	
	673	1	count	
	240	1	count	
	109	1	count	
	150	1	count	
	189	1	count	
	164	1	count	
	122	1	count	
	227	1	count	
	95	1	count	
	1070	1	count	
228. Bird survival	50	5	numeric(3), binary(2) [survival]	BIRD.DAT
229. Trailing digits in data	4000	1	nominal	TRAILING.DAT
	50	1	nominal	
	141	1	nominal	
	100	1	nominal	
	125	1	nominal	
230. Silica in meteors	22	1	percentage	SILICA.DAT
231. Husbands and wives	200	5	rep(2) numeric(2), numeric	HUSBANDS.DAT
232. Presurgical stress	19	2	rep(2) numeric	STRESS.DAT
233. Bladder cancer study	82	13	binary, rep(12) count	BLADDER.DAT
234. Interspike waiting times	1	100	rep(100) numeric [time series]	WAITING.DAT
235. Traffic flow	349	1	numeric	TRAFFIC.DAT
236. Waiting times for planned pregnancies	678	2	binary, count	PLANNED.DAT

Title	Data structure		Filename
237. Iowa soil samples	18	3	numeric(3)	SAMPLES.DAT
238. Greenland turbot	68	7	numeric, rep(6) count	TURBOT.DAT
239. Insect data	30	4	nominal, numeric(3)	INSECT.DAT
	6	3	numeric(3)	
240. Skin resistance	16	5	rep(5) numeric	RESIST.DAT
241. Energy maintenance in sheep	64	2	numeric(2)	ENERGY.DAT
242. Lengths of lives of rats	195	2	binary, numeric	RATLIVES.DAT
243. Crying babies	18	3	county(2), binary	CRYING.DAT
244. Plant competition data	147	4	nominal, numeric(3) [spatial]	PLANT.DAT
245. Product sales	1	300	rep(300) count [time series]	PRODUCT.DAT
246. Timber data	50	3	numeric(3)	TIMBER.DAT
247. Parasite data	70	4	binary, numeric, count(2)	PARASITE.DAT
248. Azimuth data	18	1	numeric	AZIMUTH.DAT
249. Perpendicular distance models	164	1	numeric	PERPEND.DAT
	68	1	numeric	
	74	1	numeric	
250. Japanese black pines	204	2	numeric(2)	PINES.DAT
251. Treatments for locally unresectable gastric cancer	90	3	binary(2), numeric [survival]	GASTRIC.DAT
252. Strength of beams	10	3	numeric(3)	BEAMS.DAT
253. Salinity values	30	2	nominal, numeric	SALINITY.DAT
254. Epileptic seizures and chemotherapy	59	7	binary, rep(4) count, count(2)	CHEMO.DAT
255. Period between earthquakes	1	62	rep(62) count [time series]	QUAKES.DAT
256. Alveolar-bronchiolar adenomas	23	2	proportion, count	ALVEOLAR.DAT
257. Woodlice	37	1	count [spatial]	WOODLICE.DAT
	24	1	count [spatial]	
258. Drilling times	6	161	binary, rep(80) numeric(2) [time series]	DRILLING.DAT
259. Failure data	1	27	rep(9) (rep(2) numeric, count)	FAILURE.DAT

Title	Data structure		Filename
260. Spatial presence-absence data	24 ×	24	binary [spatial]	SPATIAL.DAT
261. Neuralgia data	18	5	binary(3), numeric(2)	NEURAL.DAT
262. Temperatures in America	56	3	minimum, numeric(2)	USTEMP.DAT
263. Weldon's dice data – (currently) the last word	26306	1	count	WELDON.DAT
	7006	1	count	
	4096	1	count	
	4096	1	count	
264. Putting computers to the test	16	13	numeric(13)	TEST.DAT
265. Assessing the effects of toxicity	40	2	categorical, numeric	TOXICITY.DAT
266. Testing homogeneity	11	2	binary, count	HOMOGEN.DAT
267. Random digits	6	10	counts(10)	DIGITS.DAT
268. Ice cream consumption	1	120	rep(30) numeric(4) [time series]	ICECREAM.DAT
269. Hospital data	12	4	numeric(3), count	HOSPITAL.DAT
270. Aquifer data	85	3	numeric(3)	AQUIFER.DAT
271. Windmill data	25	2	numeric(2)	WINDMILL.DAT
272. Examination times and scores	134	2	numeric(2)	EXAM.DAT
273. Gamma irradiation	12	2	numeric(2)	GAMMA.DAT
274. Ramus heights	20	4	rep(4) numeric	RAMUS.DAT
275. A soil experiment	21	3	nominal(2), numeric	SOIL.DAT
276. Rabbit foetuses	24	3	binary, count(3)	RABBIT.DAT
277. Blood fat concentration	371	3	binary, numeric(2)	BLOODFAT.DAT
278. Snowfall data	63	1	numeric	SNOWFALL.DAT
279. Counting alpha-particles	2608	1	count	COUNTING.DAT
280. Eruptions of the Old Faithful geyser	299	1	numeric	GEYSER.DAT
281. Starch films	94	3	nominal(2), numeric(2)	STARCH.DAT
282. Sheep diet	16	7	nominal(2), numeric, rep(4) numeric [latin square]	SHEEP.DAT

Title	Data structure			Filename
283. Horse-kicks	196	1	count	KICKS.DAT
	122	1	count	
	196	20	rep(20) count	
284. Pneumonia risk in smokers with chickenpox	7	2	rep(2) numeric	SMOKERS.DAT
285. Anorexia data	63	3	nominal, rep(2) numeric	ANOREXIA.DAT
286. Breast development of Turkish girls	318	2	categorical(2)	TURKISH.DAT
287. Treatment of enuresis	29	3	binary, rep(2) count	ENURESIS.DAT
288. Births in the USA	53	3	nominal, rep(2) percentage	BIRTHS.DAT
289. Flying bomb hits on London during World War II	576	1	count	BOMB.DAT
290. Brain and body weights of animals	28	2	numeric(2)	BRAIN.DAT
291. Severe Ideopathic Respiratory Distress Syndrome	50	2	binary, numeric	DISTRESS.DAT
292. Lowest temperatures for US cities	22	4	rep(4) numeric	LOWEST.DAT
293. Plasma citrate concentrations	10	5	rep(5) numeric	PLASMA.DAT
294. Voting in congress	15 × 15		[dissimilarity matrix]	CONGRESS.DAT
295. Cortisol levels in psychotics	71	2	nominal, numeric	CORTISOL.DAT
296. An historic data set: crime and drinking	1426	2	binary, nominal	HISTORIC.DAT
297. Depression in adolescents	465	4	numeric, binary(3)	DEPRESS.DAT
298. English and Greek teachers	32	2	rep(2) numeric	TEACHERS.DAT
299. Fun runners	33	2	nominal, numeric	FUN.DAT
300. Olympic jumping events	1	84	rep(21) numeric(4) [time series]	JUMPING.DAT
301. Blood lactic acid	20	2	numeric(2)	LACTIC.DAT
302. Skin cancers	15	4	binary, numeric, count(2)	SKIN.DAT
303. Mental status of school-age children	1600	2	categorical(2)	MENTAL.DAT
304. Diabetic mice	57	2	nominal, numeric	DIABETIC.DAT
305. Absorption of ions by potato	5	3	numeric(3)	POTATO.DAT

Title	Data structure			Filename
306. Survival of cancer patients	25	3	binary(2), numeric	SURVIVAL.DAT
307. Major earthquakes since 1900: fatalities and intensities	1	40	rep(20) (numeric, count) [time series]	EARTH.DAT
308. Presidents, popes and monarchs	66	2	nominal, numeric [survival]	MONARCH.DAT
309. Cervical cancer	366	2	binary(2)	CERVICAL.DAT
310. Sickle cell disease	41	2	nominal, numeric	SICKLE.DAT
311. Men's olympic sprint times	1	100	rep(20) maxima(5)	SPRINT.DAT
312. Women's olympic swimming	1	28	rep(7) maxima(14)	SWIMMING.DAT
313. A thrombosis study	24	2	binary, numeric	THROMBOS.DAT
314. Deaths in USA in 1966	1	12	rep(12) count [time series]	DEATH.DAT
315. Comparison of experimental method	315	2	binary(2)	COMPARE.DAT
316. Bat-to-prey detection distances	11	3	numeric(2), binary	BAT.DAT
317. Sexual performance of castrated mice	3	7	count(7)	CASTRATE.DAT
	3	6	count(6)	
318. The cloud point of a liquid	19	2	percentage, numeric	LIQUID.DAT
319. Romano-British pottery	26	6	nominal, percentage(5)	POTTERY.DAT
320. A split plot field trial	72	4	nominal(2), numeric(2) [split plot]	SPLIT.DAT
321. Olympic triple jump distances	21	1	maximum	TRIPLE.DAT
322. Species of flea beetle	74	3	nominal, numeric(2)	FLEA.DAT
323. Survival times of cancer patients	64	2	nominal, numeric [survival]	PATIENT.DAT
324. Birthweights of Poland China pigs	56	2	nominal, numeric	POLAND.DAT
325. Performance of various computer CPUs	209	8	numeric(6), count(2)	COMPUTER.DAT
326. Monthly deaths from lung diseases in the UK	1	72	count	MONTHLY.DAT
327. Convictions for drunkenness	947	2	numeric, binary	DRUNK.DAT
328. Ear infections in swimmers	287	5	binary(3), ordinal, count	EAR.DAT
329. A perception experiment	24	3	numeric(2), binary	PERCEPT.DAT
330. Boiling points in the Alps	17	2	numeric(2)	ALPS.DAT

Title	Data structure		Filename
331. Remission times of leukaemia patients	42	3	REMISS.DAT
332. Foster feeding rats of different genotype	61	3	FOSTER.DAT
333. Record times for Scottish hill races	35	3	HILLRACE.DAT
334. Hardness of timber	36	2	HARDNESS.DAT
335. Dimensions of jellyfish	46	3	JELLY.DAT
336. Malaria parasites	50000	1	MALARIA.DAT
337. The New York Choral Society	107	2	CHORAL.DAT
338. Testing crash helmets	1	266	HELMETS.DAT
339. The speed of light	100	3	LIGHT.DAT
340. Murder-suicides through aircraft crashes	17	2	AIRCRAFT.DAT
341. Average monthly temperatures for Nottingham, 1920–1939	1	240	NOTTS.DAT
342. Insurance premiums	23	4	PREMIUM.DAT
343. Testing authorship	19220	2	AUTHOR.DAT
344. The rise of the planet Venus	14	3	VENUS.DAT
345. Pulse rates of Peruvian Indians	39	1	PULSE.DAT
346. Vocabulary of children	10	2	VOCAB.DAT
347. Educating cats	13	2	CATS.DAT
348. Shocking rats	16	2	SHOCKING.DAT
349. Smoking habits	20	3	HABITS.DAT
350. Mothers of schizophrenics	40	2	MOTHERS.DAT
351. Unemployment and suicide rates	11	2	UNEMPLOY.DAT
352. Main US budget items for 1978 and 1979	8	2	BUDGET.DAT
353. Damage to ships	81	5	SHIPS.DAT
354. Methadone treatment of heroin addicts	266	5	HEROIN.DAT
355. Smoking and motherhood	6851	4	SMOKEMUM.DAT

Data structure descriptions (right-hand column):

Title	Data structure description
331	numeric, binary(2)
332	nominal(2), numeric
333	numeric(3)
334	numeric(2)
335	binary, numeric(2)
336	count
337	nominal, numeric
338	rep(133) numeric(2)
339	numeric, nominal(2) [randomized block]
340	numeric, count
341	rep(240) numeric [time series]
342	binary(2), numeric(2)
343	binary, count
344	numeric(3)
345	numeric
346	numeric(2)
347	count, numeric
348	count, numeric
349	categorical, nominal, proportion
350	binary, count
351	percentage, rate
352	rep(2) numeric
353	nominal, categorical, binary, numeric, count
354	binary(3), numeric(2)
355	binary(4)

Title	Data structure			Filename
356. Crime rates in the USA	50	7	rate(7)	CRIME.DAT
357. Olympic decathlon, 1988	34	11	maxima(11)	OLYMPIC.DAT
358. Measurements of dog mandibles	7	6	numeric(6)	MANDIBLE.DAT
359. Children with Down's syndrome	7	3	numeric, count(2)	CHILDREN.DAT
360. Protein consumption in European countries	25	9	numeric(9)	PROTEIN.DAT
361. Male Egyptian skulls	150	5	numeric(5)	SKULLS.DAT
362. Olympic women's heptathlon 1988	25	8	numeric(8)	HEPTATH.DAT
363. Percentages employed in different industries in Europe	26	9	percentage(9)	EUROPE.DAT
364. Ages at death of English rulers	42	1	numeric	RULERS.DAT
365. Deaths in coal mining disasters	166	1	count	MINING.DAT
366. A study in obesity	583	2	categorical, binary	OBESITY.DAT
367. The star cluster CYG OB1	47	2	numeric(2)	STAR.DAT
368. Frequency of breast self-examination	1216	2	numeric, categorical	SELFEXAM.DAT
369. Annual wages of production line workers in the USA	30	1	numeric	WORKERS.DAT
370. One way to kill a cat	41	2	numeric(2)	KILL.DAT
371. Viral lesions on tobacco leaves	8	2	rep(2) count	TOBACCO.DAT
372. Finger ridges of identical twins	12	2	rep(2) count	TWINS.DAT
373. Statures of brother and sister	11	2	rep(2) numeric	STATURES.DAT
374. Candidates in the 1992 British Election	11	10	count(10)	BRITISH.DAT
375. Papers presented at the first ten ICPRs	10	10	rep(10) count	PAPERS.DAT
376. Body mass index data	20	3	numeric(3)	BODYMASS.DAT
377. Citation rates	26	1	count	CITATION.DAT
378. A World Heavyweight Boxing Championship match	1	36	binary, rep(5) count(7)	BOXING.DAT
379. Assessment of conference papers	30	8	rank(8)	CONF.DAT
380. Clinical trial for sprains	46	4	rep(4) categorical	SPRAINS.DAT

Title	Data structure			Filename
381. Jeans sales in the UK	1	72	rep(72) numeric [time series]	JEANS.DAT
382. The thirteen most common psychiatric disorders in the UK	22	13	numeric(13)	DISORDER.DAT
383. Randomized experiment on tomato yield	11	3	binary, numeric, categorical	TOMYIELD.DAT
384. Screening for psychiatric disorder	40	26	binary(26)	SCREEN.DAT
385. Salaries in different occupations	144	2	binary, numeric	SALARIES.DAT
386. Whisky prices	42	2	binary, numeric	WHISKY.DAT
387. 1992 London Marathon	100	2	binary, numeric	MARATHON.DAT
388. Predicting cervical cancer	30	2	categorical(2)	PREDICT.DAT
389. Treatments for head and neck cancer	96	3	binary, numeric	HEADNECK.DAT
390. Tau particle decay	13	2	numeric(2)	TAU.DAT
391. Factors influencing depression	419	3	binary(3)	FACTORS.DAT
	419	3	binary(3)	
392. Vocabulary scores of university students	54	1	numeric	UNIV.DAT
393. Two groups drug comparison	5	5	binary, rep(4) numeric	GROUPS.DAT
394. Blood glucose levels	6	60	rep(60) numeric	GLUCOSE.DAT
395. Rat body weights with diet supplements	16	12	nominal, rep(11) numeric	RATBODY.DAT
396. Diet supplements	15	7	nominal, rep(6) numeric	DIET.DAT
397. Visual acuity and lens strength	7	8	rep(4) numeric(2)	VISUAL.DAT
398. Plasma ascorbic acid levels in hospital patients	12	7	rep(7) numeric	ASCORBIC.DAT
399. Salsolinol excretion rates	14	5	binary, rep(4) numeric	EXCRETE.DAT
400. Response profiles to three drugs	46	3	binary(3)	RESPONSE.DAT
401. Patterns of psychotropic drug consumption	5773	5	binary(3), numeric(2)	PSYCHO.DAT
402. Wool data	27	4	binary(3), numeric	WOOL.DAT
403. Survival times of animals	48	3	nominal(2), numeric	ANIMALS.DAT
404. Effectiveness of slimming clinics	34	3	binary(2), numeric	SLIMMING.DAT
405. Prevalence of vertebral fractures	1005	2	binary, numeric	FRACTURE.DAT

Title	Data structure			Filename
Title				**Filename**
406. Agreement between two examination markers	29	2	categorical(2)	MARKERS.DAT
407. Subjective estimates of lengths of pieces of string	15	4	numeric(4)	STRING.DAT
408. Comparison of two kitchen scales	19	2	numeric(2)	SCALES.DAT
409. Performance in mathematical degrees by sex in England and Scotland	7115	3	binary(2), categorical	DEGREES.DAT
410. Chemical process	24	3	numeric(2), percentage	CHEMICAL.DAT
411. Carcinoma of the uterine cervix	118	7	categorical(7)	UTERINE.DAT
412. Rainfall in Minneapolis/St Paul	1	30	numeric [time series]	MINN.DAT
413. Book condition by strength of paper	840	2	categorical(2)	BOOK.DAT
414. Oxford and Cambridge boat race crews	18	2	binary, numeric	BOATRACE.DAT
415. Comparisons between 48 major cities in 1991	48	3	numeric(3)	CITIES.DAT
416. Height and resting pulse measurements for a sample of hospital patients	50	2	numeric(2)	RESTING.DAT
417. Coalminers' breathing difficulties	17232	3	binary(2), numeric	COALMINE.DAT
418. Toxicity of potassium cyanate on trout eggs	48	4	numeric, binary, count(2)	TROUT.DAT
419. Asthma death, corticosteroid use, and fenoterol use	585	3	binary(3)	ASTHMA.DAT
420. Seed germination	1255	3	binary(3)	SEED.DAT
421. Depression and friendship amongst children	49 / 49 / 49	3	binary(3), binary(3), binary(3)	FRIENDS.DAT
422. Social class of fathers and sons	845	2	categorical(2)	SOCIAL.DAT
423. Survival data for patients with chronic active hepatitis	44	2	numeric, binary [survival]	CHRONIC.DAT
424. Survival times of leukaemia patients	33	3	binary, numeric(2) [survival]	LEUKWBC.DAT

Title	Data structure			Filename
425. Dissimilarity ratings of World War II politicians	12 ×	12	[dissimilarity matrix]	POLITICS.DAT
426. Rate of formation of a chemical impurity	6	3	numeric(3)	IMPURITY.DAT
427. Road distances between major UK towns	23 ×	23	[distance matrix]	ROAD.DAT
428. Penicillin manufacture	20	4	categorical, nominal(2), numeric [randomised blocks]	PENICILL.DAT
429. Early detection of autism	2	13	percentage(13)	AUTISM.DAT
430. Adolescent attempted suicide rates	20	4	numeric, binary, counts(2)	ADOLESCE.DAT
431. Lung cancer cases and person-years experience by age and period of diagnosis	140	5	numeric(2), binary(2), count	DIAGNOSE.DAT
432. Yields of wheat	13	4	numeric(4)	HARVEST.DAT
433. Estimating numbers of unobserved species	14	10	count(10)	SPECIES.DAT
434. Distance and direction travelled by blue periwinkles	31	2	numeric(2)	PERIWINK.DAT
435. Birth season and psychosis	12	3	counts(2), percentage	SEASON.DAT
436. Psychotropic drug use by sex, age, and physical illness	1307	4	binary(3), categorical	PHYSICAL.DAT
437. Mouse lymphoma mutation	24	3	binary, numeric(2)	MOUSE.DAT
438. Stopping distance	63	2	numeric(2)	STOPPING.DAT
439. Decontaminants for *M. bovis*	129	3	nominal, numeric(2)	MBOVIS.DAT
440. Breast cancer and ambulatory status over time	37	7	rep(7) categorical	AMBULAT.DAT
441. Numbers of icebergs sighted	1	24	rep(12) count(2)	ICEBERG.DAT
442. Ratings of synchronized swimming	40	5	numeric(5)	SYNCHRO.DAT

Title	Data structure		Filename
443. Numbers of revertant colonies of TA98 Salmonella	3	12	rep(6) (numeric, count)	COLONIES.DAT
444. Landsat multi-spectral scanner data	3 12 × 12 20	12 12 5	numeric(12) [correlation matrix] nominal, numeric(2), count(2)	LANDSAT.DAT
445. Effectiveness of insecticides	20	5		EFFECT.DAT
446. Strengths of glass fibres	63	1	numeric(1)	GLASS.DAT
447. Psychoactive drug use	13 × 13	13	[correlation matrix]	DRUGUSE.DAT
448. Estimated abdominal disease diagnosis	50	4	nominal, probability(3)	ABDOMEN.DAT
449. Statistics examination results	10	14	percentage(14)	STATS.DAT
450. Eight variables measured on female psychiatric patients	118	8	numeric(3), categorical(2), binary(3)	FEMALE.DAT
451. Suicide thoughts	295	2	categorical(2)	THOUGHTS.DAT
452. Similarities between eight offences	8 × 8	8	[similarity matrix]	OFFENCES.DAT
453. Language abilities	12 × 12	12	[proximity matrix]	LANGUAGE.DAT
454. Heat evolved by setting cement	13	5	percentage(4), numeric	SETTING.DAT
455. Quality control in bread baking	29	10	categorical(6), numeric(4)	BREAD.DAT
456. Death rates from heart disease among doctors	20	4	binary(2), count, numeric	DISEASE.DAT
457. Renal transplant data	148	3	numeric, binary, count	RENAL.DAT
458. Strengths of welds in high density polyethylene	34	5	numeric(5)	WELDS.DAT
459. Subjective health assessment in 5 regions	5855	2	nominal, categorical	ASSESS.DAT
460. AIDS incidence data for the USA	1	26	rep(26) count [time series]	AIDS.DAT
461. Stock recruitment of Skeena River sockeye salmon	1	28	rep(28) count(2) [time series]	SALMON.DAT
462. Tumour response to chemotherapy	299	3	binary(2), categorical	TUMOUR.DAT
463. Cervical cancer deaths in four European countries	4	3	numeric, count(2)	COUNTRY.DAT

Title	Data structure		Filename
464. Performance of a credit scoring instrument	22	5	numeric, percentage(4)	CREDIT.DAT
465. Weights of mice	13	7	rep(7) numeric	MICE.DAT
466. Poisson random variate generation	60	4	binary, nominal, numeric(2)	POISSON.DAT
467. Survival times of gastric cancer patients	90	3	binary(2), numeric [survival]	TIME.DAT
468. Time for assays to register HIV positivity	47	6	binary(2), rep(4) counts	HIV.DAT
469. Mean daily SIDS deaths by temperature	10	4	numeric(3), count	MEAN.DAT
470. Number of SIDS deaths per day	242	2	count, numeric	SIDS.DAT
471. Predicting benefit from group psychotherapy	64	3	binary, rep(2) categorical	BENEFIT.DAT
472. African statisticians in 1977	48	3	count(3)	AFRICAN.DAT
473. Railway accidents on British Rail, 1970-83	1	84	rep(14) (count(5), numeric) [time series]	RAILWAY.DAT
474. Australian higher education students, 1981 to 1988	8	24	rep(8) (numeric, binary, count)	EDUCATE.DAT
475. Contents of books on multivariate analysis	20	7	count(7)	CONTENTS.DAT
476. Age of menarche in Polish girls	25	3	numeric, count, proportion	POLISH.DAT
477. Level of Lake Victoria Nyanza and numbers of sunspots	1	40	rep(20) (numeric, count) [time series]	LAKE.DAT
478. Surgical deaths	40	5	numeric, binary(2), count(2)	SURGICAL.DAT
479. Mean monthly temperature at various altitudes	14	24	rep(12) numeric(2)	ALTITUDE.DAT
480. Aircraft air-conditioning systems failures	12	($\leq$) 30	rep(30) numeric	SYSTEMS.DAT
481. Female age distribution in Mexico in 1960	75	2	numeric, count	MEXICO.DAT

Title	Data structure			Filename
482. Volume of hens' eggs	12	3	numeric(3)	EGGS.DAT
483. Smoking and health	36	4	numeric, nominal, count(2)	HEALTH.DAT
484. Herniorrhaphy	32	9	numeric(2), binary(2), categorical(5)	HERNIOR.DAT
485. Birthweight by gestational age	23	4	numeric(3), count	AGE.DAT
486. Underdispersed word count data	100	1	count	UNDER.DAT
	100	1	count	
487. Red core disease in strawberries	144	5	categorical(2), nominal(2), count [randomized blocks]	REDCORE.DAT
488. Failure times for a piece of electronic equipment	107	1	numeric [survival]	EQUIP.DAT
489. Survival time of patients with lymphocytic non-Hodgkin's lymphoma	38	3	binary(2), numeric [survival]	LYMPHOMA.DAT
490. Prestige, income, education, and suicide rates for 36 occupations	36	4	categorical, rate, numeric(2)	PRESTIGE.DAT
491. Natural scientists and engineers in the US	1	16	rep(8) count(2)	ENGINEER.DAT
492. Correlation between nine statements about pain	9 × 9		[correlation matrix]	PAIN.DAT
493. Manufacture of truck leaf springs	16	7	binary(5), numeric(2)	TRUCK.DAT
494. Housing satisfaction around Montevideo, Minnesota	35	4	binary, count(3)	HOUSING.DAT
495. Car changing patterns	16 × 16		[transition matrix]	CAR.DAT
496. *In vitro* chromosome aberration assays	9	4	binary, numeric, count(2)	INVITRO.DAT
497. Visits to dentist	10218	1	count	VISITS.DAT
498. Myocardial infarction and obesity, smoking and vasectomy	36	4	binary(4)	MYOCARD.DAT
499. Specific heat of water at various temperatures	6	12	rep(6) numeric(2)	HEAT.DAT

Title	Data structure			Filename
500. Numbers of packets of cereal purchased over 13 weeks	2000	1	count	CEREAL.DAT
501. Accelerated life tests on electrical insulation	40	3	numeric(2), binary [survival]	ELECTRIC.DAT
502. Cardiac oxygen consumption and mechanical function	11	6	binary, numeric(5)	OXYGEN.DAT
503. Coffee consumption and coronary heart disease	1718	2	binary(2)	CONSUME.DAT
	151	3	binary(3)	
504. Strength of sheet metal	32	6	binary(5), numeric	METAL.DAT
505. Mortality from testicular cancer in England and Wales	13	22	rep(11) (numeric, rate)	TESTICLE.DAT
506. Fathers' and sons' occupations	14	14	counts(14)	FATHERS.DAT
507. Numbers of purchases of toilet tissue	1816	2	nominal, count	TISSUE.DAT
508. Haematology of paint sprayers	103	6	numeric(6)	PAINT.DAT
509. Measurement of protein content in ground wheat	24	7	percentage, number(6)	GROUND.DAT
510. Winning times in men's running events	17	5	maxima(4), numeric	WINNING.DAT

SUBJECT INDEX

All references are to table numbers.